MULTIPHASE SCIENCE AND TECHNOLOGY

AF606156

MULTIPHASE SCIENCE AND TECHNOLOGY

Editorial Advisory Board

D. AZIZ
Department of Chemical Engineering
The University of Calgary
Calgary T2N 1N4, Canada

S. BANERJEE
Chemical and Nuclear Engineering
Department
University of California
Santa Barbara, California 93106, USA

H. BRAUER
Technische Universität Berlin
Institut für Chemieingenieurtechnik
1 Berlin 10, RFA

J. C. CHARPENTIER
CPIC/ENSIC
54042 Nancy Cedex, France

J. G. COLLIER
Central Electricity Generating Board
Generation and Construction Division
Barnwood, Gloucester GL4 7RS, England

M. CUMO
Comitato per l'Energia Nucleare
Centro Studi Nucleari della Casaccia
Departimento Reattori Termici
Rome, Italy

M. GIOT
Département Thermodynamique
et Turbomachines
B-1348 Louvain la Neuve, Belgium

D. HASSON
Technion
Department of Chemical Engineering
Haifa, Israel

M. HIRATA
University of Tokyo
Department of Mechanical Engineering
Bunkyo-Ku
Tokyo 113, Japan

A. KONORSKI
Instytut Maszyn Przeplywowych
80 - 952 Gdansk, Poland

K. R. LAWTHER
A.A.E.C. Research Establishment
Private mail bag, Sutherland 2232
NSW Australia

F. MARBLE
Jet Propulsion Center
California Institute of Technology
Pasadena, CA 91125, USA

F. MAYINGER
Lehrstule A. für Thermodynamik
Technische Universitat München
8000 München 2
Arcistrasse 21
München, FRG

V. NAKORYAKOV
Institute of Thermophysics
Novosibirsk 63 0090, USSR

G. OOMS
Shell Research B.V.
Badhuisweg 3
Postbus 3003
1003AA Amsterdam, The Netherlands

F. RESCH
Institut de Mécanique Statistique de la
Turbulence
13003 Marseille, France

L. E. SCRIVEN
Department of Chemical Engineering
and Materials Science
University of Minnesota
Minneapolis, MN 55455, USA

G. V. TSIKLAURI
Institute of High Temperature
Krasnokas Armenaya
Moscow, USSR

M. WICKS
Shell Oil Company
P. O. Box 3105
Houston, TX 77001, USA

24 0081616 4

£66 00

CENTRAL LIBRARY
Imperial College, London SW7 2AZ

THREE WEEK LOAN
To be returned by last date stamped

30 SEP 1996
10 JAN 1997
- 8 OCT 1997
26 MAY 1998
- 1 JUN 1999
- 6 NOV 2000
11 OCT 2001
30 AUG 2005
23 OCT 2009
01 FEB 2010
14 APR 2010
10 MAY 2010
02 JUN 2010

MULTIPHASE SCIENCE AND TECHNOLOGY

Volume 5

Edited by

G. F. Hewitt

Thermal Hydraulics Division
U.K. Atomic Energy Authority
Harwell Laboratory, England
and
Department of Chemical Engineering and Chemical Technology
Imperial College of Science and Technology
London, England

J. M. Delhaye

Commissariat à l'Energie Atomique
Centre d'Etudes Nucléaires de Grenoble
Service d'Etudes Thermohydrauliques
Grenoble, France

N. Zuber

Division of Reactor Safety Research
U.S. Nuclear Regulatory Commission
Washington, D.C., U.S.A.

HEMISPHERE PUBLISHING CORPORATION
A member of the Taylor & Francis Group

New York   Washington   Philadelphia   London

66.04 MUL

MULTIPHASE SCIENCE AND TECHNOLOGY, Volume 5

Copyright © 1990 by Hemisphere Publishing Corporation. All rights reserved. Printed in the United States of America. Except as permitted under the United States Copyright Act of 1976, no part of this publication may be reproduced or distributed in any form or by any means, or stored in a data base or retrieval system, without the prior written permission of the publisher.

1 2 3 4 5 6 7 8 9 0 BCBC 8 9 8 7 6 5 4 3 2 1 0 9

Library of Congress Cataloging in Publication Data

Multiphase science and technology.

Includes bibliographies and indexes.
1. Vapor-liquid equilibrium—Addresses, essays, lectures. 2. Ebullition—Addresses, essays, lectures. 3. Cooling—Addresses, essays, lectures.
I. Hewitt, G. F. (Geoffrey Frederick) II. Delhaye, J. M., date. III. Zuber, N.
TP156.E65 660.2'96 88-655096
ISBN 0-89116-650-5 (v. 5)
ISSN 0-276-1459

CONTENTS

Preface

Multiphase Science and Technology has the overall objective of providing authoritative and archival information on important subject areas in multiphase systems. So far, the material has arisen from two main sources, namely articles commissioned for the series and material arising from the International Workshops on Two-Phase Phase Flow Fundamentals. These Workshops (organised under the auspices of the US Department of Energy) were set up to provide a critical assessment and evaluation of the state-of-the-art in prediction of two-phase flow systems. The present Volume contains four chapters, the first three being based on invited lectures at the Second International Workshop, though with extensive review and revision; these three chapters are concerned respectively with the two-fluid model, with multiphase instrumentation and experimental techniques and with computational methods. The final chapter in the volume is a commissioned one and relates to inertial coupling and macroscopic properties of suspensions.

Chapter 1 ("Two-fluid model for two-phase flow" by M. Ishii) gives a detailed discussion of the two-fluid model formulation of two-phase flow. Specifically, the full three-dimensional model is obtained from a time averaging of the local instantaneous formulation. The author concludes that the formulation of the constitutive and closure relationships for the two-fluid model is not very satisfactory. The importance of the existence of small fluid particles in all flow regimes is stressed and recommended correlations for drop size and droplet entrainment in annular flow are presented. The problems of modelling with turbulent transfer processes in two-phase flows is also reviewed. This article provides a useful and realistic assessment of the current state-of-the-art on the two-fluid model.

The second chapter ("Computational methods for multiphase flow" by W. Wulff) discusses the numerical solution of the governing equations in multiphase flow. For most practical systems, solutions of the balance equations for multiphase flows have to be obtained by numerical means. This paper gives a survey of numerical methods for solving the partially or completely discretised equation. Issues of computing accuracy and economy are discussed. The numerical methods used in the major computer codes for multiphase flow analysis are reviewed and it is shown that further work is needed on quantification of computing errors and improvement of computing economy. This authoritative article should be of major archival interest for all those involved in numerical simulations of multiphase flow systems.

The third chapter ("Multiphase instrumentation and experimental techniques" by J. M. Delhaye) gives an appraisal of the state-of-the-art of measuring techniques in gas-liquid two-phase flow and discusses current needs in instrumentation for research and industrial applications. Specific techniques reviewed include probes for local void fraction measurement, new developments on impedance devices, measurement of time dependent relative velocity of bubbles using a moving camera technique, the use of high speed video systems etc. The philosophy of instrumentation development is also reviewed. This article should provide a readable introduction to two-phase instrumentation for those entering or already in the field, in particular giving a realistic appraisal of the objectives of instrumentation systems.

The final chapter ("Inertial coupling in two-phase flow: Macroscopic properties of suspensions in an inviscid fluid" by G. B. Wallis) is a major new work which addresses the problem of the macroscopic properties of a flow suspension of spheres. Classical solutions for the motion of a single sphere are reviewed and extended and self consistent models are derived for suspension. A major result of the work is the recognition that Maxwell's equations for electrical conductivity of a suspension of spheres can be used to derive the added mass of the same array as well as the added kinetic energy due to relative motion. The applications of the formulations are explored in a number of simple cases including slug flow and flow through a nozzle. This chapter breaks considerable new ground and in itself should contribute significantly to the ongoing debate about this area.

It is believed that the contents of this volume will be of great interest to all those working on multiphase flow systems, both from the experimental and analytical viewpoints.

The Editors
17 April

Contents of Volume 1

Contents of Volume 2

Contents of Volume 3

Contents of Volume 4

CHAPTER 1

TWO-FLUID MODEL FOR TWO-PHASE FLOW

M. Ishii

School of Nuclear Engineering, Purdue University, West Lafayette, Indiana 47907

ABSTRACT

The two-fluid model formulation is discussed in detail. The emphasis of the paper is on the three-dimensional formulation and the closure issues. The origin of the interfacial and turbulent transfer terms in the averaged formulation is explained and their original mathematical forms are examined. The interfacial transfer of mass, momentum, and energy is proportional to the interfacial area and driving force. This is not a postulate but a result of the careful examination of the mathematical form of the exact interfacial terms. These two effects are considered separately. Since all the interfacial transfer terms involve the interfacial area concentration, the accurate modeling of the local interfacial area concentration is the first step to be taken for development of reliable two-fluid model closure relations. The interfacial momentum interaction has been studied in terms of the standard drag, lift, virtual mass, and Basset forces. Available analytical and semi-empirical correlations and closure relations are reviewed and existing shortcomings are pointed out. The other major area of importance is the modeling of turbulent transfer in two-phase flow. The two-phase turbulence cannot be understood without understanding the interfacial drag and lift forces accurately. There are some indications that the mixing length type model may not be sufficient to describe the three-dimensional turbulent and flow structures. Although it is a very difficult challenge, the two-phase flow turbulence should be investigated both experimentally and analytically with long time-scale research.

1. INTRODUCTION

The most important characteristic of two-phase flow is the existence of an interface separating phases and associated discontinuities of properties across the phase interface. Because of the deformable nature of a gas-liquid interface, a limitless number of interface configurations are possible. Therefore, at first glance it might appear that various two-phase systems or phenomena have very little in common. Actually, the contrary is true. Many two-phase systems have a common structure or topography of an interface. It is recalled that single-phase flows can be classified according to the internal flow geometry into laminar, transitional and turbulent flows; whereas two-phase flows can be classified into several different flow regimes or flow patterns according to the geometry of an interface. The simplest classification is to use three flow regimes - separated flows, transitional or mixed flows and dispersed flows. However, more detailed classifications have been proposed by a number of researchers (Dukler and Taitel, 1977a, 1977b; Griffith and Wallis, 1961; Bergles et al., 1968; Mishima and Ishii, 1984; Baker, 194; Hoogendoorn, 1959; Govier and Omer, 1962; Mandhane et al., 1974; Weisman and Kang, 1981).

The various transfer mechanisms between a two-phase mixture and a surrounding wall, as well as between two-phases strongly depend on the two-phase flow regimes. However, in view of the topographical and structural classification, it might be expected that many of the two-phase systems would exhibit a large number of steady-state and dynamic similarities. This indeed is the fundamental assumption in developing various two-phase flow models and closure relations. It is assumed that if a firm understanding of the thermo-fluid dynamic characteristics of two-phase flow is attained, these results can be applied to predict the operations performance of systems and components in a variety of technologies. Such generalized analyses have been used in the field of single phase flow. The thermofluid dynamic characteristics of laminar and of turbulent flows were studied, and the results applied to various technologies. In the field of two-phase flow, however, the opposite approach has most often been followed. The tendency has been to analyze the thermal or fluid dynamic problems of a particular system, component, or process. Consequently, a broad understanding of the thermofluid dynamic behavior of two-phase systems has not been attained yet, nor is a generalized method available at present to analyze and predict the performance of these systems.

The design of engineering systems and the ability to predict their performance depend on the availability of experimental data and of conceptual models and predictive tools that can be used to describe a physical process with a required degree of accuracy and reliability. From a scientific, as well as from a practical point of view, it is essential that the various characteristics and properties of such conceptual models and predictive tools be formulated clearly and rationally, and supported by experimental data. This requires specially designed experiments that must be conducted in conjunction with and in support of analytical investigations.

It is well established in continuum mechanics that the conceptual models and predictive tools for single-phase flow of a gas or liquid are formulated in terms of field equations that describe the conservation laws of mass, momentum, energy, charge, etc. These field equations are then complemented by appropriate constitutive equations such as the constitutive equations of state, stress, chemical reactions, etc., which specify the thermodynamic, transport, and chemical properties of a given constituent material. It is to be expected, therefore, that the conceptual models that describe the steady-state and dynamic characteristics of structured multiphase or multicomponent media should also be formulated in terms of the appropriate field equations and constitutive or closure relations. However, the derivation of such equations for the flow of structured media is considerably more complicated than for single-phase flow.

To appreciate the difficulties in deriving balance equations for structured media, one must recall that in continuum mechanics the field theories are constructed on integral balances of mass, momentum, and energy. Thus, if the variables in the region of integration are continuously differentiable and the Jacobian transformation between material and spatial coordinates exists, then the Eulerian differential balance equations can be obtained by using Leibnitz's rule or more specifically the Reynolds transport theorem, which allows interchanging differential and integral operations. In two-phase flow the presence of an interface introduces great difficulties in the mathematical and physical formulation of the problem. From the mathematical point of view, two-phase flow can be considered as a field that is subdivided into single-phase regions, with moving boundaries separating the constituent phases. The differential balance holds for each subregion, but it cannot be applied to the set of these subregions in the normal sense without violating the above conditions of continuity. Because of this, physical problems of two-phase flow are represented by macroscopic field equations and closure relations using a continuum formulation.

From the point of view of physics, the difficulties encountered in deriving the field equations and closure relations appropriate to multiphase flow systems stem from the presence of an interface and the fact that both the steady and dynamic characteristics of multiphase flows depend on the structure of the flow. For example, the steady-state and the dynamic characteristics of dispersed two-phase flow systems depend on the collective dynamics of solid particles, bubbles, or droplets interacting with each other and with the surrounding continuous phase; whereas, in the case of separated flows these characteristics depend on the structure and dynamics of the interface. To determine the collective interaction of particles, bubbles, or droplets - the dynamics of the interface - it is necessary to first describe the local properties of the flow and then obtain a macroscopic description by means of appropriate averaging procedures. For dispersed flows, for example, it may be necessary to model the rates of nucleation, evaporation, or condensation, and motion and disintegration of droplets or bubbles as well as the collisions and coalescence of several droplets or bubbles. For separated flow, the structure and the dynamics of the interface (such as the wave form, propagation velocity and entrainment) greatly influence the rates of mass, energy, and momentum transfer as well as the stability of the system.

To derive the field equations and closure relations appropriate to two-phase flow, it is necessary to describe the local characteristics of the flow from which the macroscopic properties should be obtained by means of an appropriate averaging procedure. Since closure relations are for the averaged fields, the experimental observations and data as well as accurate modeling which takes into account the scale effects are essential to the development of a reliable two-phase flow formulation.

In view of the great importance of two-phase flow analyses in various industries, considerable efforts have been made to develop accurate two-phase flow formulations. In particular, in relation to safety analyses of nuclear reactors, some progress in understanding and predicting complicated transient two-phase flow phenomena has been made in three different fronts, namely, development of accurate models, improvement in numerical techniques, and detailed experiments in small and large scales. Recently, comprehensive reviews of the state of the arts from several different aspects have been presented at the First International Workshop on Two-phase Flow Fundamentals at The National Bureau of Standards (Kestin and Podowski, 1985). Early developments in this area are reviewed in standard text books on two-phase flow (Tong, 1975; Brodkey, 1967; Wallis, 1969; Hewitt and Hall-Taylor, 1970; Collier, 1982; Govier and Aziz, 1972; Hsu and Graham, 1976; Lahey and Moody, 1977; Delhaye et al. 1981). However, the fundamental difficulties persist (Kestin and Podowski, 1985; Ishii, 1975, 1981; Wallis, 1980; Ishii and Mishima, 1984)

1980; Ishii and Mishima, 1984) and the existing models have a number of shortcomings and limitations.

In most two-phase flow formulations, physical problems are represented by macroscopic field and closure relations (Ishii, 1975; Vernier and Delhaye, 1968; Kocamustafaogullari, 1971; Drew and Wood, 1985; Nigmatulin, 1979). Two-phase flow models can be developed from several different methods. However, various averaging methods applied to the local instantaneous formulation give the most rigorous mathematical foundation (Ishii, 1975, 1981; Vernier and Delhaye, 1968). Because of the development of detailed averaging methods, the focus of the two-phase flow modeling efforts has shifted from the field equations to the constitutive equations and closure relations.

The development of the closure relations for the macroscopic two-phase flow formulation presents one of the most difficult problems ever encountered in fluid mechanics and heat transfer. The essential difficulties are of two types. First, the transport mechanisms involved in two-phase flow phenomena are very complicated due to the existence and motion of a phase interface. Often several mechanisms are coupled. Second, an understanding of scale effects is very difficult to establish due to several factors such as;

1) existence of both the internal scale (interfacial length scale corresponding to bubble size etc.) and external scale (hydraulic diameter),

2) a large number of variables involved,

3) unavailability of incompleteness of detailed two-phase flow instrumentation techniques, and

4) lack of experimental data over sufficient parameter ranges in terms of system scales, properties, and operating conditions.

Because of the macroscopic nature of the two-phase flow formulation, the development of the closure relations is forced to be strongly empirical. This implies that, in many cases, little data base exists to extend existing correlations. Therefore, a prediction of two-phase flow in new or hypothetical situations has often been very difficult and unreliable. It is important that a researcher knows the limitations of a model used and the ranges of data base for the closure relations applied. Unfortunately, this aspect has been frequently neglected by practicing engineers.

It is recalled that most of the reliable closure relations are flow regime dependent. This requires that the flow regime criteria to be accurate, however, even the experimental determinations of the flow regimes have been quite subjective until recently. Several researchers (Jones and Zuber, 1975; Hubbard and Dukler, 1966; Vince and Lahey, 1980) have proposed objective methods based on statistical analyses of data. However, the concept of the flow regimes is based on global observations and phenomena, therefore, their identification based on one parameter, such as the void fraction or pressure drop, can still be subjective. This is particularly true in a geometrically complicated system. Often the flow regime depends on the size of the volume used to define them. Another problem is that flow regimes are sensitive to the initial and boundary conditions. Thus the use of the flow regime transition criteria developed for a fully developed flow can often be wrong. Furthermore, the flow regime itself is not a mathematical entity in the two-phase formulation. One of the important parameters characterizing the interfacial geometry or two-phase flow regimes is the interfacial area concentration. It has been pointed out that the interfacial area concentration is particularly important in a two-fuid model formulation (Ishii, 1975; Ishii and Mishima, 1981). However, sufficient research has not been carried out in relation to the interfacial area concentration.

Until quite recently, various mixture models have been used extensively because of its simplicity, both in terms of the field equations and necessary closure relations. In view of the limited data base presently available and difficulties associated with detailed measurements in two-phase flow, an advanced mixture model such as the drift-flux model (Ishii and Zuber, 1979; Chawla and Ishii, 1978; Ohkawa and Lahey, 1983) is probably the most reliable and accurate analytical tool for standard two-phase flow problems. However, more detailed formulation based on a two-fluid model (Ishii, 1975; Vernier and Delhaye, 1968; Kocamustafaogullari, 1971; Drew and Wood, 1985; Nigmatulin, 1979; Ishii and Mishima, 1981; Chawla and Ishii, 1980; Bouré, 1985) is also available. The two-fluid is formulated by considering each phase separately in terms of two sets of field equations for mass, momentum, and energy for each phase in the averaged fields. In the macroscopic field equations, the interaction terms that couple the transport of mass, momentum, and energy of each phase appear after proper averging. These terms represent the transfer of mass, momentum, and energy at a phase interface and satisfy the macroscopic jump conditions as additional balance equations at the interface. Because the number of necessary closure relations are considerably higher for the two-fluid model than for the drift-flux model, much more detailed experimental data are necessary to develop satisfactory closure relations. However, sufficient experimental information for developing reliable closure relations for the two-fluid is often nost available. The present state of the art in two-phase flow measurements implies that considerable uncertainties exist in the closure relations for the two-fluid model. In spite of these shortcomings, however, there are a number of situations where the two-fluid model is more advantageous than the drift-flux model. For example, the two-fluid model may be necessary to model a transient where two phases are weakly coupled both dynamically and thermally. In general, the two-fluid model is more advantageous than the mixture model for;

1) mechanically non-equilibrium flows such as developing flows, entrance flows, sudden mixing, co-current to counter current flow transition, and flow regime transitions;

2) flow phenomena where dynamical interactions between phases are important such as interfacial wave and compressible wave propagation pheomena; and

3) thermally non-equilibrium phenomena.

In the present paper, the two-fluid model formulation, field equations and closure relations are discussed in detail,

2. LOCAL INSTANT FORMULATION

In a microscopic sense, a two-phase flow system consists of a number of single phase regions bounded by moving interfaces. Therefore, mathematically it is possible to formulate a two-phase flow problem by considering a field which is subdivided into single phase regions with moving boundaries. The standard differential balance equations hold for each subregion. At the phase interfaces, appropriate jump and boundary conditions are used to match the solutions of these differential equations. This type of formulation is a direct extension of the single phase flow formulation and a two-phase flow problem is solved for the local instant variables. This is called a local instant formulation of two-phase flow in order to distinguish it from macroscopic formulations.

Such a formulation would result in a moving multi-boundary problem with positions of interfaces being unknown. For most two-phase problems, the mathematical difficulties encountered by using this local instant formulation can be considerable and in many cases they are insurmountable. Thus, in general, direct applications of

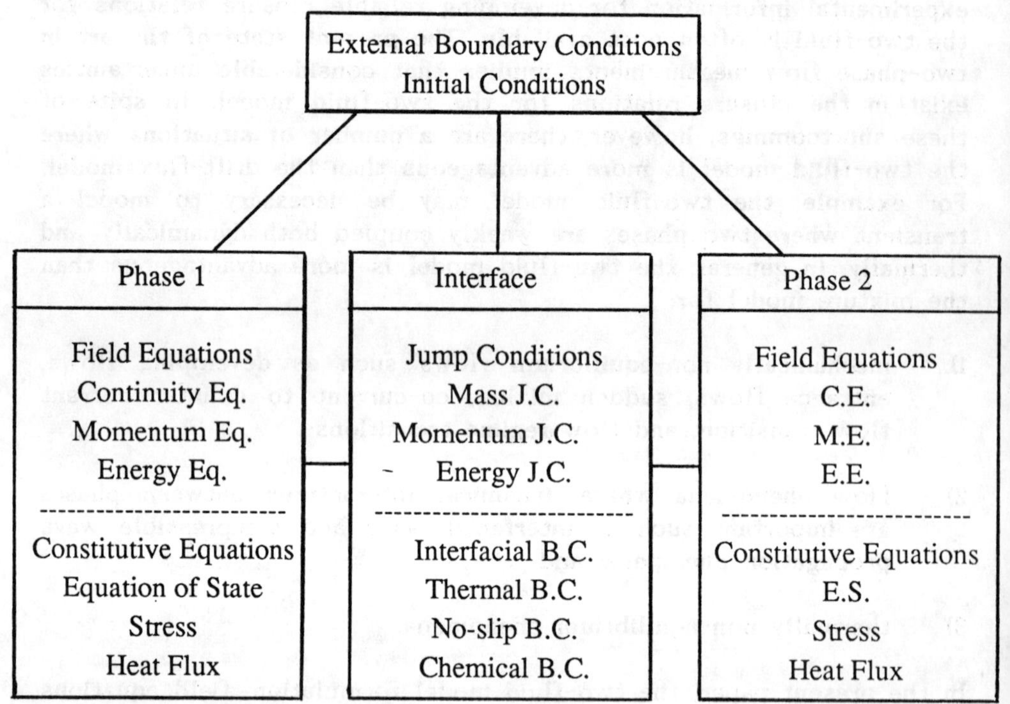

Figure 1 Local instant formulation.

the local instant formulation to practical two-phase flow problems have a very limited value. However, there are two fundamentally important benefits in the local instant formulation. First, it is the fundamental base of all the macroscopic two-phase flow models obtained from averaging. Second, it can be applied to study basic phenomena in very simple geometry. The analytical results obtained there can be a very useful guide for understanding various two-phase flow transfer mechanisms and developing general closure relations. The local instant formulation is illustrated in Fig. 1.

A detailed discussion of the local instant formulation has been given by Ishii (1975). In what follows, a brief summary is given. The differential form of the general balance equation is

$$\frac{\partial \rho_k \psi_k}{\partial t} + \nabla \cdot (\mathbf{v}_k \rho_k \psi_k) = - \nabla \cdot \mathbf{J}_k + \rho_k \varphi_k \tag{1}$$

where ρ is the density, $\boldsymbol{v}$ is the velocity, and ψ is the quantity conserved, $\boldsymbol{J}$ is the efflux and φ is the body source. The subscript k denotes k-phase. This balance equation holds within the subregion occupied by the k-phase.

The general balance equation can be applied for the conservation of mass, momentum, and energy separately. Then the continuity equation is obtained by setting $\psi_k = 1$, $J_k = 0$, and $\varphi_k = 0$ in Eq. (1).

$$\frac{\partial \rho_k}{\partial t} + \nabla \cdot (\rho_k \boldsymbol{v}_k) = 0 \tag{2}$$

The momentum equation can be obtained by introducing the surface stress tensor $\boldsymbol{T}_k = \tau_k - p_k \boldsymbol{I}$ and setting $\psi_k = \boldsymbol{v}_k$, $\boldsymbol{J}_k = -\boldsymbol{T}_k$ and $\varphi_k = \boldsymbol{g}$ in Eq. (1). Here τ is the shear stress, p the pressure, and $\boldsymbol{g}$ the gravity. Thus

$$\frac{\partial \rho_k \boldsymbol{v}_k}{\partial t} + \nabla \cdot (\rho_k \boldsymbol{v}_k \boldsymbol{v}_k) = - \nabla p_k + \nabla \cdot \tau_k + \rho_k \boldsymbol{g} \tag{3}$$

The conservation of energy is given for the total energy, thus by setting $\psi_k = U_k + v_k^2/2$, $J_k = q_k - \boldsymbol{T}_k \cdot \boldsymbol{v}_k$ and $\varphi_k = \boldsymbol{g} \cdot \boldsymbol{v}_k + Q_k/\rho_k$ in Eq. (1), the total energy balance can be obtained. Here u is the internal energy,. q the heat flux, and Q_k is the internal heat generation. Thus

$$\frac{\partial \rho_k (u_k + v_k^2/2)}{\partial t} + \nabla \cdot \left[\rho_k \, (u_k + v_k^2/2) \boldsymbol{v}_k\right]$$

$$= - \nabla \cdot \boldsymbol{q}_k + \nabla \cdot (\boldsymbol{T}_k \cdot \boldsymbol{v}_k) + \rho_k \boldsymbol{g} \cdot \boldsymbol{v}_k + Q_k \tag{4}$$

The conservation of the angular momentum for non-polar fluids implies that

$$\boldsymbol{T}_k = \boldsymbol{T}_k^{+}$$

where $\boldsymbol{T}_k^{+}$ denotes the transposed stress tensor.

The continuity, momentum, and energy equations describe the basic conservation laws for non-polar fluids. In order to solve these balance equations, it is necessary to specify the fluxes and body sources by various constitutive relations as well as to give the fundamental equation of state. It is also noted that there are several important transformations of the above equations. For two-phase flow analysis, the enthalpy energy equation is particularly useful. It can be obtained by eliminating the kinetic energy terms in the total energy equation in view of the momentum equation then introducing the enthalpy defined by

$$i_k = u_k + p_k/\rho_k \tag{6}$$

Thus,

$$\frac{\partial \rho_k i_k}{\partial t} + \nabla \cdot (\rho_k i_k \mathbf{v}_k) = - \nabla \cdot \mathbf{q}_k + \tau_k : \nabla \mathbf{v}_k + Q_k + \left[\frac{\partial p_k}{\partial t} + \mathbf{v}_k \cdot \nabla p_k\right] \tag{7}$$

The standard differential balance equations given above can be applied to each phase with different constitutive relations up to a phase interface. At the interface a particular form of the balance equations should be used in order to account for the special characteristic of the interface. By considering the interface as a singular surface across which the density, energy, and velocity suffer a jump discontinuity, the jump conditions can be derived (Ishii, 1975).

The mass jump condition is given by

$$\sum_{k=1}^{2} \rho_k \mathbf{n}_k \cdot (\mathbf{v}_k - \mathbf{v}_i) = 0 \tag{8}$$

where $\mathbf{n}_k$ is the outward unit normal vector for the k-phase and $\mathbf{v}_i$ is the interface velocity with the tangential velocity equal to the fluid particle velocity. This equation states that the mass fluxes balance at the interface and there is no mass capacity at the interface.

The momentum jump condition is given by

$$\sum_{k=1}^{2} \left\{\rho_k \boldsymbol{n}_k \cdot (\boldsymbol{v}_k - \boldsymbol{v}_i)\boldsymbol{v}_k - \boldsymbol{n}_k \cdot \boldsymbol{T}_k\right\} + (\boldsymbol{t}_\alpha a^{\alpha\beta}\sigma)_{,\beta} = 0 \tag{9}$$

where σ, $\boldsymbol{t}_\alpha$, $a^{\alpha\beta}$, and $(\ )_{,\beta}$ is the surface tension, hybrid tensor between the space and surface coordinates, metric tensor of the surface, and surface covariant derivative, respectively. This equation expresses the balance between the momentum fluxes from the bulk fluids and the surface force.

The energy jump condition is given by

$$- T_i \left\{\frac{d_s}{dt}\left(\frac{d\sigma}{dT_i}\right) + \frac{d\sigma}{dT_i} \nabla_s \cdot \boldsymbol{v}_i\right\} = (\boldsymbol{t}_\alpha a^{\alpha\beta}\sigma)_{,\beta} \cdot \boldsymbol{v}_i + \sum_{k=1}^{2} \left\{\rho_k \boldsymbol{n}_k \cdot (\boldsymbol{v}_k - \boldsymbol{v}_i)\left(u_k + \frac{v_k^2}{2}\right) + \boldsymbol{n}_k \cdot (- \boldsymbol{T}_k \cdot \boldsymbol{v}_k + \boldsymbol{q}_k)\right\} \tag{10}$$

Here only the surface energy associated with the surface tension is considered and the thermal equation of state for the interface $\sigma = \sigma(T_i)$ has been used. The subscript s denotes the differential operators associated with the surface coordinates. In general, the surface energy and the surface work terms are small compared with the latent heat transfer. Thus it may be simplified to

$$\sum_{k=1}^{2} \left\{\rho_k \boldsymbol{n}_k \cdot (\boldsymbol{v}_k - \boldsymbol{v}_i)\left(u_k + \frac{v_k^2}{2}\right) + \boldsymbol{n}_k \cdot (- \boldsymbol{T}_k \cdot \boldsymbol{v}_k + \boldsymbol{q}_k)\right\} = 0 \tag{11}$$

It is important to note that the above three jump conditions do not constitute sufficient matching conditions at the interface. Consequently, they should be supplemented by various boundary conditions which restrict the kinematic, dynamic, and thermal relations between phases. Details of the derivation of these have been given by Ishii (1975). The simplest conditions can be obtained by assuming the entropy production at the interface to be zero. These are given below. For viscous fluids, the no-slip condition is obtained. This is given as

$$v_{1t} = v_{2t} = v_{it} \tag{12}$$

where t denotes a tangential component at the interface. The thermal boundary condition is given by the thermal equilibrium condition at the interface, thus

$$T_{1i} = T_{2i} = T_i \tag{13}$$

which can also be considered as the definition of T_i. The chemical boundary condition for the interface with mass transfer is given by

$$g_1 - g_2 = \frac{\{\rho_1 n_i \cdot (v_1 - v_2)\}^2}{2} \left(\frac{1}{\rho_1^2} - \frac{1}{\rho_2^2}\right) + \left(\frac{\tau_{nn1}}{\rho_1} - \frac{\tau_{nn2}}{\rho_2}\right) \tag{14}$$

where g_k is the Gibbs free energy and τ_{nnk} is the normal stress at the interface. The last equation restricts the pressure at each side of the interface with respect to the saturation pressure (Ishii, 1975). The right-hand side of Eq. (14) may be neglected except for cases with very violent phase changes. Then the result can be simplified to the following condition.

$$p_{ki} - p^{sat}(T_i) = 2\, H_{fg}\, \sigma \frac{\rho_k}{|\Delta\rho|} \tag{15}$$

where p_{ki} is the interfacial pressure for the k-phase, $p_{sat}(T_i)$ is the saturation pressure corresponding to T_i, H_{fg} is the surface mean curvature looking from liquid to vapor, and $\Delta\rho$ is the density difference. Since the mean curvature of the liquid phase H_{fg} is positive for a droplet and negative for a bubble, the phase pressures at the interface are both over the saturation pressure for a droplet flow, and they are both under it for a bubbly flow. For many practical problems of two-phase flow, even Eq. (15) may be too complicated to use. In such cases, the complete thermostatic equilibrium condition given below is used.

$$p_{gi} = p_{fi} = p_{sat}(T_i) \tag{16}$$

3. FOUNDATION OF TWO-PHASE FLOW MODELS

It can be said that overwhelming difficulties encountered in the local instant formulation of two-phase flows stem from:

1) existence of the deformable moving interfaces with their motion being unknown, and

2) existence of the fluctuations of variables due to turbulences and to the motions of the interfaces.

The first effect causes complicated coupling between the field equations of each phase and the interfacial conditions, whereas the second effect inevitably introduces statistical characteristics originating from the instability of the Navier-Stokes equations and of the interfacial waves. Since these difficulties exist in almost all two-phase flow systems, an application of the local instant formulation to obtain a solution is severely limited.

As most two-phase flows observed in practical engineering systems have extremely complicated interfacial geometries and motions, it is not possible to solve for local instant motions of the fluid particles. Such microscopic details of the fluid motions and of other variables are rarely needed for an engineering problem, but rather macroscopic aspects of the flow are much more important.

In the past, various approaches have been taken (Ishii and Mishima, 1984), however, the method based on the averaging of the local instant formulation gives the most rational approach. By averaging the fields, the details of the local instant formulation are partially eliminated. Only the macroscopic and statistical effects appear in the averaged equations. Instead of considering local instant transfers at an interface, collective interactions between phases are modeled in a macroscopic two-phase flow formulation.

The averaging procedure is basically low pass filtering which eliminates unwanted high frequency signals from local instant fluctuations. A collective contribution of interfaces which affect the macroscopic processes should be taken into account in the formulation through various closure relations.

Averaging procedures which can be applied to two-phase systems can be classified into three main groups based on the basic physical concepts used to formulate dynamical problems (Ishii, 1975, 1981). These are the Eulerian, Lagrangian, and Boltzmann statistical averages. They can be further divided into sub-groups based on a variable with which averaging is defined. The above classifications and the definitions of various averaging are given below:

i) *Eulerian Average - Eulerian Mean Value*

Function: $F = F(t,x)$

Time mean value (temporal): $\frac{1}{\Delta t} \int_{\Delta t} F(t,x)dt$

Spatial mean value: $\frac{1}{\Delta R} \int_{\Delta R} F(t,x)dR(x)$

Volume: $\frac{1}{\Delta v} \int_{\Delta v} F(t,x)dv$

Area: $\frac{1}{\Delta a} \int_{\Delta a} F(t,x)da$

Line: $\frac{1}{\Delta c} \int_{\Delta c} F(t,x)dc$

Statistical Mean value: $\frac{1}{N} \sum_{n=1}^{N} F_n(t,x)$

Mixed mean value: combination of above operations

ii) *Lagrangian Average - Lagrangian Mean Value*

Function: $F = F(t,X);\ X = X(x,t)$

Time mean value (temporal): $\frac{1}{\Delta t} \int_{\Delta t} F(t,X)dt$

Statistical mean value: $\frac{1}{N} \sum_{n=1}^{N} F_n(t,x)$

iii) *Boltzmann Statistical Average*

Molecular density function: $f = f(x,\xi,t)$
(particle)

Transport properties: $\psi(t,x) = \frac{\int \psi(\xi) f d\xi}{\int f d\xi}$

Here it is noted that x and X are the spatial and the material coordinates, respectively, whereas ξ is the phase velocity, i.e., the velocity of individual particles. Furthermore, the true time or statistical averaging is defined by taking the limit $\Delta t \rightarrow \infty$ or $N \rightarrow \infty$, which is only conceptual and cannot be realized.

The basic concept underlining the Eulerian averaging is the time-space description of physical phenomena. In a so-called Eulerian description, the time and space coordinates are taken as independent variables and various dependent variables express their changes with respect to these coordinates. Since the standard field equations of continuum mechanics adapt to this description, it is natural to consider averaging with respect to these independent variables, i.e., the time and the spatial averaging. furthermore, these averaging are basically an integral operator, thus they have an effect of smoothing out instant or local variations within a domain of integration.

The Lagrangian mean values are directly related to the Lagrangian description of mechanics. As the particle coordinate X displaces the spatial variable x of the Eulerian description, this averaging is naturally fitted to a study of the dynamics of a particle. Thus if our interest is on the behavior of an individual particle rather than on the collective mechanics of a group of particles, the Lagrangian average is useful.

In order to study two-phase flow systems, many of the above averaging methods have been used by various researchers. The applications of averaging can be divided into two main categories:

1) to define properties and then to correlate experimental data, and

2) to obtain macroscopic field equations and closure relations which can be used to predict macroscopic two-phase phenomena.

The most elementary use is to define mean properties and motions which include various kinds of concentrations, density, velocity, and energy of each phase or of a mixture. These properly defined mean values then can be used for various experimental purposes and for developments of empirical correlations. The choice of averaging and instrumentations are closely coupled, since in general measured quantities represent some kinds of mean values themselves.

The most important and widely used group is the Eulerian averaging. This can be explained by the traditional and effective use of the Eulerian description of thermo-fluid mechanics and its close relation to experimental observations and instrumentations. In particular, the Eulerian time (or statistical) and Eulerian area averaging procedures form the theoretical foundation of rigorous formulations of two-phase flow problems (Ishii, 1975, 1977; Vernier and Delhaye, 1968; Kocamustafaogullari, 1971; Drew and Wood, 1985; Nigmatulin, 1984; Zuber and Findlay, 1965; Ishii and Zuber, 1979; Chawla and Ishii, 1978, 1980; Bouré, 1985; Teleto, 1945; Frankl, 1953; Kalinin, 1970; Sha and Slattery, 1980; Ishii.

The Eulerian time (or statistical) averaging, which has been used extensively for analyzing a single-phase turbulent flow, is the foundation of a fully three-dimensional two-phase flow model (Ishii, 1975; Vernier and Delhaye, 1968; Ishii and Mishima, 1984). The model is useful where a quasi-one-dimensional approach is too inaccurate; for example, flows in an entrance region or in a vessel. Except these special cases, the Eulerian area averaging over a cross section of a flow area gives the most practical two-phase flow formulation (Zuber and Findlay, 1965; Ishii, 1977; Chawla and Ishii, 1980). By area averaging, the resulting field equations become quasi-one-dimensional. Therefore, the information on changes of variables in the direction normal to the main flow within a flow channel is basically lost. Thus the transfer of momentum and energy between the wall and fluid should be specified by empirical correlations or by simplified models. When the area averaging is applied to a subchannel within a rod bundle, a cross flow momentum equation (Chawla and Ishii, 1978, 1980; Rowe, 1967; Amsden and Harlow, 1978; Thurgood, 1978) should be included in the formulation. Furthermore, it is noted that generally the average of a product of variables is different from a product of the averages. This difference is defined as a covariance and it plays a very important role in the averaged formulation. For example, in the time averaged equations (Ishii, 1975) the well-known turbulent transfer terms, such as the Reynolds stress, correspond to the covariance arising from the convective fluxes. In the area averaged formulation, the important effect of phase and velocity distributions on the relative velocity was pointed out and several efforts to model this effect have been made (Zuber and Findlay, 1965; Ishii, 1977; Bankoff, 1960; Ishii and Mishima, 1984). The volume averaging or mixed averaging have also been applied to develop a three-dimensional model (Drew, 1971). It is also noted that most of the earlier work on two-phase flow models (Brodkey, 1967; Hewitt and Hall-Taylor, 1970; Kutateladze, 1952; Levy, 1960; Isbin et al. 1954) have a form of the one-dimensional formulation although they are obtained from a simple control volume analysis rather than from rigorous averaging methods.

Besides these different approaches for developing a model, there are two fundamentally different formulations of the macroscopic field equations for two-phase flow systems: namely, the two-fluid model and the mixture model. The two-fluid model (Ishii, 1975; Vernier and Delhaye, 1968; Kocamustafaogullari, 1971) is formulated by considering each phase separately in terms of two sets of conservation equations governing the balance of mass, momentum, and energy of each phase. Previous studies have indicated that unless phase momentum interaction terms are accurately modeled, the advantage of the two-fluid model over the mixture model disappears and in certain cases numerical instabilities result. At least two transient forces, i.e., the virtual mass and Basset forces, exist in addition to the drag and interfacial shear forces. Since these derivative terms can influence the stability of the system of differential equations, careful evaluation of these terms is necessary.

4. THREE-DIMENSIONAL TWO-FLUID MODEL

4.1 Averaged Properties

The importance of the Eulerian time averaging in studying a single-phase turbulent flow is well-known. Since the most useful information in analyzing standard fluid flow systems are the time mean values rather than the local instant responses of the fluid particles, its use both in experimental and analytical purposes is indispensable in turbulent flow studies. For example, mean velocity, temperature, and pressure or the heat transfer coefficient and the friction factor are the important mean values routinely required in standard problems. Furthermore, commonly used experimental methods and measurements are well suited for the application of time averages. Thus a single-phase turbulent flow has been studied in great depth by using the time or statistical averaged field equations with the closure relations expressed by mean values. Although these equations which are based on time averages may not give answers to the fundamental origin, structure, and transport mechanisms of turbulent flow, their applications to engineering systems are widely accepted as almost a unique means of solving problems.

In discussing the importance of the Eulerian time averaging applied to two-phase flow, it is recalled that in many two-phase flows the local instant fluctuations of variables are caused not only by turbulences but also by rapidly moving and deforming interfaces. Confronted by these complicated flow fluctuations, the solutions from the local instant formulation are inaccessible, thus in order to derive appropriate field and closure relations, it is necessary to apply some averaging procedure to the original formulation. In view of the preceding discussion on the importance and usefulness of the time average in a single phase turbulent flow analysis, it is both natural and logical that we also apply the time averaging to the two-phase flows to develop a fully three-dimensional formulation.

It is expected that the averaged field equations distinctly exhibit macroscopic phenomena of the system from hopelessly complicated interfacial and turbulent fluctuations, since they enter the formulation only statistically. There are two notable consequences in the time averaging when it is applied to two-phase flow:

1) smoothing out of turbulent fluctuations in a similar sense as in single phase flow, and

2) bringing two phases, which are alternately occupying a volume element, into two continua simultaneously existing at same point with a properly defined statistical expectation for each phase.

Furthermore, it can be recognized that the closure relations appearing in the averaged formulation should be expressed through the time mean values. These closure relations can be developed from simple mechanistic or phenomenological modeling of two-phase transport phenomena together with various experimental data which are commonly expressed by the mean values.

In what follows, the three-dimensional two-phase flow formulation based on time averaging is discussed. First it is assumed that the occupant of any particular point is alternating randomly between the two phases and that the time averaged functions are sufficiently smooth in the new coordinates, i.e., the time coordinate has a minimum scale of Δt denoting the time interval of averaging under which a time differential operator has no physical importance.

The averaging is carried out for the fluid properties and field equations in order to treat them as a mixture of continua. First, by taking a fixed time interval Δt of the averaging it is assumed that it be large enough to smooth out the local variations of properties yet small compared to the macroscopic time constant of the unsteadiness of the bulk fluid. This assumption is identical to that made in analyzing turbulent single-phase flow. After choosing any particular reference point and time $(\boldsymbol{x}_o, t_o)$, there are definite times, t_1, t_2, ..., t_j, ... referring to the interfaces which pass the point x_o from time $(t_o - \Delta t/2)$ to $(t_o + \Delta t/2)$. In averaging various properties and field equations, it becomes necessary to distinguish three states (phase 1, phase 2, and interface), therefore state density functions M_1, M_2, M_s which alternate between 1 and 0 are assigned. These variables become 1 only when the state denoted by a subscript occupies a point, otherwise they are zero. By denoting the total time interval by $[\Delta t]$ and the total time interval without $[\Delta t]_s$ by $[\Delta t]_T$ the local void fraction of the kth-phase is defined by

$$\alpha_k \equiv \frac{1}{\Delta t}\int_{[\Delta t]_T} M_k dt = \frac{\Delta t_k}{\Delta t} \tag{17}$$

Here Δt_k is the time occupied by the kth-phase and $[\Delta t]_s$ is the time interval associated with an interface.

Similarly, various time averages and weighted mean values are defined as follows.

$$\bar{F} = \frac{1}{\Delta t}\int_{[\Delta t]_T} F\, dt \qquad \text{: mixture average} \tag{18}$$

$$\bar{F}_k = \frac{1}{\Delta t}\int_{[\Delta t]_k} F_k\, dt \qquad \text{: kth-phase average} \tag{19}$$

$$\psi_m = \bar{\psi} = \overline{\rho\psi}/\bar{\rho} \qquad \text{: mixture weighted mean value} \tag{20}$$

$$\rho_m = \bar{\rho} \qquad \text{: mixture density} \tag{21}$$

$$\bar{\bar{F}}_k = \bar{F}_k/\alpha_k \qquad \text{: kth-phase mean value} \tag{22}$$

$$\bar{\psi}_k = \overline{\rho_k\psi_k}/\bar{\rho}_k \qquad \text{: kth-phase weighted mean value} \tag{23}$$

From the above definitions of mean values, it is straightforward to obtain the following important relations.

1) Density:

$$\rho_m = \sum_{k=1}^{2} \alpha_k \bar{\bar{\rho}}_k \tag{24}$$

2) Internal Energy: $u_m = \sum_k \alpha_k \bar{\bar{\rho}}_k \bar{u}_k / \rho_m$ (25)

3) Pressure: $p_m = \sum_k \alpha_k \bar{\bar{p}}_k$ (26)

4) Enthalpy: $i_m = \sum_k \alpha_k \bar{\bar{\rho}}_k \bar{i}_k / \rho_m$ (27)

Then it can be shown that

$$i_m = u_m + p_m/\rho_m \quad \text{and} \quad \bar{i}_k = \bar{u}_k + \bar{\bar{p}}_k/\bar{\bar{\rho}}_k \tag{28}$$

5) Velocity: It is necessary to introduce two properly defined mean velocity fields in the formulation in order to take into account the effects of the relative motions between phases. However, there are several velocity fields which are useful in analyzing various aspects of a two-phase flow problem. A selection of velocity fields for a particular problem depends on characteristics and nature of the flow as well as on the forms of available closure relations. In what follows, the velocity fields which are important in studying various aspects of two-phase flow systems are presented.

The mixture center of mass velocity is defined by

$$v_m \equiv \sum_k \alpha_k \bar{\bar{\rho}}_k \bar{v}_k / \rho_m \tag{29}$$

The relative velocity between the phases is given by

$$v_r \equiv \bar{v}_1 - \bar{v}_2 \tag{30}$$

And the volumetric fluxes of each phase are given by

$$j_k \equiv \alpha_k \bar{v}_k \tag{31}$$

which can be considered as the velocity when one of the phases superficially occupies the entire time interval Δt with the total amount of the flow fixed. Accordingly the mixture volumetric flux, i.e., the velocity of the center of volume, is defined by

$$j = \sum_k j_k = \sum_k \alpha_k \bar{v}_k \tag{32}$$

If the relative velocity between the phases exists, the velocities v_m and j are not equal because of the difference of the densities of the phases.

The diffusion velocities, i.e., the velocities with respect to the mass center of the mixture, are defined by

$$V_{km} \equiv \bar{v}_k - v_m \tag{33}$$

which are frequently used in the analyses of heterogeneous chemically reacting systems.

In a two-phase flow system, the drift velocities of phases, i.e., the velocities with respect to the center of volume, are important, because the closure relations for these velocities in the drift-flux formulation are relatively simple for several flow regimes (Wallis, 1969; Zuber and Findlay, 1965; Ishii, 1977). By definition,

$$V_{kj} \equiv \bar{v}_k - j \tag{34}$$

Several important relations between the above velocities can be obtained directly from the definitions. For example,

$$V_{1j} = \alpha_2 v_r \quad \text{and} \quad V_{2j} = -\alpha_1 v_r \tag{35}$$

and j and v_m are related by

$$j = v_m - \alpha_1 \alpha_2 \frac{(\bar{\bar{\rho}}_1 - \bar{\bar{\rho}}_2)}{\rho_m} v_r \tag{36}$$

$$= v_m - \alpha_1 \frac{(\bar{\bar{\rho}}_1 - \bar{\bar{\rho}}_2)}{\rho_m} V_{1j} \quad (37)$$

4.2 Basic Concept of Two-Fluid Model

The two-fluid model is formulated by considering each phase separately. Thus, the model is expressed in terms of two sets of conservation equations governing the balance of mass, momentum, and energy in each phase. However, since the averaged fields of one phase are not independent of the other phase, the interaction terms appear in these balance equations. The terms denoted by Γ_k, M_k, and E_k are the mass, momentum, and energy transfers to the kth-phase from the interfaces. As these quantities also should obey the balance laws at the interfaces, interfacial transfer conditions from the local jump conditions can be obtained. Consequently, six differential field equations with three interfacial transfer conditions govern the macroscopic two-phase flow systems. In the two-fluid model formulation, the transfer processes of each phase are expressed by their own balance equations, thus it is anticipated that the model can predict more detailed changes and phase interactions than the mixture (drift flux) model. However, this means that the two-fluid model is more complicated not only in terms of the number of field equations involved but also in terms of the necessary constitutive and closure equations. It is evident that these equations should be accurate for the model to be useful. This is particularly true with the interaction terms Γ_k, M_k, and E_k, since without these interfacial exchanges in the field equations, the two phases are essentially independent. These interaction terms decide the degree of coupling between the phases, thus the transfer processes in each phase are greatly influenced by these terms.

The real importance of the two-fluid model is that it can take into account the dynamic interactions between phases. This is accomplished by using the momentum equations for each phase and two independent velocity fields in the formulation. Thus it is expected that two-fluid model can be useful to the analyses of wave propagations and of the flow regime changes. Particularly if the two phases are weakly coupled such that the waves can propagate in each phase with different velocities, the two-fluid model should be used to study these phenomena. The second application to the analysis of flow regimes can be explained by the fact that the changes of flow regimes occur mainly due to the instabilities of the interfaces. The stability and dynamics of the interfaces are governed by the interfacial transfer of momentum.

The three-dimensional two-fluid model can be obtained from the time averaging of the local instant formulation (Ishii, 1975). The other averaging methods such as the combination of volume-time averaging are also possible, however, the resulting form of the field equations do not differ significantly. The differences exist in terms of the physical meaning of the averaged quantities. This is quite an important point to remember in using different formulations.

The following diagram summarizes the derivation of the three-dimensional two-fluid model. It is noted that the two-fluid model formulation retains the basic structure of the local instant formulation. However, because of the time averaging, the phases are considered to co-exist at each point and time such that two sets of the field equations can be solved simultaneously over the entire flow field.

4.3 Two-fluid Model Field Equations

The two-fluid model is characterized by two independent velocity fields which specify the motions of each phase. The most natural choice of velocity fields is obviously the mass weighted mean phase velocity of each phase. The two-fluid model developed by Ishii (1975) is presented below. The suitable form of the continuity equations to be used in the model should be

$$\frac{\partial \alpha_k \bar{\bar{\rho}}_k}{\partial t} + \nabla \cdot (\alpha_k \bar{\bar{\rho}}_k \bar{v}_k) = \Gamma_k \tag{38}$$

with the interfacial mass transfer condition

$$\sum_{k=1}^{2} \Gamma_k = 0 \tag{39}$$

Here

$$\Gamma_k \equiv -\frac{1}{\Delta t} \sum_j \left\{ \frac{1}{v_{ni}} \, n_k \cdot \rho_k (v_k - v_i) \right\} \tag{40}$$

The summation j is taken over the number of interfaces passing the point within Δt. Thus the term denoted by Γ_k represents the rate of

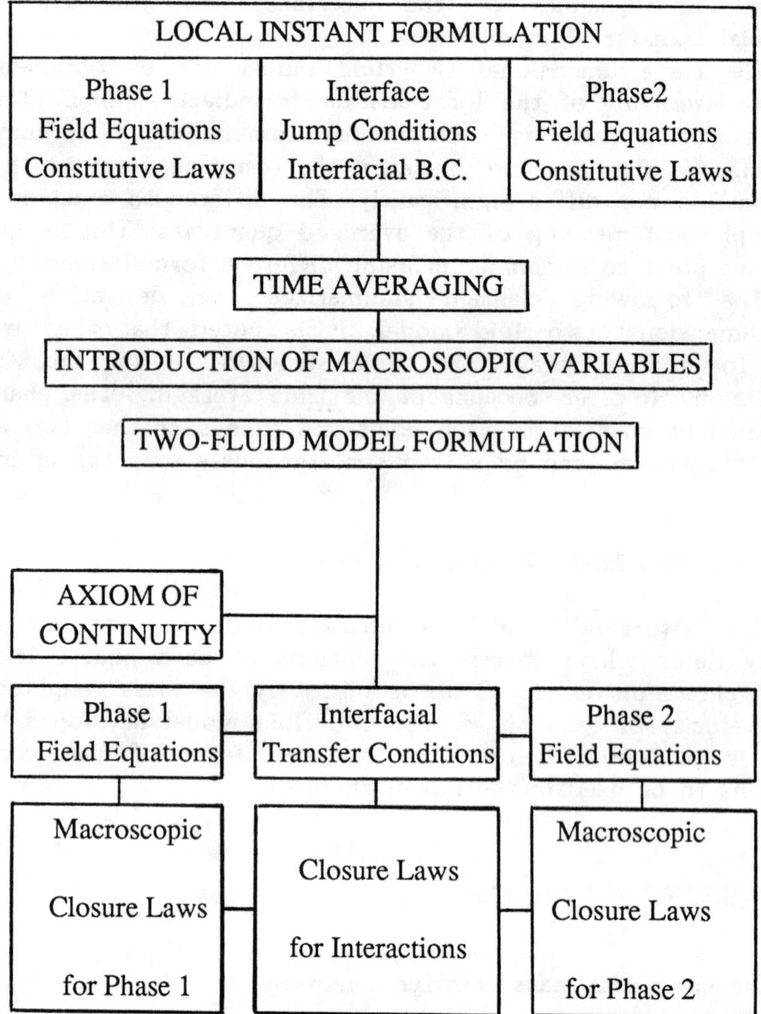

mass transfer to the k-phase from the phase changes at the interfaces. It appears because the local continuity equation has been integrated in time to obtain a macroscopic field equation. These two equations are the general statements of the conservation of mass in the macroscopic field, thus they hold regardless of the mechanism of phase changes.

In the two-fluid model formulation, the conservation of momentum is expressed by two momentum equations with the interfacial momentum transfer condition. The appropriate field equations should be expressed by the center of mass velocity of each phase $\bar{v}_k$, thus the two momentum equations are given by

$$\frac{\partial \alpha_k \bar{\bar{\rho}} \bar{v}_k}{\partial t} + \nabla \cdot (\alpha_k \bar{\bar{\rho}}_k \bar{v}_k \bar{v}_k) = - \nabla(\alpha_k \bar{\bar{p}}_k) + \nabla \cdot [\alpha_k (\bar{\bar{\tau}}_k + \tau_k^T)]$$

$$+ \alpha_k \bar{\bar{\rho}}_k g + M_k \tag{41}$$

The interfacial transfer condition has the form

$$\sum_{k=1}^{2} M_k = M_m \tag{42}$$

The exact forms of M_k, M_m, and τ_k^T are given by

$$M_k = - \frac{1}{\Delta t} \sum_j \left\{ \frac{1}{v_{ni}} n_k \cdot [\rho_k (v_k - v_i) v_k - T_k] \right\} \tag{43}$$

$$M_m = \frac{1}{\Delta t} \sum_j \left\{ \frac{1}{v_{ni}} (t_\alpha a^{\alpha\beta} \sigma)_{,\beta} \right\} \tag{44}$$

$$\tau_k^T = - \overline{\overline{\rho_k v'_k v'_k}} \tag{45}$$

here the interfacial momentum source term M_m arises from the surface tension force (Ishii, 1975).

The most fundamental form of the conservation of energy is expressed by the balance of the total energy. For the two fluid model formulation, two total energy equations with the interfacial energy transfer condition are obtained. Therefore,

$$\frac{\partial}{\partial t} \left[\alpha_k \bar{\bar{\rho}}_k \left(\bar{\varepsilon}_k + \frac{\bar{v}_k^2}{2} \right) \right] + \nabla \cdot \left[\alpha_k \bar{\bar{\rho}}_k \left(\bar{\varepsilon}_k + \frac{\bar{v}_k^2}{2} \right) \bar{v}_k \right] = - \nabla \cdot [\alpha_k (q_k \;\; q_k^T)]$$

$$- \nabla \cdot (\alpha_k \rho_k \bar{v}_k) + \nabla \cdot \left[\alpha_k (\bar{\bar{\tau}}_k + \tau_k^T) \, \bar{v}_k \right] + \alpha_k \rho_k g \cdot \bar{v}_k + E_k \tag{46}$$

The interfacial transfer condition couples the energy transport process of two phases, and given by

$$\sum_{k=1}^{2} E_k = E_m \tag{47}$$

Here the interfacial energy source term E_m arises from the energy associated with the surface tension (Ishii, 1975). If the effects of the surface tension on the interfacial transfer can be neglected then M_m and E_m can be neglected., In order to write the total energy equation in the form given above, the effective energy $\bar{\varepsilon}_k$ has been defined as

$$\bar{\varepsilon}_k \equiv \bar{u}_k + \frac{\overline{(v_k')^2}}{2} \tag{48}$$

Thus the effective energy consists of the standard thermal energy and turbulent kinetic energy.

For two-phase flow analyses, the enthalpy energy equation is important and it is frequently used to solve problems involving phase changes. The enthalpy energy equation can be obtained by eliminating the kinetic energy associated with the mean velocity between the momentum and energy equation. Thus by introducing the effective enthalpy $\bar{H}_k$ given by

$$\bar{H}_k = \bar{i}_k + \frac{\overline{(v_k')^2}}{2} = \bar{\varepsilon}_k + \bar{\bar{p}}_k/\bar{\bar{\rho}}_k \tag{49}$$

the enthalpy energy equation becomes

$$\frac{\partial \alpha_k \bar{\bar{\rho}}_k \bar{H}_k}{\partial t} + \nabla \cdot \left(\alpha_k \bar{\bar{\rho}}_k \bar{H}_k \bar{v}_k\right) = - \nabla \cdot (\alpha_k \bar{\bar{q}}_k) - \nabla \cdot (\alpha_k q_k^T)$$

$$+ \frac{D_k}{Dt} (\alpha_k \bar{\bar{p}}_k) + \alpha_k (\bar{\bar{\tau}}_k + \tau_k^T) : \nabla \bar{v}_k$$

$$+ \left(\frac{v_k^2}{2} \Gamma_k - M_k \cdot \bar{v}_k + E_k\right) \tag{50}$$

The exact forms of q_k^T and E_k are given by

$$q_k^T = \overline{\overline{\rho_k \left(u_k' + \frac{v_k'^2}{2}\right) v_k'}} - \overline{\overline{T_k \cdot v_k'}} \tag{51}$$

$$E_k = -\frac{1}{\Delta t} \sum_j \left\{ \frac{1}{v_{ni}} n_k \cdot \left[\rho_k (v_k - v_i) \left(u_k + \frac{v_k^2}{2} \right) - T_k \cdot v_k + q_k \right] \right\} \tag{52}$$

It is noted here that the definition of the turbulent heat flux has been changed slightly from the one originally given by Ishii (1975) to include the turbulent work term in it. The last three terms of the right-hand side of Eq. (50) given by

$$\Lambda_k \equiv \frac{\bar{v}_k^2}{2} \Gamma_k - M_k \cdot \bar{v}_k + E_k \tag{53}$$

can be considered as the interfacial transfer contribution to the phase enthalpy change. In many practical heat transfer problems, it is convenient to use the thermal energy equation instead of the total energy equation. This is particularly true for low-speed two-phase flows with heat additions where the phase change and thermal terms dominate the energy transfer.

4.4 Simplified Form for Interfacial Transfer

The interfacial momentum transfer term given by Eq. (43) can be divided into several physically meaningful components (Ishii, 1975; Ishii and Mishima, 1981). Thus

$$M_k = \Gamma_k v_{ki} + \bar{\bar{p}}_{ki} \nabla \alpha_k + M_{ik} - \nabla \alpha_k \cdot \tau_i \tag{51}$$

The generalized interfacial force M_{ik} represents the combined statistical effects of the pressure and shear stress deviations from the interfacial mean values. Therefore, M_{ik} should include the skin and form drag forces, transient forces, lift force, and particle collision force. The macroscopic momentum jump condition becomes

$$\sum_k \mathbf{M}_{ik} = 0 \tag{52}$$

Furthermore, from the normal component of the jump condition

$$\bar{\bar{p}}_{gi} - \bar{\bar{p}}_{fi} \simeq -2\, H_{fg}\, \sigma \tag{53}$$

where the effect of the normal stress and thrust force due to phase changes has been neglected (Ishii, 1975).

The contribution of the interfacial transfer to the enthalpy change given by Λ_k consists of the thermal and mechanical energy transfer terms. It can be expressed as

$$\Lambda_k = \left(\Gamma_k \bar{H}_{ki} + a_i q''_{ki}\right) - \bar{\bar{p}}_{ki} \frac{D_k \alpha_k}{Dt} + (\mathbf{M}_{ik} - \nabla\alpha_k \cdot \tau_i) \cdot (\mathbf{v}_{ki} - \mathbf{v}_k) + W^T_{ki} \tag{54}$$

where W^T_{ki} is the turbulent flux of work due to the interfacial force. a_i is the interfacial area concentration. If the mechanical terms can be neglected in comparison with the thermal terms, then

$$\Lambda_k \simeq \left(\Gamma_k \bar{H}_{ki} + a_i q''_{ki}\right) \tag{55}$$

And the macroscopic energy jump condition becomes

$$\sum_k \left(\Gamma_k \bar{H}_{ki} + a_i q''_{ki}\right) \simeq 0 \tag{56}$$

5. INTERFACIAL TRANSFER

5.1 Interfacial Area Concentration

One of the important parameters for the interfacial transfer is the interfacial area concentration (Ishii, 1975; Ishii and Mishima, 1981; Kataoka et al. 1984, 1985, 1986). For a time averaged system it is given by

$$a_i = \frac{1}{L_s} = \frac{1}{\Delta t} \sum_j \left(\frac{1}{v_{ni}} \right)_j \tag{57}$$

L_s denotes the interfacial length scale. The interfacial area concentration also has the physical meaning of the interfacial area per unit volume, thus

$$a_i = \frac{\text{interfacial area}}{\text{mixture volume}} \tag{58}$$

From the field equations for the two-fluid model it can be seen that several interfacial transfer terms appear on the right-hand side of the equations. As these interfacial transfer terms also should obey the balance laws at an interface, the interfacial transfer conditions could be obtained from an average of the local jump conditions (Ishii, 1975). In summary, they are given by

$$\left.\begin{aligned} &\sum_k \Gamma_k = 0 \\ &\sum_k M_{ik} = 0 \\ &\sum (\Gamma_k \bar{H}_{ki} + q''_{ki}\, a_i) \simeq 0 \end{aligned}\right\} \tag{59}$$

The interfacial force term may be expressed as a linear combination of several important physical forces (Ishii and Mishima, 1984; Ishii and Zuber, 1979; Ishii and Chawla, 1979; Zuber, 1964, Drew and Lahey, 1979) as

$$M_{ik} = M_{ik}^D + M_{ik}^V + M_{ik}^B + M_{ik}^L + M_{ik}^C \tag{60}$$

where the superscripts D, V, B, L, and C stand for the steady-state drag, virtual mass, Basset, lift, and collision forces, respectively.

The standard drag force acting on a particle under steady-state conditions can be given in terms of the drag coefficient C_D based on the relative velocity as

$$F_D = -\frac{1}{2} C_D \rho_c v_r |v_r| A_d \tag{61}$$

where A_d is the projected area of a typical particle and v_r is the relative velocity given by $v_r = v_d - v_c$; here the subscripts d and c denote the dispersed and continuous phases. Hence the portion of M_{id} represented by the drag force becomes

$$M_{id}^D = \alpha_d F_D / B_d = -\left(\alpha_d \frac{A_d}{B_d}\right) \frac{C_D}{2} \rho_c v_r |v_r| \tag{62}$$

Here B_d is the volume of a typical particle.

Now for a dispersed flow system, several important length scales (Ishii and Mishima 1981) can be defined as follows:

Sauter mean radius: $$r_{sm} \equiv \frac{3B_d}{A_i} \tag{63}$$

Drag radius: $$r_D \equiv \frac{3B_d}{4A_d} \tag{64}$$

Equivalent radius: $$r_v \equiv \left(\frac{3}{4\pi} B_d\right)^{1/3} \tag{65}$$

Surface radius: $$r_s \equiv \left(\frac{A_i}{r4}\right)^{1/2} \tag{66}$$

where A_i is the surface area of a typical particle. For spherical particles the above defined radii are all equivalent. For a system with a distribution of particle sizes it is necessary to take corresponding mean values for these parameters (Kataoka et al. 1983). The number density N_d of a dispersed phase is given by

$$N_d = \alpha_d / B_d \tag{67}$$

and the interfacial area concentration a_i by

$$a_i = N_d A_i \tag{68}$$

Using the above definitions the area concentration can be expressed in a number of forms. For example,

$$a_i = \frac{3\alpha_d}{r_{sm}} = \frac{3\alpha_d}{r_v}\left(\frac{r_s}{r_v}\right)^2 = \frac{3\alpha_d}{r_D}\left(\frac{r_D}{r_{sm}}\right) \tag{69}$$

Therefore, the drag term becomes

$$\alpha_d F_D/B_d = -a_i\left[\frac{C_D}{4}\left(\frac{r_{sm}}{r_D}\right)\frac{\rho_c v_r|v_r|}{2}\right] \tag{70}$$

This indicates that the drag force per unit volume of mixture is proportional to the interfacial area concentration and drag coefficient. The ratio of the Sauter mean radius to the drag radius appears on the right-hand side of Eq. (70) as a shape factor. It is noted that for non-spherical particles or for a system with a particle size distribution, the shape factor may not be one. However, for a spherical particle, the shape factor, r_{sm}/r_D, is one regardless of the particle size distribution as shown by Ishii and Kataoka (1982) and Kataoka et al. (1983).

By introducing the mean mass transfer per unit area defined by

$$\Gamma_k \equiv a_i m_k \tag{71}$$

the interfacial energy transfer term can be rewritten as

$$\Gamma_k \bar{H}_{ki} + \frac{q''_{ki}}{L_s} = a_i (m_k \bar{H}_{ki} + q''_{ki}) \tag{72}$$

The above shows that the interfacial area concentration is the first order geometrical effect of the interface on the interfacial transfer. The interfacial transfer terms are now expressed as a product of the interfacial area and driving force as,

$$\textit{Interfacial Transfer} \sim a_i \times \textit{Driving Force} \tag{73}$$

It is essential to make a conceptual distinction between the effects of these two parameters (Ishii and Mishima, 1981; Kataoka et al. 1986). The interfacial transfer of mass, momentum, and energy increases with an increasing interfacial area concentration toward the mechanical and thermal equilibrium.

It should be emphasized here that the form given by Eq. (73) is not arbitrarily chosen. A detailed examination of the interfacial transfer in terms of the local instant variables in the averaged balance equations shows that it contains the weighting function proportional to the interfacial area concentration (Ishii, 1975). This can be easily seen from the definition of the interfacial area concentration given by Eq. (57) and the exact forms of the interfacial transfer terms given by Eqs. (40), (43), and (49). Thus expressing the interfacial transfer as a product of the interfacial area concentration and driving force is also mathematically proper.

The geometrical relations developed for the interfacial area show the importance of the existence and size of small fluid particles for all flow regimes (Ishii and Mishima, 1981). This implies that the number density of small bubbles are important in bubbly, slug, and churn turbulent flows, whereas in annular-mist or dispersed droplet flows, the droplet size distribution and amount of entrainment are important in determining the interfacial area.

The cross-sectionally area averaged interfacial area of a slug or churn-turbulent flow (Ishii and Mishima, 1981) can be given approximately by

$$a_i = \frac{4.5}{D}\,\frac{\langle\alpha\rangle - \alpha_{gs}}{1 - \alpha_{gs}} + \frac{3\alpha_{gs}}{r_{sm}}\,\frac{1 - \langle\alpha\rangle}{1 - \alpha_{gs}} \tag{74}$$

where $\langle\alpha\rangle$. α_{gs}, and D are the average overall void fraction, average void fraction in the liquid slug section, and hydraulic diameter, respectively. Hence, the first term on the right-hand side of Eq. (74) represents the contribution of large slug or churn flow bubbles, whereas the second term represents the effect of small bubbles in the liquid slug section. For a bubbly flow $\langle\alpha\rangle = \alpha_{gs}$ and the first term drops out, and one obtains the standard form of $a_i = 3\langle\alpha\rangle/r_{sm}$.

For annular or annular mist flow the interfacial area is given approximately by

$$a_i = \frac{4C_{an}}{D} \sqrt{\frac{<\alpha>}{1 - \alpha_{fd}}} + \frac{<\alpha>}{1 - \alpha_{fd}} \frac{3\alpha_{fd}}{r_{sm}} \tag{75}$$

where $<\alpha>$, α_{fd}, and C_{an} denote the mean void fraction, the liquid-drop fraction in the gas core alone, and the roughness parameter due to waves in the film $(C_{an} \geq 1)$, respectively. The first term on the right-hand side represents the interfacial area from the wavy interface, whereas the second term is the contribution due to entrained droplets.

The inception of droplet entrainment occurs (Ishii and Kataoka, 1982; Ishii and Grolmes, 1975) when the following inequality is satisfied.

$$\frac{\mu_f j_g}{\sigma} \sqrt{\frac{\rho_g}{\rho_f}} \geq 11.78\, N_\mu^{0.8}\, Re_f^{-1/3} \quad \text{for } Re_f < 1635$$

$$\geq N_\mu^{0.8} \quad \text{for } Re_f \geq 1635 \tag{76}$$

where the viscosity number is defined by $N_\mu = \mu_f/(\rho_f\sigma \sqrt{\sigma/g\Delta\rho})^{1/2}$ and the liquid Reynolds number by $Re_f = \rho_f j_f D/\mu_f$.

The entrainment fraction in annular flow is defined by

$$E \equiv j_{fe}/j_f \tag{77}$$

whre j_{fe} and j_f are the droplet volumetric flux and total liquid volumetric flux, respectively. Then the droplet void fraction in the core alone can be given approximately by

$$\alpha_{fd} \simeq E \frac{j_f}{j_g} \tag{78}$$

if the relative velocity between the gas core flow and droplets is small. By knowing the value of E the interfacial area can be calculated from Eqs. (75) and (78).

A correlation for the amount of entrained liquid has been developed from a simple model (Ishii and Mishima, 1982). The entrained fraction reaches a quasi-equilibrium value at a point far removed from the tube entrance when the entrainment and deposition processes attain a quasi-equilibrium condition. In this region, E can be given by the following correlation

$$E = \tanh(7.25 \times 10^{-7}\ We^{1.25}\ Re_f^{0.25}) \tag{79}$$

where the entrainment Weber number We and total liquid Reynolds number is defined by

$$We \equiv \frac{\rho_g j_g^2 D}{\sigma}\left(\frac{\Delta\rho}{\rho_f}\right)^{1/3}$$

and (80)

$$Re_f \equiv \rho_f j_f D/\mu_f$$

The above correlation has been compared to many experimental data for air-water systems in the ranges of $1 < p < 4$ bar, $0.95 < D < 3.2$ cm, $370 < Re_f < 6400$ and $j_g < 100$ m/s, and the result has been satisfactory (Ishii and Mishima, 1982).

Correlations for the droplet size and size distribution for annular-mist flow have been developed by Kataoka et al. (1983) and Ishii and Kataoka (1982). According to their correlations,

$$r_{sm} = 0.00394\ \frac{\sigma}{\rho_g j_g^2}\ Re_g^{2/3}\left(\frac{\rho_g}{\rho_f}\right)^{-1/3}\left(\frac{\mu_g}{\mu_f}\right)^{2/3} \tag{81}$$

where r_{sm} is the Sauter mean radius. Re_g is the gas Reynolds number defined by $Re_g \equiv \rho_g j_g D/\mu_g$. This correlation indicates the existence of a large number of small droplets. These droplets are much smaller than the size predicted by the standard Weber number criterion based on the relative velocity between the droplet and gas.

5.2 Drag Force

The drag force for a single particle depends on the nature of a particle, i.e., solid particle, drop or bubble, shape, size and flow around them. The drag force in a multiparticle system is further complicated by the existence of other particles. The multiparticle drag correlation has been developed by considering the similarity between single and multiparticle systems (Ishii and Zuber, 1979). In the analysis, following flow regimes were considered.

Solid-particle system { Viscous regime
Newton's regime

Fluid-particle system { Viscous regime (Undistorted-particle regime)
Distorted-particle regime
Churn-turbulent-flow regime
Slug-flow regime

In the viscous regime, distortions of fluid particles are negligible. Therefore, for this regime, solid and fluid particle systems are considered together. The other flow regimes are analyzed separately because of significant differences in the flow around the particles and the motions of the interfaces.

Table 1 summarizes the drag coefficient in various flow regimes. Since these correlations are obtained from the assumed similarity hypothesis, their validity should be tested against experimental data. In a multiparticle system, a drag force cannot be measured directly under normal conditions. Therefore, considerable care should be taken in these comparisons.

The comparison of the theoretical predictions to over 1000 experimental data in terms of the relative velocity indicated that satisfactory agreements could be obtained at wide ranges of the particle concentration and Reynolds number. For spherical solid-particle systems the data from the Stokes regime up to the Newton's regime within the concentration range of 0-0.55 were examined. For fluid-particle systems the distorted-particle and churn-turbulent regimes were extensively studied because of their practical importance. The applicability of the correlation at up to the highest concentration range for spherical-solid-particle systems was accomplished by the introduction of the maximum packing in the mixture-viscosity relation. This was a definite improvement over the existing correlations. It is also noted that the model was sufficient up to the foam or dense-packing regime with the concentration ranging from 0.5 to 0.95 for both bubbly and droplet flows. These comparisons indicated that the postulated drag-similarity law based on the mixture-viscosity concept was appropriate. Therefore, the drag law governing the motions of bubbles, drops and particles in various dispersed two-phase flows can be explained by a unified and consistent model developed under the study.

The above correlations show that the drag coefficient increases with the increasing volumetric concentration of fluid particles in the viscous and distorted particle regimes. This is due to the effect of the other particles on the drag force through the deformation of the flow field. Since the particles are more rigid than the fluid against deformations, the particles will impose a system of forces that will react upon the fluid. As a result of additional stresses, the original particle sees an increase in the resistance to its motion which appears to it as an increase in the drag coefficient. On the other hand, in the churn-turbulent or slug flow regimes, the drag coefficient decreases as the particle volumetric concentration increases. Therefore, the effect of α_d on C_D is opposite of that in the viscous or distorted particle regimes. This peculiar trend can be explained by the entrainment of other particles in a wake of larger particles. This entrainment promotes the channeling of the dispersed phase without increasing the drag force. As the volumetric concentration increases, the interaction among fluid particles increases in the direction of reducing the drag force.

The above correlations for the drag coefficient for multiparticle systems have been developed from the steady-state and adiabatic formulation. It was postulated that the transient effect on the momentum-exchange term could be taken into account by an essentially linear closure relation in the form given by Eq. (60). Therefore, it was indirectly assumed that the standard-drag coefficient developed in the analysis could also be used under transient conditions. The additional interfacial forces due to the inertia effect and development of a boundary layer in transient flow are considered separately.

The phase change at the particle surface contributes to the interfacial momentum transfer in two different ways. There is a direct effect of momentum carried by the mass undergoing phase change, as can be seen from the momentum equation and Eq. (51). The standard drag may be affected by the changing particle size or shape due to phase change and modification of the boundary layer around the particles by additional mass flux normal to the surface. However, the effects of heat transfer and phase changes are considered as secondary. These effects appear only indirectly through the local variables such as the void fraction, particle sizes and component velocities. To assess the significance of the phase change effect on the drag coefficient, further experimental and analytical studies are required.

Table 1

Local Drag Coefficients in Multiparticle System (Ishii and Mishima, 1981)

	Fluid particle system			Solid particle system
	Bubble in liquid	Drop in liquid	Drop in gas	
Viscosity model	$\frac{\mu_m}{\mu_c} = \left(1 - \frac{\alpha_d}{\alpha_{dm}}\right)^{-2.5\alpha_{dm}\mu^*}$, $\mu^* \equiv \frac{\mu_d + 0.4\mu_c}{\mu_d + \mu_c}$			
Max. packing α_{dm}	~1	~1	0.62~	~0.62
μ^*	0.4	~0.7	1	1
μ_m/μ_c	$(1-\alpha_d)^{-1}$	$(1-\alpha_d)^{-1.75}$	$(1-\alpha_d)^{-2.5}$	$(1 - \frac{\alpha_d}{0.62})^{-1.55}$
Stokes regime C_D	$C_D = 24/N_{Re}$, where $N_{Re} = 2r_d\rho_c v_r/\mu_m$			
Viscous regime C_D	$C_D = 24(1 + 0.1\,N_{Re}^{0.75})/N_{Re}$			
Newton's regime C_D	--		$C_D = 0.45\left\{\frac{1 + 17.67[f(\alpha_d)]^{6/7}}{18.67f(\alpha_d)}\right\}^2$	
Distorted particle regime C_D	$C_D = \frac{4}{3} r_d \sqrt{\frac{g\Delta\rho}{\sigma}} \left\{\frac{1 + 17.67[f(\alpha_d)]^{6/7}}{18.67f(\alpha_d)}\right\}^2$			where
	$f(\alpha_d) = (1-\alpha_d)^{1.5}$,	$= (1-\alpha_d)^{2.25}$,	$= (1-\alpha_d)^3$	$f(\alpha_d) = \sqrt{1 - \alpha_d}\left(\frac{\mu_c}{\mu_m}\right)$
Churn-turbulent flow regime C_D	$C_D = \frac{3}{8}(1-\alpha_d)^2$			
Slug flow C_D	$C_D = 9.8\,(1-\alpha_d)^3$			

5.3 Lift Force

The drag force plays a very important role in determining the overall axial relative motion of the phases and cross sectionally averaged void fraction. In contrast the lift force may be important in determining the transverse void distribution, since it can generate a force perpendicular to the main relative motion. The lift force is generated by particle rotation and shear flow in the continuous phase. Saffman (1965) showed that at very low Reynolds number the lift force in a simple shear flow is given by

$$F_L = 6.46\, \mu_c\, r_d\, v_r \sqrt{\frac{r_d^2\, \rho_c}{\mu_c} \frac{dv_c}{dy}} \tag{82}$$

for a spherical particle. At higher Reynolds number, a significant lift force can be generated by rotation of a particle. However, the lift force due to a shear flow at moderate to higher Reynolds numbers is not well understood even for a single particle case (Clift et al. 1978). The formation of the wake and unsteady particle motion complicate both experimental measurements and theoretical analyses of the lift force. Therefore, the exact nature of the life force in two-phase flow is not known, where the effects of the multiparticle and overlapping wakes further complicate the situation. Drew and Lahey (1979) postulated an interfacial force given by

$$F_L = C_L\, (\bar{v}_c - \bar{v}_d) \cdot (\nabla \bar{v}_c)^+ \tag{83}$$

It can be seen that this force acts in the direction perpendicular to the relative velocity. Thus it is considered to be a lift force. The force due to the relative motion and deformation is then generalized as

$$\frac{M_{ic}^L B_d}{\alpha_d} = F_L = C_L\, (\bar{v}_c - \bar{v}_d) \cdot \left[\nabla \bar{v}_c + (\nabla \bar{v}_c)^+\right] \tag{84}$$

5.4 Virtual Mass Force

The generalized interfacial force has been modeled as a linear combination of several forces in Eq. (60). One of the transient forces is known as the virtual mass force. This force arises due to the force required to accelerate the apparent mass of surrounding phase when the relative velocity changes. This increased resistance can be easily understood by the change in the total kinetic energy of fluid around an accelerating particle. The well-known result for an accelerating sphere in inviscid fluid (Milne-Thomson, 1968) is given by

$$\frac{M_{id}^{V} B_d}{\alpha_d} = F_V = -\frac{B_d}{2}\rho_c \frac{d\mathbf{v}_r}{dt} \tag{85}$$

Here F_V is the virtual mass force acting on a particle.

The correction factor for the virtual mass force at higher Reynolds number was obtained empirically (Odar and Hamilton, 1964), thus

$$F_V = -\left\{2.1 - \frac{0.132\, C_a^2}{1 + 0.12\, C_a^2}\right\} \frac{B_d}{2}\rho_c \frac{d\mathbf{v}_r}{dt} \tag{86}$$

where the acceleration modulus C_a is defined by

$$C_a = \frac{2r_d}{v_r^2}\frac{dv_r}{dt} \tag{87}$$

The above expression reduces to the theoretical result given by Eq. (85) as $\mathbf{v}_r \rightarrow 0$. At the other limit of $\mathbf{v}_r \rightarrow \infty$, the virtual mass force becomes approximately twice that given by Eq. (85). However, it is important to note that the correction factor is bounded by 1 and 2.1.

The effect of the virtual mass force is much larger for a bubble in liquid than for a drop in gas due to the higher liquid density surrounding the bubble. The above result was extended to two-phase flow by including the effect of the particle concentration by Zuber (1964). This was done by using the existing solution for an accelerating sphere inside another sphere. It is given by

$$\alpha_d F_V / B_d = -\frac{1}{2}\alpha_d \frac{1 + 2\alpha_d}{1 - \alpha_d}\rho_c \frac{D_d}{Dt}(\bar{\mathbf{v}}_d - \bar{\mathbf{v}}_c) \tag{88}$$

Drew and Lahey (1979) studied a necessary condition for the requirement of the frame-indifference of the constitutive equations. They determined that the virtual mass force F_V should satisfy

$$F_v \sim \left[\frac{D_c \bar{v}_d}{Dt} - \frac{D_d \bar{v}_c}{Dt} - (1 - \lambda)\, v_r \cdot \nabla v_r\right] \tag{89}$$

However, currently Drew and Lahey (1987) consider that it is possible to partition the lift and virtual mass forces so that both are individually objective, but this is not necessary nor natural. The objectivity should be met for the combined force of the virtual mass and lift forces. This was demonstrated by the analytical result for a single bubble accelerating in a shear flow.

In view of Zuber's study (1964) on the effect of concentration and the above frame-indifference condition, a new form for F_v is proposed (Ishii and Mishima 1984). Due to the acceleration of the particles relative to the fluid, the acceleration drag arises. This should be proportional to the induced mass $\rho_c B_d^*$ and the frame-indifferent relative-acceleration vector. Hence,

$$F_v = -\rho_c B_d^* \left[\frac{D_c \bar{v}_d}{Dt} - \frac{D_d \bar{v}_c}{Dt} - (1 - \lambda)\, v_r \cdot \nabla v_r\right] \tag{90}$$

If it can be assumed that λ is constant, its value can be obtained by taking the limit $\alpha_d \to 0$ and comparing the result to Eq. (85). Then the value of λ should be 2.

The effect of the concentration on B_d^* can be taken into account by the method used by Zuber (1964). Thus, from the solution for the induced mass for a sphere moving within an outer sphere, B_d^* may, be approximated by

$$B_d^* \doteq \frac{1}{2} B_d \frac{1 + 2\alpha_d}{1 - \alpha_d} \tag{91}$$

where α_d is the volumetric fraction of the dispersed phase.

Under the assumption of λ = constant, the expression for the virtual mass force (Ishii and Mishima, 1981).

$$\alpha_d F_v / B_d = -\frac{1}{2} \alpha_d \frac{1 + 2\alpha_d}{1 - \alpha_d} \rho_c \left(\frac{D_d v_r}{Dt} - v_r \cdot \nabla \bar{v}_c\right) \tag{92}$$

The above equation indicates that the virtual mass force F_v per article increases considerably with increasing particle concentration.

This relation implies that the effect of concentration on dynamic coupling can be scaled by a factor of $(1 + 2\alpha_d)/(1 - \alpha_d)$. Mokeyev (1977) used an electrohydrodynamic analog method to determine the velocity potential through an electric field potential and obtained an empirical function $B_d^*/B_d = 0.5 + 2.1\alpha_d$. The theoretical result of Eq. (92) compares favorably with this correlation.

For a slug flow, the virtual mass force can be developed from a simple potential flow analysis (Ishii and Mishima, 1981). By considering a train of spherical-edged cylindrical bubbles of diameter D_b and length L_b, the virtual mass force becomes

$$\alpha_d F_v/B_d = -5\left[0.66\alpha_d + 0.27\left(\frac{L_b - D_b}{L}\right)\right]\rho_c\frac{\partial \mathbf{v}_r}{\partial t} \qquad (93)$$

where a simplification has been made from an approximation $\alpha_b \simeq 0.8$ where α_b is the void fraction at the slug bubble section. For a limiting case of a train of spherical bubbles, $L_b = D_b$, the above equation reduces to

$$\alpha_d F_v/B_d = -3.3\alpha_d\rho_c\frac{\partial \mathbf{v}_r}{\partial t} \qquad (94)$$

On the other hand, if $L_b >> D_b$, L_b/L can be approximated by α_d/α_b. The virtual mass force for a slug flow given by Eq. (92) is expressed in terms of the relative acceleration in the absence of a large convective acceleration. However, if the convective acceleration cannot be neglected, a special convective derivative in the form of Eq. (92) may be more appropriate. Thus for a general case,

$$\alpha_d F_v/B_d = -5\left[0.66\alpha_d + 0.27\left(\frac{L_b - D_b}{L}\right)\right]\rho_c\left(\frac{D_d \mathbf{v}_r}{Dt} - \mathbf{v}_r\cdot\nabla\bar{\mathbf{v}}\right)_c \qquad (95)$$

Due to a similarity in flow geometries, the virtual mass force for a churn turbulent flow may be approximated by the solution for a slug flow given above. In a liquid dispersed flow, the virtual mass force becomes considerably smaller than that in a vapor dispersed flow. This decrease is caused by a change in the continuous phase density to be used. By changing ρ_c from ρ_f to ρ_g, the virtual mass force for a droplet flow becomes insignificant compared with that in a bubbly flow. This also indicates that the virtual mass force should be reduced considerably in annular and annular dispeersed flow.

In most cases, the actual value of the virtual mass force is not large comparing with other terms such as the steady-state drag force. However, in rapid transient or high convection flow it can be important. Because of the time and space derivatives in the expression of the virtual mass force, the mathematical characteristic of the differential balance equations is significantly affected by this term (Drew et al. 1979). This implies that the numerical stability of the mathematical formulation may be strongly influenced by this term.

5.5 Basset Force

The Basset force is the effect of acceleration on the viscous drag and the boundary layer development. This term takes into account the history effect of the change of the viscous force, and for a simple case it is expressed by a time integral. For unsteady motion of a spherical particle at very low Reynolds number, the Basset history effect is given by (Basset, 1961; Landau and Lifshitz, 1959)

$$F_B = 6\, r_d^2\, \pi \rho_c \mu_c \int_{-\infty}^{t} \left(\frac{dv_r}{dt}\right) \frac{d\xi}{\sqrt{\pi \rho_c \mu_c (t-\xi)}} \tag{96}$$

The form of this integral results from the momentum diffusion from the particle.

The Basset force at higher Reynolds number was studied empirically by Odar and Hamilton (1964) for a rigid spherical particle. The result is given by

$$F_B = \left[0.48 + \frac{0.52\, C_a^3}{(1 + C_a)^3}\right] 6\, r_d^2 \sqrt{\pi \rho_c \mu_c} \int_{-\infty}^{t} \left(\frac{dv_r}{dt}\right) \frac{d\xi}{\sqrt{t - \xi}} \tag{97}$$

where C_a is the acceleration modulus defined by Eq. (87). It can be seen that as $v_r \to 0$ the acceleration modulus $C_a \to \infty$. Therefore, the above expression reduces to the creeping flow result given by Eq. (96). At the other limit of $C_a \to 0$ which can occur at very high velocities, the Basset force reduces to approximately one half of that given by Eq. (96). Hence the correction factor for higher Reynolds number flow is bounded by 0.48 and 1. This suggests that the form of the solution for low Reynolds numbers given by Eq. (96) basically describes the overall physics of this effect at all Reynolds numbers.

Based on the above observations, the theoretical result of Eq. (96) was extended to multiparticle two-phase flow (Zuber, 1964; Ishii and Mishima, 1981) as

$$M_{id}^{B} = \frac{\alpha_d F_B}{B_d} = \frac{9}{2}\frac{\alpha_d}{r_d}\sqrt{\frac{\rho_c \mu_m}{r_d}} \int_{-\infty}^{t} \frac{D_d \mathbf{v}_r}{D\xi} \frac{d\xi}{\sqrt{t-\xi}} \tag{98}$$

In place of the continuous phase velocity μ_c, the mixture viscosity μ_m is used in the above expression due to the following reason. When a single particle moves through a dispersed two-phase mixture, it imparts a motion to the continuous phase. However, as the fluid moves, its deformation causes the translational and rotational movements of other particles in the neighborhood. Since the particles are more rigid than the fluid against deformations, the original particle sees an increase in the resistance to its motion which appears as if it arises from an increased viscosity. Consequently, in analyzing the motion of the suspended particles, mixture viscosity should be used. The effect of the particle collisions may be indirectly reflected in the mixture viscosity through the void fraction. The viscosity model (Ishii and Zuber, 1979) is based on the extension of the linear correlation for the mixture viscosity for fluid particles along the power relation for solid particles based on the maximum packing α_{dm}. It is given by

$$\frac{\mu_m}{\mu_c} = \left(1 - \frac{\alpha_d}{\alpha_{dm}}\right)^{-2.5\alpha_{dm}(\mu_d + 0.4\mu_c)/(\mu_d+\mu_c)} \tag{99}$$

The maximum packing α_{dm} for solid particle systems ranges from 0.5 to 0.74. However, α_{dm} = 0.62 suffices for most of the practical cases. For a bubbly flow, theoretical α_{dm} can be much higher because of the deformation of bubbles. In the absence of turbulent motions and particle coalescences, the void fraction in a fluid particle system can be as high as 0.95. By taking α_{dm} to be unity, it is possible to include these foam or dense packing regimes in the analysis. Therefore, for fluid particle systems, we take α_{dm} = 1. The use of the mixture viscosity model is consistent with the correlation for the steady state drag force discussed in Section 5.2.

Drew and Wood (1985) proposed a model with a derivative which is objective and with a viscosity of a continuous phase. However, it is not certain whether this derivative describes the physics correctly.

5.6 Interfacial Energy Transfer

The macroscopic interfacial balance conditions for mass and energy are given by Eq. (59). They indicate that specifying the interfacial energy transfer at both sides of the interface is equivalent to specifying the closure relation for the mass transfer Γ_k. This simplification holds only when the mechanical terms associated with interfacial energy transfer can be neglected in comparison with the latent heat transfer. However, for most practical problems this relation greatly simplifies the closure relation development for interfacial energy transfer.

First, the enthalpy at the interface can be assumed to be that which corresponds to the saturation value. Thus,

$$\bar{H}_{ki} = H_k^{sat}(T_i) \tag{100}$$

The simplest closure relation for the heat flux at the interface is given in terms of a heat transfer coefficient as

$$q''_{ki} = h_{ki}(T_i - T_k) \tag{101}$$

There are several correlations for the heat transfer coefficient for a single rigid sphere [68] without phase changes. For a moderate Reynolds number range of 100 to 4000, it is given in terms of the Nusselt number as

$$Nu = \frac{2\pi_d h_{ci}}{k_c} = 1 + 0.677\, N_{Re}^{0.47} \tag{102}$$

where N_{Re} is the particle Reynolds number.

The bubbles which are growing or collapsing experience considerable changes in the thermal boundary layer thickness due to changing volume. In such a case the heat transfer coefficient for a solid sphere becomes quite inaccurate. There are a large number of publications focussing on the dynamics of a single bubble. For this type of problems, the local instant formulation has been used and detailed analyses have been carried out. However, there is a considerable gap between the understanding of the simple case and general two-phase flow. Yadigaroglu and Bensalem (1985) reviewed various approaches used for modeling thermal non-equilibrium and interfacial mass generation in two-phase flow. Some of the typical results are given below.

Ruckenstein (1959) obtained results for a bubble translating in a liquid as

$$h_{fi} = \rho_f C_{pf} \left[\frac{2}{\pi} \frac{v_r}{r_d} \frac{1}{\lambda_f} \right]^{1/2} \tag{102}$$

where λ_f is the thermal diffusivity. Wolfert, et al. (1978) modified the thermal diffusivity in the above expression by adding the turbulent diffusivity as

$$h_{fi} = \rho_f C_{pf} \left[\frac{2}{\pi} \frac{v_r}{r_d} \frac{1}{\lambda_f (1 + \varepsilon_f)} \right]^{1/2} \tag{103}$$

Houdayer, et al. (1978) expressed the interfacial heat transfer in terms of a relaxation time constant τ_k as

$$a_i q''_{ki} = \alpha_k \rho_k C_{pk} \frac{(T_k - T_i)}{\tau_k} \tag{104}$$

The time constant was correlated against experimental data for flashing flow. This approach, therefore does not reflect the effect of the interfacial area explicitly.

The interfacial area concentration plays a very important role for the interfacial heat transfer. However, satisfactory closure relations for this term do not exist for most two-phase flow regimes. This is particularly so for flow with a phase change. The modeling of the interfacial area concentration for boiling flow, for example, is complicated by the wall nucleation source. In order to understand the development of interfacial structures it is necessary to know the nucleation site density as well as the bubble departure frequency (Kocamustafaogullari and Ishii, 1983; Riznic and Ishii, 1986). The mechanistic approach first suggested by Zuber (1964) which takes into account these effects may be necessary for the fundamental understanding of highly non-equilibrium phenomena such as flashing.

6. STRESS IN AVERAGED FIELD

It was shown that the stress in the averaged field consists of two components. These are the average viscous stress and turbulent stress. Except in highly compressible flow situations, the stress tensor of the linearly viscous fluid of Navier-Stokes may be approximated by

$$\tau_k = \mu_k \left[\nabla v_k + (\nabla v_k)^+ \right] \tag{105}$$

Then the averaged viscous stress becomes (Ishii, 1975)

$$\bar{\bar{\tau}}_k = \mu_k \left[\nabla \bar{v}_k + (\nabla \bar{v}_k)^+ \right] + \frac{\mu_k}{\alpha_k} \frac{1}{\Delta t} \Sigma \frac{1}{v_{ni}} \left[n_k (v_k - \bar{v}_k) + (v_k - \bar{v}_k)\, n_k) \right] \tag{106}$$

Thus the viscous stress is associated with two different deformation tensors. The bulk deformation tensor is defined by

$$D_{ki} \equiv \frac{1}{2} \left[\nabla \bar{v}_k + (\nabla \bar{v}_k)^+ \right] \tag{107}$$

and the interfacial deformation tensor by

$$D_{ki} \equiv \frac{1}{2\alpha_k} \frac{1}{\Delta t} \Sigma \frac{1}{v_{ni}} \left[n_k (v_k - \bar{v}_k) + (v_k - \bar{v}_k)\, n_k \right] \tag{108}$$

By using these definitions

$$\bar{\bar{\tau}}_k = 2\mu_k \left[D_{kb} + D_{ki} \right] \tag{109}$$

This indicates that the effects of the interfacial motion appears in the average viscous stress tensor also.

It can be seen that the characteristic of the interfacial deformation tensor is complex due to various mechanisms affecting it, however, Ishii (1975) showed that under special conditions it may be approximated by a simple form. For a dispersed flow $D_{di} \simeq 0$. Furthermore, if the effect of phase change or turbulence on the interfacial velocity is not large, then

$$D_{ci} = - \frac{a_c^i}{2\alpha_c} \left\{ (\nabla \alpha_c)(\bar{v}_d - \bar{v}_c) + (\bar{v}_d - \bar{v}_c)(\nabla \alpha_c) \right\} \tag{110}$$

The constant a_c^i is the mobility of phase c and has the order of magnitude of 1.

The second component of the averaged stress tensor is the Reynolds stress τ_k^T arising from the turbulent fluctuations of the velocity. Since most two-phase flows of interest are turbulent, any realistic model must account for the turbulent structure of the flows. This is, however, very difficult to accomplish in view of the fact that the theoretical description of single-phase turbulence is far from complete. Only limited experimental data are available for the understanding of the nature of two-phase flow turbulence (Serizawa et al. 1975a,b,c; Sato et al. 1981; Lance and Bataille, 1982; Theofanous and Sullivan, 1982; Marie, 1983; Wang et al. 1985; Wang, 1985; Fabre, 1983).

By defining the total deformation tensor as

$$\boldsymbol{D}_k \equiv \boldsymbol{D}_{kb} + \boldsymbol{D}_{ki} \tag{111}$$

Ishii (1975) postulates that

$$\tau_k^T = a_{ko}\boldsymbol{I} + a_{k1}\boldsymbol{D}_k + a_{k2}\boldsymbol{D}_k \cdot \boldsymbol{D}_k \tag{112}$$

which is a generalization of Prandtl's mixing length theory. It is further simplified by considering only the second term.

$$\tau_k^T = 2\mu_k^T \boldsymbol{D}_k \tag{113}$$

where μ_k^T is the eddy viscosity and depends on the scalar invariants. The final result is given by

$$\tau_k^T = 2\mu_k^{T*} \rho_k l^2 \sqrt{2\boldsymbol{D}_k : \boldsymbol{D}_k}\, \boldsymbol{D}_k \tag{114}$$

μ_k^{T*} is a scalar nondimensional function dependent on the void fraction and interfacial area concentration. Drew and Lahey (1987) pointed out some deficiencies of the above approach and recommended

$$\tau_k^T = a_{ko}\boldsymbol{I} + a_{k1}\boldsymbol{D}_k + b_1 \bar{\boldsymbol{v}}_r \bar{\boldsymbol{v}}_r - b_2 \nabla\alpha_1 \nabla\alpha_2 \tag{115}$$

Some models for dispersed two-phase flow are based on the assumption that the dispersed flow turbulent structure is proportional to that of the continuous phase. This assumption may be more applicable to finely dispersed bubbly flow than for droplet or solid particulate flow (Nigmatulin, 1979; Drew and Wood, 1985; Wang, 1985). Then the continuous phase turbulent stress is given by three terms as

$$\tau_k^T = a_c \alpha_d \rho_c \bar{v}_r^2 I + b_c \alpha_d \rho_c \bar{v}_r \bar{v}_r + \mu_c^T [\nabla \bar{v}_c + \nabla \bar{v}_c^+] \tag{116}$$

For bubbly flow

$$\tau_g^T = \frac{\rho_g}{\rho_f} \tau_f^T \tag{117}$$

is often assumed. Drew and Lahey (1982) take a different approach. The ratio for two stresses is related to the void fraction and normal velocity correlations. Thus the empirical value of the ratio can be found if detailed turbulence measurements are made. Wang et al. (1985) modified the model to take into account the lift forces acting on the bubbles in a shear flow. For this purpose the lift force per unit volume M_{di}^L in the following form has been used.

$$M_{di}^L = A \rho_f \alpha (v_f - v_g) \frac{dv_f}{dr} \tag{118}$$

where A is an empirical parameter.

There are some indications that higher order analyses may be necessary to model the two-phase flow turbulence accurately even for bubbly flows (Marié, 1983; Lance et al. 1985). Marié (1983) used the turbulent energy equation for the core region together with the assumption of the logarithmic velocity profile near the wall. The solution is obtained by matching the velocity profiles at the boundary of core and wall turbulent regions. A recent work on the phase separation and turbulence by Lahey and Drew showed also that the mixing length type model may be insufficient and higher order models such as the K-ε model may be necessary to model the three dimensional turbulent effects accurately.

Detailed experimental studies under carefully controlled conditions have been carried out by Marié (1983, 1987), Lance and

Bataille, 1982, 1985). The main focus was on the turbulence in the liquid phase of a vertical, co-current air-water bubbly flow. It was found that the turbulent kinetic energy increases strongly with the void fraction. Two distinct regimes were found. At the low void fractions the hydrodynamic interactions between bubbles are negligible, whereas at relatively higher void fractions the bubbles transfer a great amount of kinetic energy to the liquid due to their mutual interactions. This type of fundamental experimental studies for two-phase flow turbulence is essential to a development of an accurate turbulence model.

Despite various analytical and experimental studies, the turbulent stresses in two-phase flow are not well understood even for simple cases such as bubbly flows. There are several reasons for this. First, the instrumentation of two-phase turbulence is quite difficult due to the intrusiveness of local measurements. Second, the void fraction profile is coupled with the turbulent structures. Thus, the two-phase turbulence problem should be solved simultaneously with the phase distribution problem.

7. SUMMARY AND CONCLUSIONS

The two-fluid model formulation of two-phase flow has been discussed. The emphasis here is on the three-dimensional formulation. The fully three-dimensional model has been obtained from a time averaging of the local instant formulation. For predicting two-phase flow transients or flow in complicated geometries, the two-fluid model is very important because of its detailed description of thermohydraulic phenomena and phase interactions. Furthermore, the model is most flexible because of the use of the separate conservation equations for each phase.

However, in the present state of the art, the formulation of the constitutive and closure relations for the two-fluid model is not very satisfactory. The major problems are the modeling of various interfacial transfer terms and turbulent transfer mechanisms. In view of this, the interfacial transfer terms for two-fluid models have been reviewed in detail here. The interfacial transfer of mass, momentum, and energy is proportional to the interfacial area and driving force. These two effects are considered separately. The geometrical effects on the interfacial transfers are taken into account primarily by the interfacial area concentration. Basically, four flow regimes; namely, dispersed (bubbly or droplet), slug, churn-turbulent, and annular flows have been modeled, and general characteristics of the prediction have been discussed. The model shows the importance of the existence and size of small fluid particles for all flow regimes.

Recommended correlations for the droplet size and amount of droplet entrainment in annular flow have been presented. The modeling of the interfacial area concentration as well as the experimental measurement techniques are quite inadequate at the present time. Since all the interfacial transfer terms involve the interfacial area concentration, the accurate modeling of the local interfacial area concentration is the first step to be taken for the development of a reliable two-fluid model.

The interfacial momentum interaction has been studied in terms of the standard-drag, lift, virtual mass, and Basset forces. Each of these are modeled separately. The standard drag correlations for dispersed, slug, and churn-turbulent flows were obtained from the postulated drag-similarity law based on the mixture viscosity. The correlations are shown to be applicable over wide ranges of the Reynolds number and the particle concentration. The virtual mass force is studied based on the requirement of the frame-indifference and simple potential flow analyses. The results for dispersed, slug, and churn-turbulent flows are also presented. It can be said that the lift, virtual mass, and Basset forces for simple multiparticle flows or more general two-phase flows are not firmly established. Further research in this area is recommended.

The other major area of interest is the modeling of turbulent transfers in two-phase flow. There are some indications that a simple mixing length type model may not be adequaste even for a simple bubbly flow. The two-phase flow turbulence problem is coupled with the phase distribution problem even in a steady-state, fully developed flow. Therefore, the turbulence problem cannot be separated from the interfacial transfer of momentum involving drag and lift forces. This makes the measurement and modeling of two-phase flow turbulence quite difficult. The intrusive measurement techniques face a problem of distortions of the interfacial geometry and particle motions in addition to the standard flow distortion problems. Since the momentum and energy transfers between the wall and mixture as well as between phases depend strongly on the nature of two-phase turbulence, it is necessary to face this difficult challenge and attack this problem with long time-scale research.

ACKNOWLEDGEMENTS

This work was performed under the auspices of the U.S. Nuclear Regulatory Commission and U.S. Department of Energy/BES. The author would like to express his appreciation to Drs. N. Zuber and R. Lee for valuable discussions on the subjectand to Dr. O. Mauley for the support of the work.

NOMENCLATURE

A_i	Interfacial area
A_d	Projected area of a particle
A	Coefficient for lift force
a	Area element
$a^{\alpha\beta}$	Surface metric tensor
a_i	Interfacial area concentration
a_c	Turbulent stress tensor coefficient
a_{k0}	Turbulent stress tensor coefficient
a_{k1}	Turbulent stress tensor coefficient
a_{k2}	Turbulent stress tensor coefficient
B_d	Volume of a typical particle
B_d^*	Induced mass volume
b_1	Turbulent stress tensor coefficient
b_c	Turbulent stress tensor coefficient
c	Line segment
C_D	Drag coefficient
C_L	Lift coefficient
C_{an}	Annular flow film roughness parameter
C_{pk}	Heat capacity of k-phase
C_a	Acceleration modulus, Eq. (87)
D	Hydraulic diameter

D_b	Slug bubble diameter
D_{kb}	Bulk deformation tensor
D_{ki}	Interface deformation tensor
D_k	Total deformation tensor
E_k	Interfacial energy transfer term
E_m	Interfacial energy source
E	Entrainment fraction
F	Function
F_D	Drag force on a particle
F_V	Virtual mass force on a particle
F_B	Basset force on a particle
F_L	Lift force on a particle
g	Gravity
g_k	Gibbs free energy
$\bar{H}_k$	Weighted mean enthalpy of *k*-phase plus turbulent kinetic energy
H_k^{sat}	Weighted mean saturation enthalpy of *k*-phase plus turbulent kinetic energy
H_{ki}	Weighted mean interface enthalpy of *k*-phase plus turbulent kinetic energy
h_{ci}	Continuous phase interfacial heat transfer coefficient
h_{fi}	Liquid phase interfacial heat transfer coefficient
i_k	Enthalpy of k-phase
i_m	Mixture enthalpy

J_k	Flux of k-phase
j_k	Volumetric flux of k-phase
j_{fe}	Entrained droplet volumetric flux
j	Total volumetric flux
L_s	Interfacial length scale
L_b	Slug bubble length
M_k	State density function of k-phase
M_s	State density function of interface
$\mathbf{M}_k$	Interfacial momentum transfer for k-phase
$\mathbf{M}_{ik}$	Generalized interfacial force for k-phase
$\mathbf{M}_{ik}^D$	Interfacial drag force per unit volume
$\mathbf{M}_{ik}^L$	Interfacial lift force per unit volume
$\mathbf{M}_{ik}^B$	Interfacial Basset force per unit volume
$\mathbf{M}_{ik}^C$	Interfacial collision force per unit volume
$\mathbf{M}_m$	Momentum source at interface
m_k	Interfacial mass transfer per unit area
$\mathbf{n}_k$	Unit outward normal vector of k-phase at interface
N_d	Number density of particle
N_μ	Liquid viscosity number
N_{Re}	Particle Reynolds number $(2r_d \rho_c v_r/\mu_m)$
Nu	Nusselt number
N	Number of samples

p_k	Pressure of k-phase
p_m	Mixture pressure
$\bar{\bar{p}}_{ki}$	Average interfacial pressure
p^{sat}	Saturation pressure
q_k	Heat flux for k-phase
Q_k	Heat source
q_k^T	Turbulent heat flux
q''_{ki}	Interfacial heat flux
r_d	Particle radius
r_v	Volume equivalent radius
r_s	Surface equivalent radius
r_D	Drag radius
r_{sm}	Sauter mean radius
Re_f	Liquid Reynolds number $(\rho_f j_f D/\mu_f)$
Re_g	Gas Reynolds number $(= \rho_g j_g D/\mu_g)$
t	Time
T_k	Temperature of k-phase
T_i	Temperature of interface
$\mathcal{T}_k$	Stress tensor of k-phase
$\mathcal{T}_k^+$	Transposed stress tensor of k-phase
t_β	Hybrid tensor between space and surface coordinates
u_k	Internal energy of k-phase

u_m	Mixture internal energy
$\bar{u}_k$	Weighted mean internal energy
$\mathbf{v}_k$	Velocity of k-phase
$\mathbf{v}_i$	Velocity of interface
$\mathbf{v}_r$	Relative velocity
v_{ni}	Normal velocity of interface
$\mathbf{v}_m$	Mixture velocity
V_{km}	Diffusion velocity
V_{kg}	Drift velocity
v	Volume element
We	Droplet Weber number
W_{ki}^T	Turbulent flux of work due to interfacial force
$\mathbf{x}$	Space coordinates
$\mathbf{X}$	Particle coordinates
y	Transverse direction

Greek Symbols

α	Void fraction
α_k	Time (void) fraction of k-phase
α_{dm}	Maximum packing of dispersed phase
α_{gs}	Average void fraction in liquid slug section
α_{fd}	Liquid droplet fraction in gas core alone
$\langle\alpha\rangle$	Area averaged void fraction

Γ_k	Interfacial mass transfer for *k*-phase
$\Delta\rho$	Density difference
Δt	Time interval of averaging
Δt_k	Total time fraction of *k*-phase
$\bar{\varepsilon}_k$	Weighted mean internal energy plus turbulent kinetic energy
ε_f	Ratio of the turbulent to molecular diffusivities
Λ_k	Interfacial energy transfer associated with enthalpy
λ	Coefficient in virtual mass correlation
λ_f	Thermal diffusivity of liquid
μ_k	Viscosity of *k*-phase
μ_m	Viscosity of mixture
μ_k^T	Eddy viscosity
μ_k^{T*}	Non-dimensional eddy viscosity coefficient
ρ_k	Density of *k*-phase
ρ_m	Mixture density
σ	Surface tension
τ_k	Viscous stress tensor for *k*-phase
$\bar{\bar{\tau}}_k$	Average viscous stress tensor for *k*-phase
τ_k^T	Turbulent stress tensor for *k*-phase
τ_i	Interfacial stress tensor
τ_{nnk}	Normal stress at interface for k-phase
τ_k	Relaxation time constant for interfacial energy transfer

φ_k	Source term
ψ_k	Property to be conserved
ψ_m	Mixture property

Subscripts

k	k-phase ($k = g$, f or c, d etc.)
f	Liquid phase
g	Gas phase
c	Continuous phase
d	Dispersed phase
1	Phase 1
2	Phase 2
i	Interface
m	Mixture
t	Tangential component

Symbols

$(\ )_{,\beta}$	Surface covariant derivative
$(\ )^{+}$	Transposed tensor
$\dfrac{d_s}{dt}$	Surface convective derivative
$\dfrac{D_d}{Dt}$	Dispersed phase convective derivative
∇_s	Surface divergence operator
$-$	Mass weighted mean value

$-$	Average
$=$	Phase average
$(\ )'$	Fluctuating component with respect to mean
$\sum_j$	Summation over interfaces passing in Δt
$\sum_k$	Summation for both phases

REFERENCES

Amsden, A. A. and Harlow, F. H. (1978) "K-TIF: A Two-Fluid Computer Program for Downcomer Flow Dynamics," LA-6994, NRC-4.

Baker, O. (1954) "Simultaneous Flow of Oil and Gas," Oil and Gas J., Vol. 53, pp. 183-195.

Bankoff, S. G. (1960) "A Variable Density single-Fluid Model for Two-Phase Flow with Particular Reference to Steam-Water Flow," J. Heat Transfer., Trans. ASME, Vol. 82, p. 265.

Basset, A. B. (1961) *A Treatise on Hydrodynamics*, Vol. 2, Ch. 22, Deighton Bell, Cambridge, p. 1888 (Dover Edition, New York).

Bergles, A. E., Roos, J. P. and Bourne, J. G. (1968) "Investigation of Boiling Flow Regimes and Critical Heat Flux," AEC Report NYO-3304-13.

Bouré,, J. (1985) "Two-Phase Flow Models: The Closure Issue," *Multiphase Science and Technology, Vol. 3,* Hewitt, G. F., Delhaye, J. M. and Zuber, N., Eds., Hemisphere Publ. Corp., Springer-Verlag, pp.3-30, 1987.

Brodkey, R. S. (1967) *The Phenomena of Fluids Motions*, Addison-Wesley Pub. Co.

Chawla, T. C. and Ishii, M. (1978) "Equations of Motion for Two-Phase Flow in a Pin Bundle of a Nuclear Reactor," Int. J. Heat and Mass Trans., vol. 21, p. 1057.

Chawla, T. C. and Ishii, M. (1980) "Two-Fluid Model of Two-Phase Flow in a Pin bundle of a Nuclear Reactor," Int. J. Heat and Mass Trans., Vol. 23, p. 991.

Clift, R., Grace, J. R. and Weber, M. E. (1978) *Bubbles, Drops and Particles*, Academic Press, New York.

Collier, J. G. (1972) *Convective Boiling and Condensation*, McGraw-Hill, New York.

Delhaye, J. M., Giot, M. and Riethmuller, M. L. (1981) *Thermohydraulics of Two-Phase Systems for Industrial Design and Nuclear Engineering,* Hemisphere Pub. Co.

Drew, D. A. (1971) "Averaged Field Equations for Two-Phase Media," Studies in Applied Mathematics, vol. 1, pp. 205-231.

Drew, D., Cheng, L. and Lahey, R. T. Jr. (1979) "The Analysis of Virtual Mass Effects in Two-Phase Flow," Int. J. Multiphase Flow, Vol. 5, pp. 233-242.

Drew, D. A. and Lahey, Jr., R. T. (1982) "Phase Distribution Mechanisms in Turbulent Low Quality Two-Phase Flow in a Circular Pipe," J. Fluid Mech., Vol. 117, p. 91.

Drew, D. A. and Lahey, Jr., R. T. (1982) "Phase Distribution Mechanisms in Turbulent Low Quality Two-Phase Flow in a Circular Pipe," J. Fluid Mech., Vol. 117, p. 91.

Drew, D. A. and Wood, R. T. (1985) "Overview and Taxonomy of Models and Methods," Int. Workshop on Two—Phase Flow Fundamentals, Gaithersburg (Sept. 22-27).

Drew, D. A. and Lahey, Jr., R. T. (1987) Private Communication and "The Virtual Mass and Lift Force on a Sphere in Rotating and Straining Invisid Flow," Int. J. Multiphase Flow, Vol. 13, No. 10, pp. 113-121.

Dukler, A. E. and Taitel, Y. (1977a) Flow Regime Transitions for Vertical Upward Gas Liquid Flow: A Preliminary Approach Through Physical Modeling," Progress Report No. 1, NUREG-0162.

Dukler, A. E. and Taitel, Y. (1977b) "Flow Regime Transition for Vertical Upward Gas Liquid Flow," Progress Report No. 2, NUREG-0163.

Fabre, J., Masbernat, L. and Suzanne, C. (1983) "New Results on the Structure of Stratified Gas-Liquid Flow," Advances in Two-Phase Flow and Heat Trans., Vol. 1, pp. 135-154, Martinus Nijhoff Pub., Boston.

Frankl, F. I. (1953) "On the Theory of Motion of Sediment Suspension," Soviet Physics, Doklady Akademiia Nauk SSSR, Vol. 92, p. 247.

Govier, G. W. and Aziz, K. (1972) *Flow of Complex Mixtures in Pipe,* Van Nostrand-Reinhold Co., New York.

Govier, G. W. and Omer, M. M. (1962) "The Horizontal Pipeline Flow of Air-Water Mixtures," Can. J. Chem. Engrg., Vol. 40, pp. 93-104.

Griffith, P. and Wallis, G. B. (1961) "Two-Phase Slug Flow," J. Heat Trans., 83C(3), p. 307.

Hewitt, G. F. and Hall-Taylor, N. S. (1970) *Annular Two-Phase Flow,* Pergamon Press.

Hoogendoorn, C. J. (1959) "Gas-Liquid Flow in Horizontal Pipes," Chem. Engrg., Sci., Vol. 9, pp. 205-217.

Houdayer, G., Pinet, B. and Vigneron, J. (1978) "HEXECO Code - A Six-equation Model in Transient Two-Phase Flow," Proc. 2nd CSNI Specialists Meeting, Paris, June 12-14.

Hsu, Y. Y. and Graham, R. (1976) *Transfer Process in Boiling and Two-Phase Systems*, McGraw-Hill, New York.

Hubbard, M. G. and Dukler, A. E. (1966) "The Characterization of Flow Regimes for Horizontal Two-Phase Flow," Proc. 1966 Heat Trans. and Fluid Mech. Inst., Stanford University Press.

Isbin, H. S., Moen, R. H. and Mosher, D. R. (1954) "Two-Phase Pressure Drop," AECU-2994, Nov.

Ishii, M. (1975) *Thermo-Fluid Dynamic Theory of Two-Phase Flow*, Eyrolles, Paris (also Scientific and Medical Publication of France, NY).

Ishii, M. (1977) "One-Dimensional Drift-Flux Model and Constitutive Equations for Relative motion Between Phases in Various Two-Phase Flow Regimes, ANL-77-47.

Ishii, M. (1981) "Foundation of Various Two-Phase Flow Models and Their Limitations," NRC-EPRI Conf. on Simulation Methods for Nuclear Power Systems, EPRIWS-81-212, pp. 3-47.

Ishii, M. and Chawla, T. C. (1979) "Local Drag Laws in Dispersed Two-Phase Flow," ANL-79-105, NUREG/CR-1230.

Ishii, M. and Grolmes, M. A. (1975) "Inception Criteria for Droplet Entrainment in Two—phase Concurrent Film Flow," AIChE J., 21(2), p. 308.

Ishii, M. and Kataoka, I. (1982) "Hydrodynamics of Annular-Dispersed Flow," Proc. ASME PVP Conf., Orlando, Florida.

Ishii, M. and Mishima, K. (1981) "Study of Two-Fluid Model and Interfacial Area," Argonne National Laboratory Report ANL-80-111.

Ishii, M. and Mishima, K. (1982) "Liquid transfer and Entrainment correlation for Droplet-Annular Flow," 7th Int. Heat Trans. Conf., Munich, Sept.

Ishii, M. and Mishima, K. (1984) "Two-Fluid Model and Hydrodynamic Constitutive Relations," Nucl. Engng., and Design, Vol. 82, pp. 107-126.

Ishii,, M. and Zuber, N. (1979) "Drag Coefficient and relative Velocity in Bubbly, Droplet or Particulate Flows," AIChE J., Vol. 25, p. 843.

Jones, O. C. and Zuber, N. (1975) "The Interrelation of Void Fraction Fluctuations and Flow Patterns in Two-Phase Flow," Int. J. of Multiphase Flow, Vol. 2.

Kalinin, A. V. (1970) "Derivation of Fluid-mechanics Equations for a Two-Phase Medium with Phase Change," Heat Tr. Soviet Research, Vol. 2, p. 83.

Kataoka, I., Ishii, M. and Mishima, K. (1983) "Generation and Size Distribution of Droplet in Annular Two-Phase Flow," J. of Fluid Engng., Vol. 105, pp. 230-238.

Kataoka, I., Ishii, M. and Serizawa, A. (1984) "Local Formulation of Interfacial Area Concentration and Its Measurements in Two-Phase Flow," NUREG/CR-4029, ANL-84-68.

Kataoka, I., Ishii, M. and Serizawa, A. (1985) "Interfacial Area Concentration in Two-Phase Flow," 23rd Nat. Heat Trans. Conf., Denver, Aug. 4-7.

Kataoka, I., Ishii, M. and Serizawa, A. (1986) "Local Formulation and Measurement of Interfacial Area Concentration in Two-Phase Flow," Int. J. Multiphase Flow, Vol. 12, pp. 505-531.

Kestin, J. and Podowski, M. (1985) "Report on International Workshop on Two-Phase Flow Fundamentals," Report prepared for DOE, Natl. Bureau of Standards, Gaithersburg, Sept. 22-27.

Kocamustafaogullari, G. (1971) "Thermo-Fluid Dynamics of Separated Two-Phase Flow," Georgia Inst. of Tech., Ph.D. thesis.

Kocamustafaogullari, G. and Ishii, M. (1983) "Interfacial Area and Nucleation Site Density in Boiling Systems," Int. J. Heat Mass Trans., Vol. 26, pp. 1377-1387.

Kutateladze, S. S. (1952) "Heat Transfer in Condensation and Boiling," Moscow, AEC-TR-3770, USAEC Technical information Service.

Lahey, R. T. and Moody, F. J. (1977) *The Thermal-Hydraulics of a Boiling Water Nuclear Reactor*, Am. Nucl. Soc.

Lance, M. and Bataille, J. (1982) "Turbulence in the Liquid Phase of a Bubbly Air-Water Flow," Proc. of the NATO Advanced Research Workshop on the Advances in Two-Phase Flow and Heat Trans., Spitzingsee, Germany, pp. 403-427.

Lance, M., Mariè, L. and Bataille, J. (1985) "Homogeneous Turbulence in Bubbly Flow," ASME Winter Annual Meeting, Miami Beach, Fl., Nov.

Landau, L. D. and Lifshitz, E. M. (1959) *Fluid Mechanics*, Pergamon, Oxford.

Levy, S. (1960) "Steam slip-Theoretical Prediction from Momentum Model," J. of Heat Trans., Series C, Vol. 82, p. 113.

Mandhane, J. M., Gregory, G. A. and Aziz, K. (1974) "A Flow Pattern Map for Gas-Liquid Flow in Horizontal Pipes," Int. J. Multiphase Flow, Vol. 1, pp. 537-553.

Mariè, J. L. (1983) "Investigation of Two-Phase Bubbly Flows Using Laser Doppler Anemometry," PCH, Vol. 4, pp. 103-118.

Mariè, J. L. (1987) "A Modeling of the Skin Friction and Heat Transfer in Turbulent Two-Component Bubbly Flows in Pipes," Int. J. Multiphase Flow, Vol. 13, No. 3, pp. 309-326.

Michiyoshi, I. and Serizawa, A. (1984) "Turbulence in Two-Phase Bubbly Flow," paper presented at the Japan-U.S. Seminar on Two-Phase Flow Dynamics, Japan.

Milne-Thomson, L. M. (1968) *Theoretical Hydrodynamics*, Macmillan Press, 5th Edition.

Mishima, K. and Ishii, M. (1984) "Flow Regime Transition Criteria for Upward Two-Phase Flow in Vertical Tubes," Int. J. Heat Mass Trans., Vol. 27, pp. 723-737.

Mokeyev, Yu. G. (1977) "Effect of Particle Concentration on Their Drag and Induced Mass," Fluid Mech. - Soviet Research 6, p. 161.

Nigmatulin, R. I. (1979) "Spatial Averaging in the Mechanics of Heterogeneous and Dispersed Systems," Int. J. Multiphase Flow, Vol. 5, pp. 353-385.

Odar, F. and Hamilton, W. S. (1964) "Forces on a Sphere Accelerating in a Viscous Fluid," J. Fluid Mech., Vol. 18, pp. 302-314.

Ohkawa, K. and Lahey, R. T. (1983) "The Development of Gamma Ray Scattering Densitometer and Its Applications to the Measurement of Two-Phase Density Distribution in an Annular Test Section," NUREG/CR-3374.

Riznic, J. and Ishii, M. (1986) "Modeling of Vapor Generation in flashing flow," NUREG/CR-4501, ANL-86-1.

Rowe, D. S. (1967) "Cross-Flow Mixing Between Parallel Flow channels During Boiling," BNWL-371 PT1.

Ruckenstein, E. (1959) "On Heat Transfer Between Vapour Bubbles in Motion and the Liquid from which They are Generated," Chem. Engrg. Sci., Vol. 10, pp. 22-30.

Sato, Y., Sadatomi, M. and Sekoguchi, K. (1981) "Momentum and Heat Transfer in Two-Phase Bubble Flow-II," Int. J. Multiphase Flow, Vol. 7, pp. 179-190.

Serizawa, A., Kataoka, I. and Michiyoshi, I. (1975a) "Turbulence Structure of Air-Water Bubbly Flow-I. Measuring Techniques," Int. J. Multiphase Flow, Vol. 2, pp. 221-233.

Serizawa, A., Kataoka, I. and Michiyoshi, I. (1975b) "Turbulence Structure of Air-Water Bubbly Flow-II. Local Properties," Int. J. Multiphase Flow, Vol. 2, pp. 235-246.

Serizawa, A., Kataoka, I. and Michiyoshi, I. (1975c) "Turbulence Structure of Air-Water Bubbly Flow-III. Transport Properties," Int. J. Multiphase Flow, Vol. 2, pp. 247-259.

Sha, W. and Slattery, J. C. (1980) "Local Volume-Time Averaged Equations of Motion for Dispersed, Turbulent, Mulitphase Flow," ANL-80-51, NUREG/CR-1491.

Saffman, P. g. (1965) "The Lift on a Small Sphere in a Slow Shear Flow," J. Fluid Mech., Vol. 22, pp. 385-400.

Teletov, S. G. (1945) "Fluid Dynamic Equations for Two-Phase Fluids," Soviet Physics, Doklady Akademii Nauk SSSR, Vol. 50, p. 99.

Theofanous, T. G. and Sullivan, J. (1982) "Turbulence in Two-Phase Dispersed Flows," J. Fluid Mech., Vol. 116, pp. 343-362.

Thurgood, M. J. et al. (1978) "Core Thermal Model Development," PNL-2653-2, NUREG/CR-0341, p. 101.

Tong, L. S. (1975) *Boiling Heat Transfer and Two-Phase Flow*, Robert E. Krieger Pub. Co.

Vernier, P. and J. M. Delhaye (1968) "General Two-Phase Flow Equations Applied to the Thermohydrodynamics of Boiling Nuclear Reactor," Energie Primaire, 4(1).

Vince, M. A. and Lahey Jr., R. T. (1980) "Flow Regime Identification and Void Fraction Measurement Techniques in Two-Phase Flow," NUREG/CR-1692.

Wallis, G. B. (1969) *One-Dimensional Two-Phase Flow*, McGraw-Hill, New York.

Wallis, G. B. (1980) "Theoretical Model of Gas-Liquid Flow," Proc. of 17th Annual Meeting of Soc. of Engng. Sci., Atlanta, p. 207.

Wang, S. K., Lahey Jr., R. T. and Jones, O. C. (1985) "Three-Dimensional Turbulence Structure Measurements in Air/Water Two-Phase Flow," Rensselaer Polytechnic Inst. Topical Report RPI/NSF-85-1.

Wang, S. L. (1985) "Three Dimensional Turbulence Structure Measurements in Air/Water Two-Phase Flow," Ph.D. Thesis, Rensselaer Polytechnic Inst., Troy, New York.

Weisman, J. and Kang, S. Y. (1981) "Flow Pattern Transitions in Vertical and Upwardly Inclined Lines," Int. J. Multiphase Flow, Vol. 7, p. 271.

Wolfert, K., Burwell, M. J. and Enix, D. (1978) "Non-Equilibrium Mass Transfer Between Liquid and Vapor Phases During Depressurization Processes in Transient Two—phase Flow," Proc. 2nd CSNI Specialists Meeting, Paris, Sept., Vol. 1, pp. 149-190.

Yadigaroglu, G. and Bensalem, A. (1985) "Interfacial Mass Generation Rate Modeling in Non-Equilibrium Two-Phase Flow," Int. Workshop on Two-Phase Flow Fundamentals, Gaithersburg, Sept. 22-27. *Multiphase Science and Technology, Vol. 3,* Hewitt, G. F., Delhaye, J. M. and Zuber, N., Eds., Hemisphere Publ. Corp., Springer-Verlag, pp.85-127, 1987.

Zuber, N. (1964) "On the Dispersed Two-Phase Flow on the Laminar Flow Regime," Chem. Engrg. Sci., Vol. 19, p. 897.

Zuber, N. and Dougherty, D. W. (1967) "Liquid Metal Challenge to the Traditional Methods of Two-Phase Flow Investigations," Sym. on Two-Phase Flow Dynamics, Vol. 1, p. 1091, EURATOM, Eindhoven, Sept.

Zuber, N. and Findlay, J. A. (1965) "Average Volumetric Concentration in Two-Phase Flow Systems," J. Heat Trans., Vol. 87, p. 453.

CHAPTER 2

MULTIPHASE INSTRUMENTATION AND EXPERIMENTAL TECHNIQUES

Jean-Marc Delhaye

Commissariat à l'Energie Atomique, Centre d'Etudes Nucléaires de Grenoble, Service d'Etudes Thermohydrauliques, France

ABSTRACT

The state of the art of measuring techniques in gas-liquid two-phase flow is commented on and typical experiments for basic research are presented. Current needs in instrumentation for research and industrial applications are discussed.

INTRODUCTION

Two-phase flow and heat transfer are important in many industrial branches such as nuclear engineering, chemical engineering, petroleum engineering, aerospace engineering, food industry and in a variety of disciplines such as oceanography or physics of clouds. Almost two decades ago, an in-depth review on two-phase flow instrumentation was edited by Letourneau and Bergles (1969) including already 368 references. Since then, interest in two-phase flow and heat transfer by scientists and engineers has continued to increase. The upcoming sophisticated computer codes to study and predict two-phase flow and heat transfer in complex geometries and/or in transient regimes has not decreased the need for measurements but rather has enhanced it. Consequently a great deal of effort has been spent on gas-liquid measuring techniques for almost thirty years now. Two categories of publications have been generated: (1) papers for the specialists in instrumentation such as the communications contained in the 746 page book edited by Delhaye and Cognet (1984), (2) books for potential users such as the Handbook of Multiphase Systems edited by Hetsroni (1982) which provides selection charts and guidelines for choosing a measuring technique.

The purpose of this paper is not to give an exhaustive survey of what has been done in two-phase flow instrumentation for the last years. Neither is it my intention to provide an exegesis of a few selected methods as I did in my keynote paper delivered at the Eighth International Heat Transfer Conference (Delhaye, 1986). What I would like to do is to give my feelings on some techniques, to bring into the limelight some others that deserve attention and to suggest a series of actions to be undertaken. To achieve these objectives I will first comment on some promising measuring methods and then give some examples of sophisticated experimental techniques. Finally, I will suggest the use of a new mathematical tool for extracting more information from the signals delivered by a sensor.

I hope that the reader will react positively to my comments by indicating new promising techniques, encountered difficulties or by rebutting my statements on what I consider as needs and priorities.

1. PROMISING MEASURING METHODS

1.1 Local Void Fraction

Optical probes are sensitive to changes in the index of refraction in the medium which surrounds their extremities (Danel & Delhaye, 1971; Figs. 1 and 2). They permit the detection of the passage of interfaces at a point and the determination of the local void fraction defined as the ratio of the residence time of the gas phase to the residence time of the two phases.

The interest given to finely dispersed bubbly or droplet flow has brought about the development of single fiber optical probes (Delhaye,

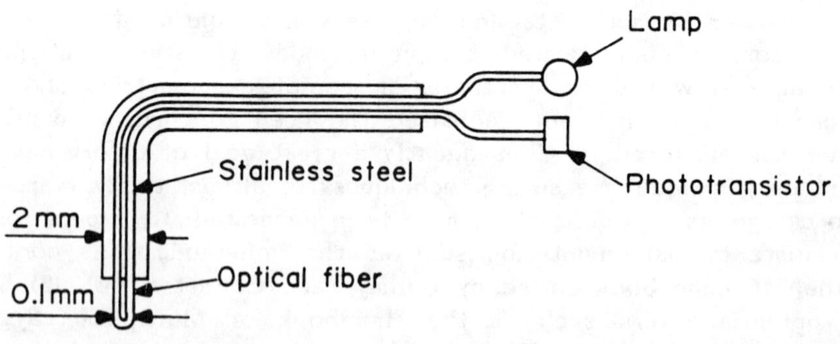

Figure 1. U-shaped fiber optical sensor (Danel & Delhaye 1971).

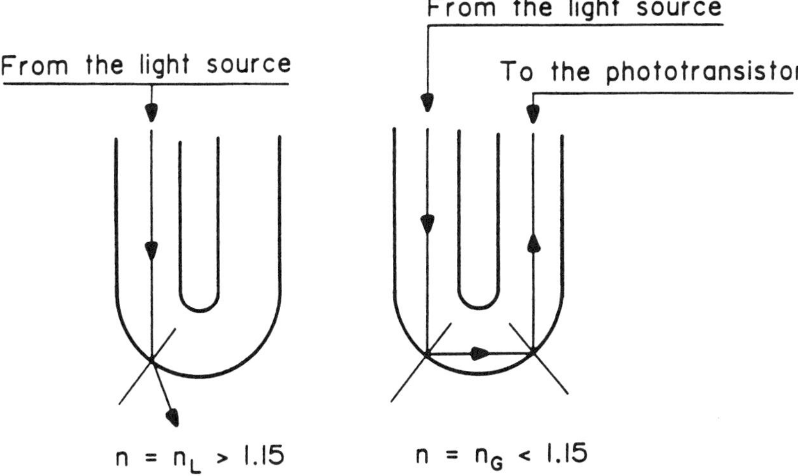

Figure 2. Active part of the U-shaped fiber optical sensor (Danel & Delhaye, 1971).

1983; Danel et al., 1984, Fig. 3). The optical fiber has a diameter of 40 μm and its extremity can be cut into a hemispherical or slender profile adapted to each individual problem. The sensitive part of the sensor can be less than 10 μm in diameter and the rise-time of the signal during the passage of an interface less than 0.4 μm.

In 1983, Delhaye et al. reported the very first reliability tests of an optical probe that was especially designed to withstand high pressures and temperatures (180 bar and 360^0C). These reliability tests were performed in an autoclave filled with water. They showed that the proposed design enabled an optical probe to be immersed in demineralized water at 180 bar and 360^0C during more than a hundred hours; the sapphire needle that was used as the sensor tip did not

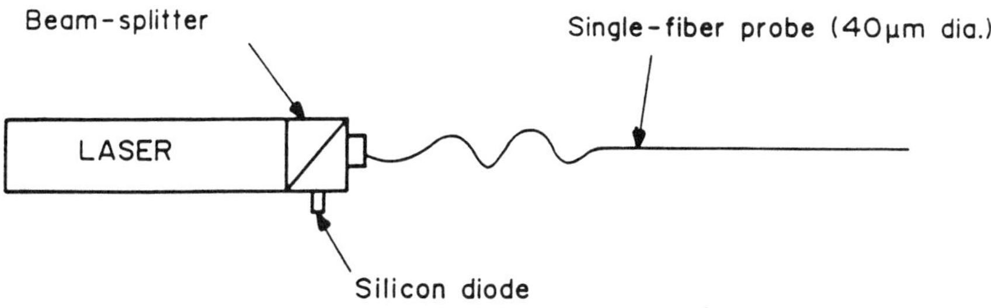

Figure 3 Single-fiber optical probe (Delhaye, 1983; Danel et al., 1984).

dissolve; the junction between the sapphire needle and the probe metallic sheath ensured a perfect seal between the inside of the probe and its environment. The optical components were not destroyed and their functions were not altered despite the high temperature surrounding. However, these probes suffered from two major drawbacks: the fabrication was not regular and the design was not appropriate for probes bent at a right-angle. Very recently a new design was tested which seems to overcome these hindrances. Moreover, this new design can be directly adapted to single fiber probe. Reliability tests are currently being performed and have already shown that such probes could withstand an immersion in demineralized water at 200 bar and 400^{0}C during more than a hundred hours.

The advantages of optical probes are numerous and well known:

(i) Contrary to the case of resistivity probes, it is not necessary that one of the phases possesses properties of known connexity.
(ii) It is not necessary that one of the phases be electrically conducting.
(iii) Their response is very rapid (less than 1 μs for single fiber probe).
(iv) They can be used over a large range of temperature and pressure due to the use of special components and design.
(v) The transparency of the fluids does not come into play.
(vi) The spatial resolution is very good (less than 10 μm for single fiber probes).

1.2 Line Void Fraction

The photon attenuation technique is the only one to measure the line void fraction defined as the ratio of the length occupied by the gas phase to the total length occupied by the two phases. It does not perturb the flow and can be made accurate enough to calibrate other techniques.

Because of the constant use of this measuring technique, Fournier and Jeandey (1984) developed a computer program simulating the behavior of an X-ray beam from its emission to its detection. This program is able to optimize the major design parameters of an experimental setup by taking into account all possible sources of uncertainty such as void fluctuations, Compton scattering, beam hardening, system drifts, positioning and statistical uncertainty. The input data of the program are: the thickness and the material of the walls traversed by the beam, the components of the two-phase flow, the characteristics of the detector (either the absorbing substance with its thickness or a detector with its efficiency). In addition, tables of absorption coefficients have been stored in the program for a wide range of chemical elements likely to be found on the beam path. The program enables the user to choose the best value for the tube high-voltage and to select filters to minimize the uncertainty.

1.3 Area Void Fraction

Impedance void meters. Impedance void meters have proved to be sufficiently adequate for measuring area void fractions in a wide range of circumstances especially in the study of void fraction waves. The most effective meters use flush-mounted electrodes at the pipe wall (Fig. 4); two versions of these meters have been used so far. In the first version a fixed electrical field is created between two electrodes whereas in the second version a rotating electrical field is created between three pairs of electrodes which are energized sequentially (Merilo et al., 1977; Snell et al., 1978). The advantage of the six electrode sensor is that its signal is less affected by the spatial distribution of the void fraction than the signal delivered by a

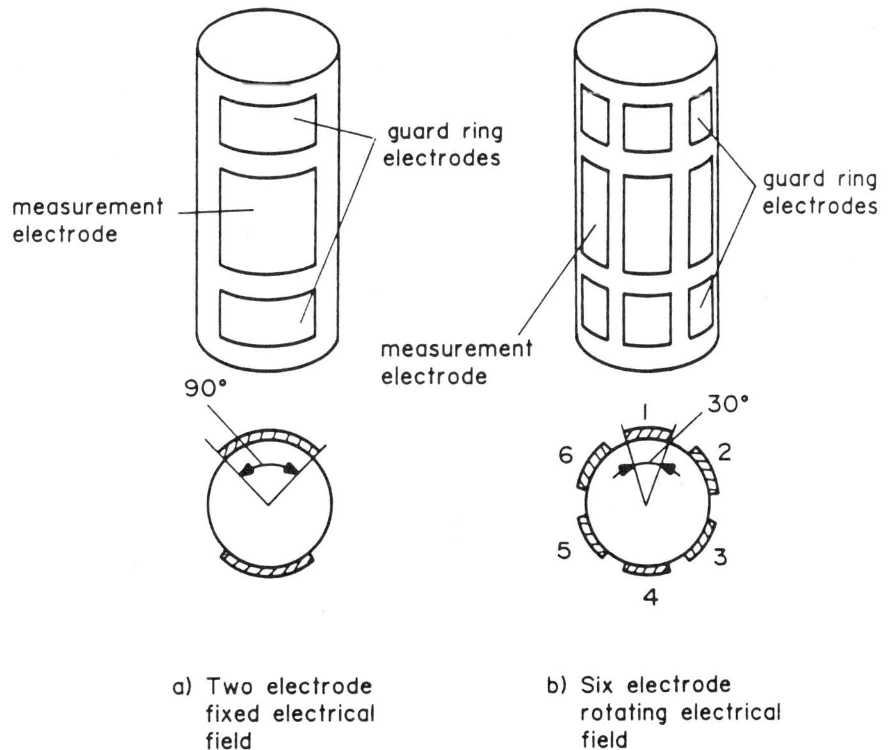

Figure 4. Impedance void meters (Delhaye et al., 1987).

two electrode sensor. An improved development of the six electrode sensor has been reported recently by Tournaire (1985, 1986, 1987). Lemonnier et al. (1988) and Delhaye et al. (1987) in connection with an experimental investigation of void fraction waves. It appears that this technique is reliable, fairly accurate and not too expensive. However, it requires walls that are not electrically conducting.

Multibeam X-ray densitometer. Several laboratories have used multibeam densitometers for measuring area void fractions in transient two-phase flow. However, the size of the standard scintillators has limited the number of beams to around ten. The recent development of medical X-ray tomography equipment has removed this constraint and void fractions measurements have been made with up to 31 beams (Jeandey, 1982). The equipment used by Jeandey (Fig. 5) consisted in a 140 kV electrostatic generator supplying the high-voltage of an X-ray tube. A flat, fan-shaped beam of X-rays covered the whole cross section of the test conduit and impinged onto a series of 31 anti-Compton collimators. Each collimator was followed by a xenon ionization chamber. The test conduit was a vertical cylindrical tube, 20 mm in diameter, followed by a divergent whose larger diameter was 60 mm. The conduit walls traversed by the X-ray beam were stainless steel, 1.5 mm thick. The pressure was 100 bar and the temperature 300^{0}C. The spatial resolution of each beam within the test section was 1 x 1 mm^2.

Neutron scattering technique. X-ray attenuation techniques are not suitable for void fraction measurements in metallic pipings when the void fraction is larger than 0.8. This is due to the effect of a small change in the liquid fraction concealed by the absorption of the X-ray beam by the metallic wall. On the contrary, neutron scattering is a very sensitive technique in this case. For a stainless-steel tube,

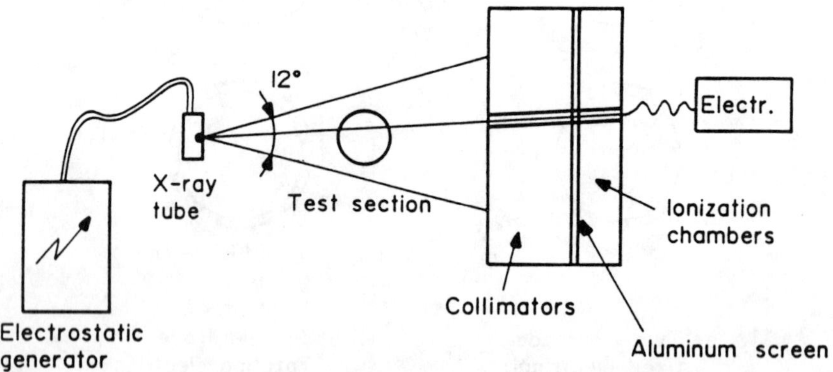

Figure 5 Multibeam X-ray densitometer (Jeandey, 1982).

12 mm in diameter, the best choice found by Freitas (1981) was the backward scattering of thermal neutrons. A neutron beam produced by a nuclear reactor was collimated in a conduit filled with helium to avoid scattering or attenuation by humidity. A bismuth shield was installed to stop the gamma rays. A boron-lined proportional detector was used to count the scattered neutrons. The contrast ratio for void fractions ranging from 0.8 to 1.0 was 2.2.

1.4 Heat Transfer Coefficient

In blowdown studies the internal wall heat flux and temperature are calculated from the measured value of the external wall temperature with a vanishing heat flux condition on this external wall. This is an inverse transient conduction calculation which is very sensitive to fluctuations in the measured temperature. Practical solutions to this problem can be found in Bonneton (1980) and a thorough presentation of this important issue is given in the book by Beck et al. (1985).

1.5 Mass Flow Rates

The measurements of the mass flow rates of both phases or of the mixture mass flow rate and quality require the use of two devices. The combination of a venturi and a turbine meter was used quite successfully in blowdown tests (Frank et al., 1977; Bonneton, 1980). However, the turbines have to be specially designed to withstand the extremely severe hydrodynamic conditions of a blowdown test. The turbines designed and manufactured in our Laboratories have the following characteristics:

Diameter: from 12 to 154 mm
Blade material: stainless steel or titanium
Rotational speed: up to 120,000 rpm for a diameter of 12 mm; up to 15,000 rpm for a diameter of 154 mm
Rotational acceleration: up to 600,000 revolutions/s^2
Maximum operating temperature: 780^{0}C (up to 900^{0}C for short periods)
Dynamic range: from 0.05 to 225 m/s for water flowing in a 46 mm turbine.

These turbines have been manufactured and sold to laboratories in Canada, Japan, Germany, Italy, Finland and France. A paper by John et al. (1984) reports on the testing of a mass flow rate measuring spool-piece involving two drag bodies, two local and one full flow turbine and a two beam gamma densitometer in transient steam-water flow. The mass flow rates determined by different combinations of the spool-piece components were compared with the reference mass flow rate measured with the KfK True Mass Flow Meter.

1.6 Interfacial Areas

Consider a two-phase mixture flowing in a pipe and a control volume limited by two cross-sectional planes and the pipe wall. The volumetric specific area is defined as the interfacial area enclosed in the control volume divided by the measure of the control volume. The volumetric specific area is then an instantaneous quantity defined over a volume. Conversely, the local specific area at a given point is defined as the sum of the reciprocals of the speeds of displacement of the interfaces passing through the point, divided by the time interval that is considered. The local specific area is then a local quantity defined over a time interval. One can show that the time-average of the volumetric specific area is identical to the volume-average of the local specific area (Delhaye, 1976; Delhaye and Achard, 1977, 1978; Ishii, 1987).

Measurements of the volumetric specific area. For bubbly flows the ultrasonic technique as described by Jones et al. (1986), Delhaye (1986) and Bensler et al. (1987) looks very promising because of its simplicity and versatility. It compares fairly well with the photographic technique or the light attenuation technique.

Measurement of the local specific area. Following the first extensive investigation on interfacial area measurements (veteau, 1981; Veteau and Charlot, 1981), the multisensor probe was thoroughly employed by Kataoka et al. (1985, 1986) in air-water bubbly and slug flows. However, this technique should still be compared systematically to a light or ultrasonic attenuation method to be entirely qualified.

1.7 Wall Shear Stress

Souhar and Cognet (1984), Cognet et al. (1984) as well as Nakoryakov et al. (1984) measured the wall shear stress in two-phase flow by means of the diffusion-controlled electrolysis technique. They used a double sensor that enables the direction of the wall shear stress to be determined. This is particularly important in slug flow where the liquid film surrounding the gas plug moves downward.

A simple method was used by Onsrud et al. (1987) to measure the wall shear stress in a wind tunnel and, according to the authors, in annular two-phase flow. Two static pressure holes of different diameters are drilled through the channel wall with their axes coplanar with the channel axis but inclined at two different angles (Fig. 6). One hole has its axis facing the upstream zone of the pipe whereas the axis of the second hole faces the downstream zone. The wall shear stress is proportional to a certain power of the pressure difference between the two static pressure holes. A systematic appraisal of this simple and cheap method would be very worthwhile.

1.8 Liquid Velocity in Bubbly Flow

Mariè (1983) showed that laser Doppler anemometry could measure the statistical properties of the turbulence of the liquid phase in a bubbly flow. The results were in good agreement with those obtained by hot film anemometry (Fig. 7). In a subsequent paper, Mariè and Lance (1984) gave the value of the void fraction beyond which laser Doppler anemometry cannot be used. This critical value is a function of the thickness of the bubbly flow traversed by the light beams (Fig. 8). It decreases from 0.3 for a thickness of 20 mm to 0.02 for a thickness of 450 mm.

2. EXAMPLES OF SOPHISTICATED EXPERIMENTAL TECHNIQUES

2.1 The Transient Drag on a Fluid Sphere

Given a sphere moving in an infinite medium whose velocity at infinity is uniform but time-dependent, the transient drag on this sphere is given by the Boussinesq-Basset-Oseen-Tchen equation if the sphere is solid, if there is no slip at the wall and if the relative Reynolds number is small. Pham Dan Tâm (1981, 1982) solved the case

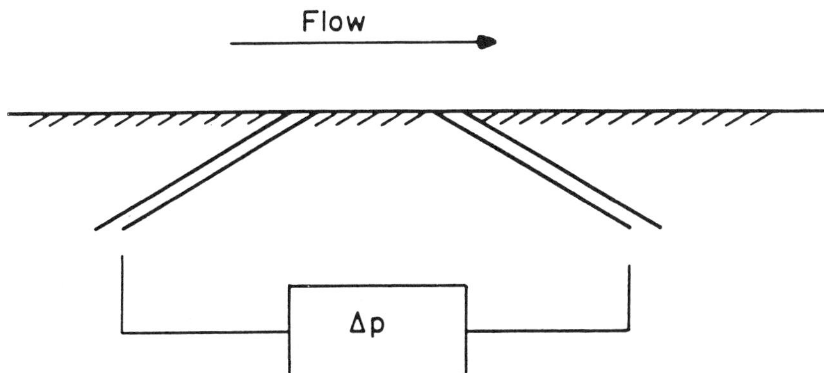

Figure 6 Wall shear stress sensor (Onsrud et al., 1987)

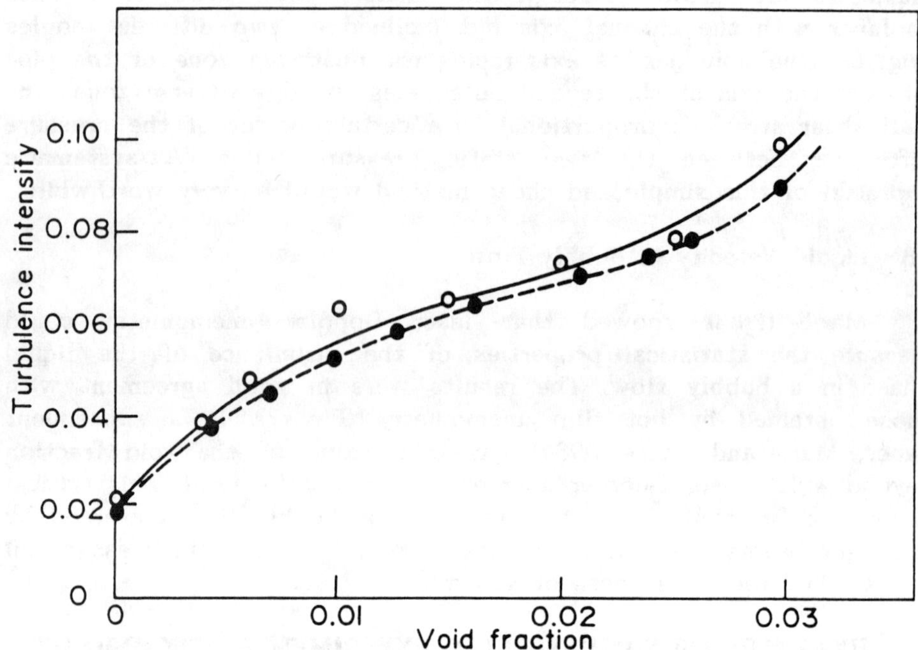

Figure 7 Turbulence itensity of the liquid phase in a bubbly flow (Mariė, 1983): 0 laser Doppler anemometry, 0 hot film anemometry; superficial liquid velocity: 0.5 m/s (Mariė, 1983).

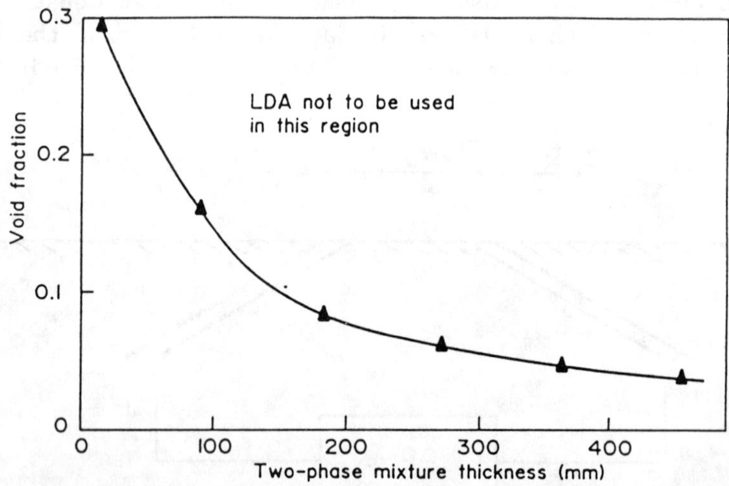

Figure 8 Applicability of laser doppler anemometry in two-phase bubbly flow (Mariė and Lance, 1984).

where the sphere is a fluid medium with a vanishing tangential stress at the wall and where the relative Reynolds number tends to approach infinity. In connection with his theoretical study, Pham Dan Tâm (1981) measured the time-dependent relative velocity between a small bubble and the surrounding fluid over a 2 m high test section, 10 cm in diameter (Fig. 9). A sinusoidal water flow rate was imposed and a bubble was released in the axis of the test section through a hypodermic needle. The velocity of the bubble was obtained by a visualization technique. A TV camera was placed on a platform which could be displaced along rails parallel to the test section. The electronic analysis of the video signal enabled the displacement of

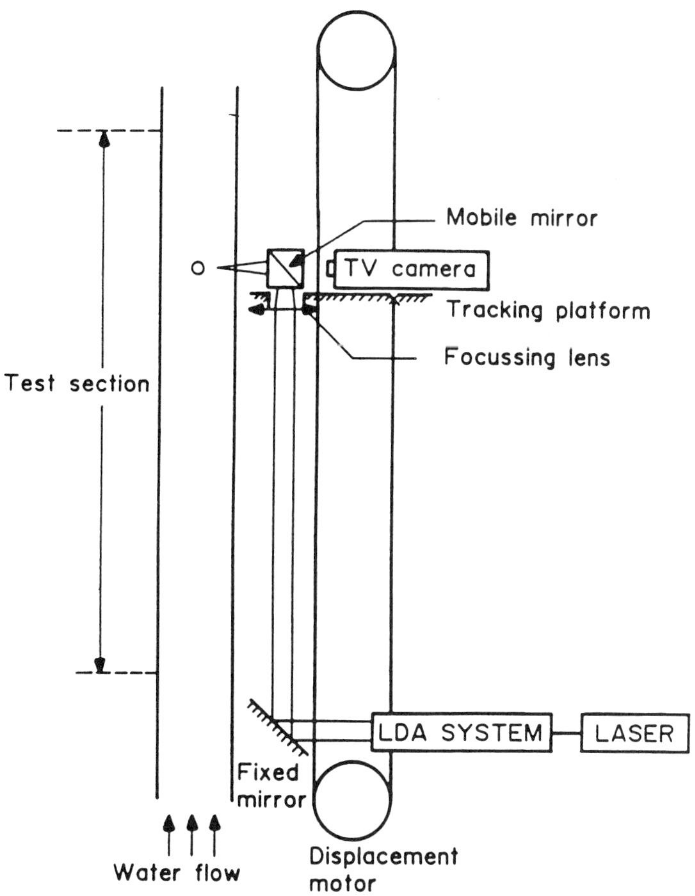

Figure 9 Experimental determination of the time-dependent relative velocity between a small bubble and the surrounding fluid (Pham Dan Tâm, 1981).

the platform to be monitored and the position of the bubble, hence its velocity, to be determined. The velocity of the water surrounding the bubble was measured by means of a laser Doppler anemometer using the backscattering mode. An argon laser and the optical unit were placed on the ground and the laser beams were focused in the test section by means of a stationary mirror and a movable mirror mounted on the platform. The experimental difficulties were extreme and are described along with all the data in Pham Dan Tâm's dissertation (1981). Typical results are shown in Figure 10.

2.2 The Bursting of a Bubble

The bursting of air bubbles at the surface of the sea plays an important role in the air-sea particulate exchanges. Wave breaking and whitecapping are responsible for the entrainment of air bubbles below the surface of the sea. After a certain soujourn time in the sea, these bubbles come back up to the free surface where they burst into the atmosphere. The bursting of such bubbles produces liquid marine aerosols which are entrained in the atmosphere. It is well known that the bursting of a bubble at a gas-liquid interface results in the release of two kinds of drops (Fig. 11). First, the liquid membrane which separates the bubble from the atmosphere ruptures and produces approximately a thousand of film drops about 10 μm of

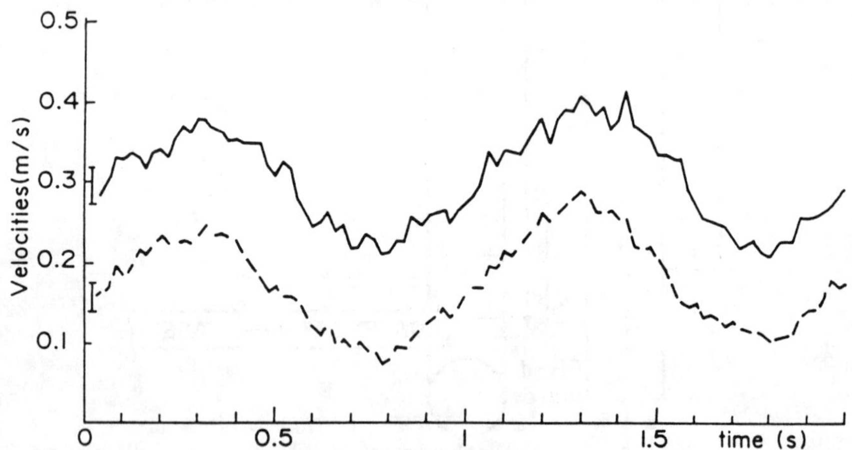

Figure 10 Time-dependent relative velocity between a small bubble and the surrounding fluid (Pham Dan Tâm, 1981); bubble diameter: 0.46 mm, ----- liquid velocity, ——— bubble velocity.

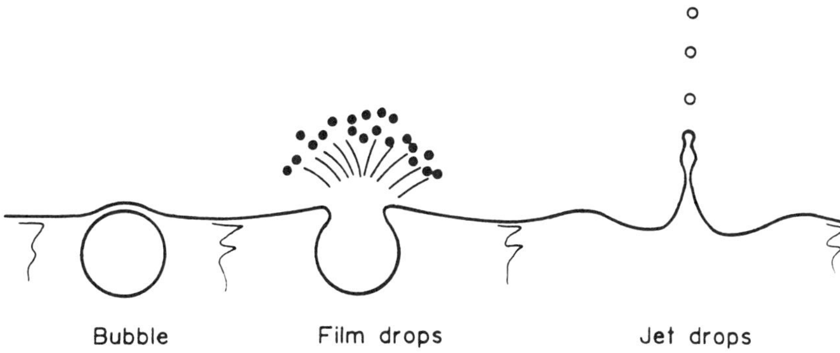

Figure 11 Bursting of a bubble at a gas-liquid interface (Resch et al., 1986).

average size which are ejected at a height of about 1 cm. Then, when the bubble collapses a liquid jet rises from the lower part of the bubble and breaks up into approximately ten jet drops about 100 μm of average size which are ejected at a hight of about 15 cm. The ejection velocity of both kinds of drops is of the order of 10 m/s. Because of their large size, the jet drops have been extensively studied for many years. However, the first experimental study concerning film drops was reported by Resch et al. (1986). A holographic Gabor technique was used (Fig. 12) with a YAG laser (105 mJ, 532 nm) whose pulse duration was 13 ns. The bursting of air bubbles of diameter ranging from 2 to 10 mm was studies in fresh water and in sea water. The synchronization problem between the bubble burst and the firing of the YAG laser was solved by using a light gate. Spectacular photographs were obtained showing the lace-typed structure of the liquid membrane just before its fragmentation.

2.3 High-Speed Video Cameras

Visualization is the first diagnostic tool that must be used when two-phase flow systems are investigated. New equipment has appeared on the market during the few years which has had an important impact on research efficiency. High-speed video cameras are easy to operate and the recorded sequences can be displayed immediately after the experiment. Frame-rates reach 200 fps for color cameras and 12,000 fps for black and white cameras. A good reference on this experimental technique is the book by Miquel (1985).

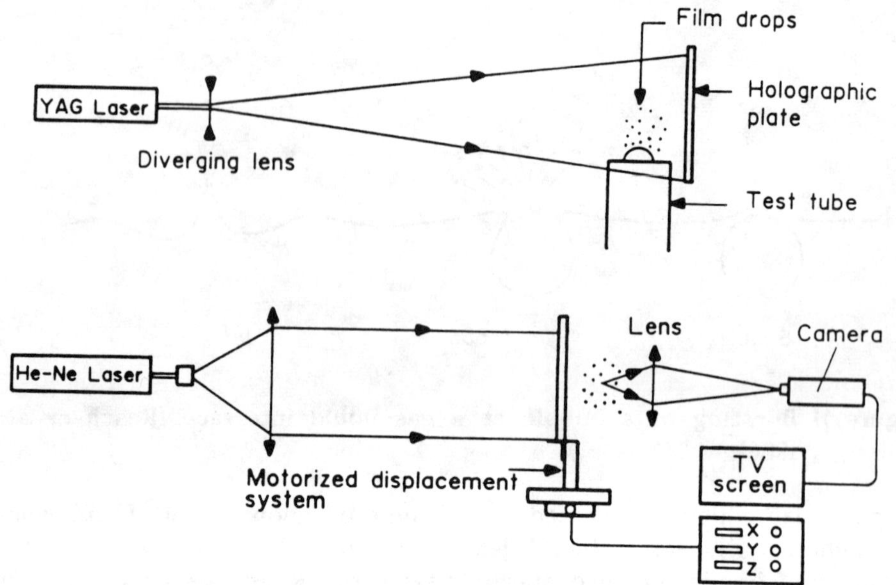

Figure 12 Holographic technique used by Resch et al. (1986) to visualize the bursting of a bubble at a gas-liquid interface.

3. RECENT ADVANCES IN DATA PROCESSING

It is worth mentioning that the second edition of the book by Bendat and Piersol appeared in 1986. This book remains a standard reference for the measurement procedures and the analysis of random data.

In a bubbly flow, the liquid velocity is a piecewise continuous function of time. Consequently, if the power density spectrum of the liquid velocity is determined directly from a simple accumulation of the different parts of the signal, high frequencies are introduced which prevent a meaningful calculation of the turbulent dissipation rate in the liquid phase. Lance (1986) suggests the use of a Laplace-Gauss weighing function to transform the actual signal into a continuous signal.

Finally, I would like to propose the systematic calculation of the Grassberger-Procaccia dimension of the attractor which governs the dynamics of any two-phase system. This dimension is relatively easy to calculate and gives a lower bound of the order of the set of differential equations which describes the two-phase system (Delhaye, 1987).

4. EXAMPLES OF INSTRUMENTATION ISSUES

The content of the previous pages of this lecture is rather optimistic. However, reality is not so nice. Scientists and engineers are constantly consulted to solve exotic problems whose solutions often require ingenuity, skill, time and money. Here are some examples of measurements which have not yet received definite solutions: area void fraction in helium 4 two-phase flow, flow rate and quality in hydrogen near critical pressures, quality in a subchannel of a LMFBR subassembly, local concentration of a noncondensable gas near a wall onto which steam condenses, position of the liquid-solid interface within a solidifying or melting drop, interfacial area in nondispersed flows, liquid film thickness close to the dryout point in a heated tube, area void fraction in 400 mm diameter pipes lying on the sea bottom, flow rate and quality in steam injection piping networks for oil recovery, nucleation rates in crystallizers, bubble detection in oil wells, detection of the onset of nucleate boiling in a rod bundle, etc.

5. NEEDS AND PRIORITIES

5.1 Avoiding Wasteful Work

Several review papers on two-phase flow instrumentation, including this one, are published every year. I think these papers do not deserve the same attention as they did a decade ago. Most of these papers contain full inventories of techniques with too many references. They will certainly be of great value to the future historians of two-phase flow, but they do not help the scientist or the engineer in choosing the best method available. Archivists and scholars waste their time and clutter journals or conference proceedings with unnecessary information.

Heavy calculations without experimental checks should be avoided. For instance, too many papers, reports and dissertations have been published on the interaction between a light beam and a small particle. On the contrary, very few papers have been published which contain valuable comparisons between different techniques of particle optical sizing.

Expenive equipment should not be developed unless there are well defined potential applications covering a wide range of parameters. I think a measuring technique should not be studied just to be studied. The final objective is not the instrument but the measured value of a parameter. Instrumentation should help the experimentalist and the theoretician. Everybody who works on instrumentation should keep this in mind.

Some techniques are worth trying, but one should be aware of the difficulties in applications from the start. Side scattering of gamma rays was used to measure the local void fraction; but the counting time is so long, around one hour, that very few people would be interested in this technique. Pulsed neutron activation raises a similar hindrance. Linear accelerators or transportable neutron sources are not common equipment. In addition, diffusion processes are not well known in two-phase flow and this may impair the advantage of an instantaneous and uniform tagging of the flow.

A sophisticated measuring technique, such as nuclear magnetic resonance, should not be developed by physicists alone, but by a joint team of physicists and specialists in two-phase flow. Several man-years have been spent on NMR applied to two-phase flows but it seems that no positive conclusions have emerged from this work.

Recently, highly efficient data acquisition systems, statistical analyzers and image processors have appeared on the market. I fear that we will be overflooded by the amount of data that this nice equipment will regurgitate.

5.2 Helping the Experimentalist in Choosing and Implementing a Measuring Technique

As said before inventories are useless, but selection charts would be of great help. The diagrams published in the Handbook of Multiphase Systems (Hetsroni, 1982) are extremely helpful. However, they should be accompanied by detailed descriptions of the selected technique, including recommendations for the experimental setup, lists of manufacturers, approximate costs, expected errors, signal processing, calibration methods and a typical actual example.

The void fraction, mass flow rate and quality measuring techniques have been well studied for many years and are used constantly all over the world. It might be the time to think of standards similar to those which exist for single phase flow measurements.

5.3 Developing New Techniques

A technique may exist, but it can be required to be adapted to extreme conditions such as high or low temperatures, high pressures, complex geometries or miniaturization. In addition, industrial environment and cost are often major constraints. Priorities cannot be given for these types of measuring techniques. Actually, these priorities depend on the country; some are concerned by oil recovery, others by nuclear energy. They also depend on the industrial branch; the needs are different in the food or automotive industries.

Sometimes the technique does not even exist. For example, interfacial areas cannot be measured in churn flow, local void fractions cannot be measured nonintrusively in transient flow and speeds of displacement of interfaces cannot be measured with good accuracy. All these measurements can be considered to be important and deserve significant effort requiring imagination, perseverance, time and money.

ACKNOWLEDGEMENTS

It is my pleasure to acknowledge the help of Mr H. Bensler for having proofread this paper.

REFERENCES

Beck, J. V., Blackwell, B. and St Clair, C. R. Jr, (1985). *Inverse Heat Conduction, Ill-Posed Problems,* John Wiley & Sons, Inc. (New York).

Bendat, J. S. and Piersol, A. G. (1986), *Random Data: Analysis and Measurement Procedures,* Wiley-Interscience, Second edition.

Bensler, H. P., Delhaye, J. M. and Favreau, C. (1987), Measurement of interfacial area in bubbly flows by means of an ultrasonic technique, *ANS Proceedings, 1987 National Heat Transfer Conference,* ANS, 240-246.

Bonneton, M. (1980), Contribution à l'ètude des coefficients de transferts de chaleur lors de la dècompression d'un tube chauffant refroidi par de l'eau, Thèse de docteur ingènieur, Institut national Polytechnique et Universitè Scientifique et Mèdicale de Grenoble.

Cognet, G., Lebouchè, M. and Souhar, M. (1984), Wall shear measurements by electrochemical probe for gas-liquid two-phase flow in vertical duct, *AIChE Journal,* Vol. 30, No. 2, 338-341.

Danel, F. and Delhaye, J. M. (1971), Sonde optique pour mesure du taux de prèsence local en ècoulement diphasique, *Mesures-Règulation-Automatisme,* Aout-Septembre, 99-101.

Danel, F., Vermeille, H, Guilloud, J. C., Jacquet, A. and Delhaye, J. M. (1984), High-speed multiple spark cameras for gas-liquid two-phase flow studies, *Measuring Techniques in Gas-Liquid Two-Phase Flows,* Delhaye, J. M. and Cognet, G., Eds., Springer-Verlag, 67-89.

Delhaye, J. M. (1976), Sur les surfaces volumiques locale et intègrale en ècoulement diphasique, *CRAS,* Sèrie A, t. 282, 243-246.

Delhaye, J. M. (1983), Two-phase pipe flow, *International Chemical Engineering,* Vol. 23, No. 3, 385-410.

Delhaye, J. M. (1986), Recent advances in two-phase flow instrumentation, *Heat Transfer 1986*, Tien, C. L., Carey, V. P. and Ferrell, J. J., Eds, Volume 1, Hemisphere Publ. Corp., 215-226.

Delhaye, J. M. (1987), Fundamentals of time-varying two-phase flow formulation. Invited lecture given at the ICHMT International Seminar on Transient Phenomena in Multiphase Flow, Dubrovnik, May 24-30.

Delhaye, J. M. and Achard, J. L. (1977), On the use of averaging operators in two-phase flow modeling, *Thermal and Hydraulic Aspects of Nuclear Reactor Safety, Vol. 1: Light Water Reactors*, Jones, O. C. and Bankoff, S. G., Eds, ASME (New-York), 289-332.

Delhaye, J. M. and Achard, J. L. (1978), On the averaging operators introduced in two-phase flow modeling, *Transient Two-Phase Flow, Proceedings of the CSNI Specialists Meeting*, August 3-4, 1976, Toronto, Banerjee, S. and Weaver, K. R., Eds., AECL, Vol. 1, 5-84.

Delhaye, J. M., Charlot, R., Danel, F. and Arnault, J. (1983), An optical probe for interface detection at high pressure (180 bar) and high temperature (360^{0}C), *Thermal-Hydraulics of Nuclear Reactors*, Volume 2, Merilo, M., Ed., ANS, 1427-1430.

Delhaye, J. M. and Cognet, G., Eds, (1984), *Measuring Techniques in Gas-Liquid Two-Phase Flows*, Springer-Verlag.

Delhaye, J. M., Favreau, C., Saiz-Jabardo, J. M. and Tournaire, A. (1987), Experimental investigation on the performance of impedance sensors with two and six electrodes for area-averaged void fraction measurements, *ANS Proceedings, 1987 National Heat Transfer Conference*, ANS, 234-239.

Fournier, Th. and Jeandey, Ch. (1984), Optimization of an experimental setup for void fraction determination by the X-ray attenuation technique, *Measuring Techniques in Gas-Liquid Two-Phase Flows*, Delhaye, J. M. and Cognet, G., Eds., Springer-Verlag, 199-228.

Frank, R., Mazars, J. and Ricque, R. (1977), Determination of mass flowrate and quality using a turbine meter and a venturi, *Heat and Fluid Flow in Water Reactor Safety*, The Institute of Mech. Engrs, London, 63-68.

Freitas, R. L. (1981) Hydrodynamique des écoulements eau-vapeur dans la zone remouillée d'un tube chauffant en renoyage, Thèse de docteur ingénieur, Institut National Polytechnique et Université Scientifique et Médicale de Grenoble.

Hetsroni, G., Ed. (1982), *Handbook of Multiphase Systems*, McGraw-Hill Book Company.

Ishii, M. (1987), Interfacial area modeling, *Multiphase Science and Technology*, Vol. 3, Hewitt, G. F., Delhaye, J. M. and Zuber, N., Eds., Hemisphere Publ. Corp. (Washington, DC), Springer-Verlag (Berlin).

Jeandey, Ch. (1982), Multibeam X-ray densitometer for flow pattern and void fraction determination in steam-water mixtures, *Measurement in Polyphase Flows*, Hedrick, T. R. and Patel, B. R., Eds, ASME G 00209, 19-28.

John, H., Reimann, J., Eisele, G., Ohlmer, E. and Schultze, W. (1984), Test of a mass flow rate measuring spool piece involving drag body, gamma densitometer, local flow turbines and full flow turbine in transient two-phase flow, *Mass Flow Measurements*, Hedrick, T. R. and Reimer, R. M., Eds., ASME, 113-121.

Jones, S. W., Amblard, A. and Favreau, C. (1986), Interaction of an ultrasonic wave with a bubbly mixture, *Experiments in Fluids*, Vol. 4, No. 6, 341-349.

Kataoka, I., Ishii, M. and Serizawa, A. (1985), Interfacial area in two-phase flow: formulation and measurement, *Multiphase Flow and Heat Transfer*, Dhir, V. K., Chen, J. C. and Jones, O. C., Eds, ASME G 00304, 131-141.

Kataoka, I., Ishii, M. and Serizawa, A. (1986), Local formulation and measurements of interfacial area concentration in two-phase flow, *Int. J. Multiphase Flow*, Vol. 12, No. 4, 505-529.

Lance, M. (1986), Etude de la turbulence dans les écoulements diphasiques dispersés, Thèse de doctorat ès sciences, Université Claude Bernard-Lyon I.

Lemonnier, H., Nakach, R., Favreau, C. and Selmer-Olsen, S. (1988), Sensitivity analysis of an impedance void meter to the void distribution in annular and bubbly flow: a theoretical study, European Two-Phase Flow Group Meeting, Brussels (Belgium).

Letourneau, B. W. and Bergles, A. E., Eds., (1969), *Two-Phase Flow Instrumentation*, Proceedings of a Session held at the 11th national Heat Transfer Conference, Minneapolis, MI, Aug. 3-6.

Marié, J. L. (1983), Investigation of two-phase bubbly flows using laser Doppler anemometry, *PhysicoChemical Hydrodynamics*, Vol. 4, No. 2, 103-118.

Marié, J. L. and Lance, M. (1984), Turbulence measurements in two-phase bubbly flows using laser Doppler anemometry, *Measuring Techniques in Gas-Liquid Two-Phase Flows*, Delhaye, J. M. and Cognet, G., Eds., Springer-Verlag, 141-148.

Merilo, M., Dechene, R. L. and Cichowlas, W. M. (1977), Void fraction measurement with a rotating field conductance gauge, *J. of Heat Transfer*, Vol. 99, 330-332.

Miquel, J. C. (1985), *L'observation en video Rapide de 50 à 25000 ts*$^{-1}$ *ou en cinéma grande vitesse*, Technique et Documentation, Lavoisier (Paris).

Onsrud, G., Persen, L. N. and Saetran, L. R. (1987), On the measurement of wall shear stress, *Experiments in Fluids*, Vol. 5, No. 1, 11-16.

Nakoryakov, V. Y., Kashinskiy, O. N. and Koz'menko, B. K. (1984), The electrochemical method for study of turbulence in two-phase flow, *Fluid Mechanics-Soviet Research,* Vol. 13, No. 3, 11-22.

Pham Dan Tâm, (1981), De la trainée instationnaire sur une petite bulle, Thèse de docteur ès sciences, Université Scientifique et Médicale et Institut National Polytechnqiue de Grenoble.

Pham Dan Tâm (1982), The unsteady drag on a spherical bubble at large Reynolds numbers, *Mechanics and Physics of Bubbles in Liquids,* Proceedings IUTAM Symposium, Pasadena, California, 15-19 June 1981, van Wijngaarden, L., Ed., *Applied Scientific Research,* Vol. 38, 247-254.

Resch, F. (1986), Marine liquid aerosol production from bursting of air bubbles, *J. Geophysical Research,* Vol. 91, No. C1, 1019-1029.

Snell, C. C., Dechene, R. L. and Newton, R. E. (1978), Two phase relative volume fraction measurement with a rotating field conductance gauge, *Measurements in Polyphase Flows,* Stock, D. E., Ed., ASME, 21-24.

Souhar, M. and Cognet, G. (1984), Wall shear stress measurement by electrochemical probes in two-phase flow - Bubble and slug regimes, *Measuring Techniques in Gas-Liquid Two-Phase Flows,* Delhaye, J. M. and Cognet, G., Eds., Springer-Verlag, 723-744.

Tournaire, A. (1985), Détection des perturbations de taux de vide dans un écoulement diphasique en conduite cylindrique. Optimisation de la méthode par sonde à impédance, CEA-R-5323.

Tournaire, A. (1986), Dependence of the instantaneous response of impedance probes on the local distribution of the void fraction in a pipe, *Int. J. Multiphase Flow,* Vol. 12, No. 6, 1019-1024.

Tournaire, A. (1987), Détection et étude des ondes de taux de vide en écoulement diphasique à bulles jusqu'à la transition bulles-bouchons, Thèse de docteur ingénieur, Institut National Polytechnique et Université Scientifique et Médicale de Grenoble.

Veteau, J. M. (1981), Contribution à l'étude des techniques de mesure de l'aire interfaciale dans les écoulements à bulles, Thèse de docteur ès sciences, Université Scientifique et Médicale et Institut National Polytechnique de Grenoble.

Veteau, J. M. and Charlot, R. (1981), Interfacial area measurements in two-phase bubbly flows: comparison between the light attenuation technique and a local method, European Two-Phase Flow Group Meeting, Eindhoven (The Netherlands), June 2-5.

CHAPTER 3

COMPUTATIONAL METHODS FOR MULTIPHASE FLOW*

Wolfgang Wulff

Brookhaven National Laboratory, Upton, New YorK, USA

SUMMARY

The balance equations of multiphase flows are classified and techniques are reviewed for rendering these partial differential into partially or completely discretized equations. Numerical methods are presented for integrating the partially, and for solving the completely discrctized, equations. The issues of computing accuracy and economy are discussed. Numerical methods used in major computer codes for multiphase flow analyses are reviewed, and it is demonstrated that quantification of computing errors and improvement of computing economy should be the objectives of future research in numerical methods for multiphase flows.

1. INTRODUCTION

Computational methods for multiphase flow analysis encompass the techniques developed in at least five branches of applied mathematics: linear algebra, analytical integration of ordinary and partial differential equations, solution of integrodifferential equations and numerical analysis. Numerical analysis in turn involves the distinct disciplines of numerical inversion of matrices, calculation of eigenvalues, numerical solution of systems of nonlinear equations, numerical interpolation and evaluation of polynomials, quadratures, numerical integration of ordinary differential equations and, most importantly, the numerical integration of partial differential equations

---

*Work performed under the auspices of the U.S. Nuclear Regulatory Commission.

by the methods of characteristics, of finite differences and of finite elements. Computing methods are, in the final analysis, the same for multiphase as for single-phase flows, yet multiphase flow computations pose far greater challenges in problem formulation and in the application of computing methods. On the other hand, multiphase flow models offer unique opportunities for simplification (cf. Sec. 4.2).

Except for applied mathematicians, computational methods are only tools which must be selected and applied in the context of modeling objectives, model formulation and available digital computer capabilities. As yet, no computing method exists which is both general and efficient for all two-phase flow analyses. There may never be a single method because universality is achieved only at the detriment of efficiency. Multiphase flow analyses are complicated. This complexity compels the two-phase flow analyst to become involved with the above wide spectrum of mathematical disciplines.

A multiphase flow is the flow of a mixture of two or more phases. A phase is a system or a portion of a system composed of any number of chemical constituents and satisfying two requirements: (1) that it is homogeneous and (2) that it has a definite boundary (Zemansky 1957, p. 566).

Clearly, the multiphase mixture is not homogeneous, and the phase boundaries or interfaces often have chaotic and stochastically changing geometries. Within the mixture, some properties vary discontinuously (density) and others have discontinuous gradients (temperature). Direct applications of the Leibniz rule for differentiating integrals and of the Gauss divergence theorems are not permitted in the derivation of local balance equations without prior averaging. The internal structure of the mixture leads to unique difficulties of measurement, of flow pattern characterization and in computing.

1.1 Purpose of Computational Methods

The aim of using computational techniques in general is to obtain specific quantitative results, either through data reduction (including regression analysis) from experiments or, more importantly here, from analytical models. The purpose of computing in multiphase flow analysis is to:

- test modeling hypotheses,
- predict steady and transient two-phase flows,
- gain insight into the behavior of multiphase flows.

Analytical closed-form solutions and even series expansions are possible only under a small number of simplifying circumstances (Gonzales-Santalo and Lahey 1971; Wulff 1978; Lahey 1981; Achard, Drew and Lahey 1985; Rizwan-Uddin and Dorning 1986). Numerical solutions by digital computers are the only available alternative for relevant multiphase flow analyses which address processes actually occurring in laboratory experiments, industrial facilities and power plants. Therefore, computational methods selected for two-phase flow studies must utilise modern digital computers effectively.

The objective of computing is to obtain, in principle with any desirable accuracy, a numerical approximation to the unique solution of a completely defined and properly posed mathematical model (Richtmyer and Morton 1967; p. 39; Courant and Hilbert 1962, p. 227), which consists of a set of partial differential and algebraic equations. For transient multiphase flows, these models constitute initial and boundary value problems and are properly posed if a unique solution exists which depends continuously on initial and boundary data. Inherent static instabilities and bifurcations require for numerical integration a problem partition into domains with unique solutions.

The objective of computing is not to discover new equations by numerical techniques, because models are to be formulated on the basis of physics. Instead, computational results can serve to test hypotheses which are implied in the *analytical* model formulation. Therefore, numerical solutions must approximate the (unknown) solution to the analytical model formulation, normally with an accuracy moderately larger than achievable instrumentation accuracy (normally one additional significant digit).

The objective of computing is not to salvage ill-posed mathematical models with numerical diffusion or other artificial gimmickry, because solutions thusly obtained are necessarily artificial, until proven otherwise. Ill-posed problems must be treated analytically first, as the inverse conduction problem (Beck et al., 1985, Chapters 4 and 5), or they must be reformulated on the basis of physics by introducing previously ignored processes, or by retaining processes which, although previously discarded as insignificant, are known to change drastically the character (classification) of the governing differential equations (see Section 2.4 below). It is equally legitimate to reformulate an ill-posed problem by eliminating irrelevant processes (acoustic phenomena in the analysis of kinematic and thermal wave propagation, as an example) and thereby rendering the problem well-posed.

A numerical integration or solution is acceptable if it is shown to approximate the analytical solution and not some test data. A comparison between computed results and test data may be necessary but is never sufficient, because such a comparison fails to reveal compensating errors, particularly the compensation of modeling deficiencies through computing errors. A numerical solution must be shown to be acceptable, based on the combination of

1) estimates of computing error bounds,
2) comparisons with analytical solutions,
3) substituting the results into the original equations.

Only those three criteria link the numerical solution uniquely to the solution of the analytical formulation.

1.2 Importance of Computational Methods

Since numerical methods are the only means for integrating the multiphase flow field equations for relevant circumstances, these methods are simply indispensable for fundamental research in, as well as for the applications of, multiphase flow analyses.

Even though some insight into multiphase flow can be derived directly from the differential equations without integration, this often requires also numerical calculations (of eigenvalues for example), and ultimately, it is necessary to integrate numerically the complicated partial or ordinary differential equations. It is impossible without computational techniques to relate theory to experiments, to make predictions or to gain insight when complicated multiphase flow is the concern. Design and safety analyses involving multiphase flow are unthinkable without computer calculations.

In the context of unresolved issues in multiphase flow however, computational problems rank far behind the issues of transfer laws (closure). One needs only to recall the long history of boundary layer analysis which led to the prediction of transfer laws in single-phase flows in simple geometries and well-defined flow regimes, along with the associated evolution of sophisticated instrumentation (schlieren photography, interferometry, hot-wire and laser anemometry, fiber deflection techniques and wall shear measurements by strain gauges), to fathom the difficulties of predicting transfer laws in multiphase flows. Here the geometry is chaotic and changes stochastically. The geometric scales are much smaller in the internal structure of multiphase flow. Any advances in the development of multiphase flow transfer laws must be preceded by the successful development of new instrumentation.

Our computational capabilities are far ahead of our abilities to predict multiphase flow regimes and their associated transfer laws for the exchange of mass, momentum and energy between the phases and for the exchange of momentum and energy between each phase and enclosing walls. These laws may be considered as part of the data base of multiphase models, just as the intrinsic constitutive laws for materials properties. Not even the most complicated multiphase model can produce more information than is contained in its data base (Wulff 1986). Existing two-phase flow thermohydraulics codes now employ detailed and complicated two-fluid models (TRAC-PF1/MOD1 1986; Ransom et al. 1985), yet they have data bases which are derived from mixture models. Thus they produce no more reliable detail than mixture models even though they absorb much more computing resources.

From a wider perspective, computational methods for digital computers belong in the range of discrete mathematics because computers are "discrete" machines, dealing with discrete points on the number line. Euler (1768) started the use of finite difference methods from calculus, which is part of continuous mathematics, dealing with continuous functions. Now, it can be expected that discrete mathematics will grow as part of computer sciences to become as important to modern science as calculus was at the time of Newton and Leibniz.

1.3 The Limitations of Computational Methods

Computational methods are unsuitable (1) to overcome the consequences from incorrectly formulated mathematical models, (2) to recover from the misapplication of a mathematical model or (3) to generate the "physics" which have been omitted from the mathematical model. It has become a serious problem of two-phase flow analyses by computers that rational inquiry into the physics of fluid and thermodynamics has been frequently abandoned in a preoccupation with numerical "fixes". Below are just one example each for the above three categories of "missions impossible".

(1) Phase separation has been analyzed by using moving regime boundaries between computational control volumes. Local balance equations have been integrated and then averaged over the control volume enclosed by moving boundaries. The second term on the right-hand side of Eq. (1.1) below, arising from Leibniz's rule, has consistently been omitted at first*

$$\iiint_{V(t)} \frac{\partial \psi}{\partial t}\, dV = \frac{d}{dt} \iiint_{V(t)} \psi dV - \iint_{A(t)} \psi \vec{v}_i \cdot d\vec{A}, \tag{1.1}$$

---

*Symbols are defined in the Nomenclature.

and afterwards, the model has often been "tuned" to produce agreement with experimental data (Snidow and Wilson, 1984; Fakory, 1985; Oh and Sursock, 1985; Baek and No, 1986; Findlay, 1961; Ahl, 1985 and many others).

(2) The two-phase flow equations are invariably misapplied to predict the position of a horizontal water or mixture level in a vertical duct. The level is understood in these cases to be the location of a sharp discontinuity in vapor volume (void) fraction.

Firstly, it must be understood that the two-phase flow equations have been derived by time averaging which obscures the location of interfaces between the phases. Secondly, the derivation of the two-phase flow equations certainly involved the use of the Gauss divergence theorem which implies continuous first-order partial derivatives and continuity of all integrands. This excludes the mixture level from the domain of the two-fluid model. The level surface is a boundary, and the two-phase flow model applies only to the open set of interior points of the mixture. Thirdly, the two-phase flow model may allow discontinuities to propagate along its singular or characteristic surfaces (see Eq. 2.25 below). In one-dimensional flows, void discontinuities move with the vapor void propagation velocity C_k, which differs markedly from the speed U of the level (Wulff 1978, Fig. 5). Using the drift flux model to compute C_k and the first principle of mass conservation at the level for computing U, one recognizes that

$$C_k = j\,\frac{\partial(\alpha C_o)}{\partial\alpha} + \frac{\partial(\alpha v_{vj})}{\partial\alpha} \neq U = \frac{j_v^+ - j_v^-}{\alpha^+ - \alpha^-} \tag{1.2}$$

and that C_k is not defined at a discontinuity. The balance equations for the continuum do not contain the information regarding the motion of a mixture level. No numerical artifice, like "level sharpener," increased bubble rise velocity, "artificial drag reduction" or "shock capturing method" will compensate for the misapplication of continuum equations to discontinuities.

On the other hand, if one were to consider the mixture level as a narrow zone of continuous density change and to use the continuum model with donor-cell differencing, then one would have to have first the instrumentation to measure and model drift velocity or drag dependencies on void fraction in the narrow zone at the level, accurately enough for differentiation, and then one would have to place at least 20 computational mesh points within this narrow zone to keep the void fraction change from decaying by numerical diffusion to less than approximately 40% (Stuhmiller and Ferguson

1979, p. 17). This approach is inacurate and impractical from a computational viewpoint. It is impossible with currently available two-phase flow instrumentation. A viable alternative is the combination of the continuum models for the low and high void two-phase mixtures below and above the level, and the last of Eqs. (1.2) in a Lagrangian reference frame for the level position (Wulff 1981).

It has been argued that existing and familiar "shock-capturing numerical methods" for single-phase gas dynamics could also be used to predict the time-dependent location of kinematic discontinuities in two-phase flow. Single-phase flow involves only three velocities at the discontinuity: up- and down-stream gas velocities and the shock velocity. Two-phase flows involve six or seven velocities, depending on whether phase change takes place at the interface. Whence there is no complete analogy between a gas dynamics shock and a kinematic "shock" in two-phase flow, and "arguments by association" are not convincing. It should be clear that any numerical method is acceptable only if it satisfies the second part of Eq. (1.2) and if it is computationally economical.

(3) The third example of frequently encountered over-extensions of computational techniques is the generation of "physics" or pseudo-physics with numerics. Whenever discrete flow parameters Y_j, space mesh size Δx, Δz, etc., or time step size Δt appear in the description of flow regime criteria (Taylor et al. 1984, Eq. 111, p. 28), level velocity (Taylor et al. 1984, Eq. 117, p. 30) or anywhere else, then the analytical model formulation is incomplete, the model is in conflict with physics (latter case above) and the results are artificial. Such modeling attempts reflect the failure of the modeler to distinguish between modeling to produce a mathematical model, and the process of solving the problem posed by the model.

Computational methods are also unsuitable to cloak intuitive remedies (Taylor et al. 1984, Eq. 443, p. 163; Ransom 1985, pp. 162-165). Solution techniques must be reducible to documented procedures which satisfy the requirements of consistency, convergence and stability, and an estimate of the error bound must be provided.

1.4 Scope of This Review

In Section 2 below are presented the fundamental forms of multiphase and mixture equations, their classification and some important relations between equation classification and physics. Section 3 presents six important principles which link modeling with model formulation and computations.

Sections 5 through 9 present the major integration methods which reduce the partial differential and algebraic equations of multiphase flow models either to ordinary differential equations or to a system of nonlinear difference equations. Sections 10 and 11 summarize the algorithms used for integrating systems of nonlinear ordinary differential equations and for solving nonlinear algebraic equations. In Section 12 I summarize important aspects of numerical stability, accuracy and convergence.

The major systems codes used for predicting transient two-phase flow in nuclear reactors are summarized in Section 13. Section 14 presents future needs in computational methods development.

This review is intended as a tutorial and concentrates on computational methods. It is based on the premise that a mathematical model has been formulated, accounting for all the relevant physics, and that a numerical solution is sought. The material for this review is drawn from published journal articles, Government reports and textbooks. The review reflects the widespread need for increased attention to standard references, particularly in the development of large two-phase flow computer codes for nuclear reactor safety analyses. This review is neither exhaustive nor all-inclusive. Instead, references are made to further details in the literature. This review is also biased toward the practical rather than theoretical aspects of computing, as it is written by an engineer and not a mathematician. Ideally, engineers should not invent their own numerical schemes as has been the case in the past. Rather they should be concerned with two-phase flow instrumentation, experiments, equipment and with mathematical modeling and then apply algorithms which are well perfected, clearly defined and documented by mathematicians. The utilization of such algorithms in the form of separate software packages or built-in firmware has been very successful (cf. Sections 13.6 and 13.8).

2. GOVERNING EQUATIONS FOR MULTIPHASE FLOWS

Two-phase flow equations have been discussed recently in detail at a special workshop (Kestin and Podowski 1985; Drew and Wood 1985). They had been derived even earlier by Delhaye (1968), Vernier and Delhaye (1968), Ishii (1971), Kocamustafaogullari (1971, area averaging), Ishii (1975, time averaging), Delhaye and Achard (1976, 1977, double time averaging), and they can be found in the book by Delhaye, Giot and Riethmuller (1981). The aim here is to establish the form of the field equations and thereby the basis for their classification and numerical solution. The discussion is limited to two-phase flows for reasons of simplicity. Multiphase flows are governed by the same types of equations and are subject to the same principles described here for two-phase flow.

The conservation or field equations for heterogeneous multiphase flows must be derived by means of rational mechanics and appropriate averaging, as in the reference cited above. Field equations must not be postulated from intuition [see e.g., Stewart and Wendroff (1984)], because they produce only postulated results which are unacceptable. Field equations derived from first principles and closure equations obtained from experiments are the principal focus where experimentalists, modelers and computational fluid dynamicists must joint to predict multiphase flow behavior by computer simulation.

2.1 The Local Instantaneous Balance Equations

Fluid flow behavior is described by the laws of mass, momentum and energy conservation. Beyond that, the entropy inequality can be used to assure compliance with the Second Law of thermodynamics, but it does not contribute directly to the prediction of flow conditions. Only if all irreversible dissipative processes and the entropy source terms (Δs) are modeled, can the entropy inequality

$$\rho \frac{Ds}{Dt} + \nabla \cdot \frac{\vec{q}}{T} = (\Delta s) > 0 \tag{2.1}$$

become a balance equation and then substitute for the energy balance. Also, the angular momentum balance contributes no new equation, it only imposes symmetry on the stress tensor Π.

The local balance equations are (Eringen 1962, p. 221; Delhaye, Giot and Riethmuller 1981, p. 101) for the k-th phase

$$\frac{\partial(\rho_k \psi_k)}{\partial t} + \nabla \cdot (\rho_k \psi_k \vec{v}_k + \vec{J}_k) = \rho_k \varphi_k , \quad k = 1,2,... \tag{2.2}$$

where ψ_k, φ_k and $\vec{J}_k$ are defined in Table 2.1 below. Equations (2.2) are valid in the open domain of that space which is entirely occupied by phase k. The definition of a phase, given in the last paragraph of the Introduction, assures continuity of the derivatives involved in Eqs. (2.2). At the phase boundary, e.g. the interface between phases 1 and 2, the conservation laws become jump conditions. They are

$$\sum_{k=1}^{2} \left(\dot{m}_k \psi_k + \hat{n}_k \cdot \vec{J}_k + \varphi_i \right) = 0, \tag{2.3}$$

where

$$\dot{m}_k = \rho_k (\vec{v}_k - \vec{v}_i) \cdot \hat{n}_k \tag{2.4}$$

and $\hat{n}_k$ is the outward normal of phase k at the interface, which moves with velocity $\vec{v}_i$.

TABLE 2.1 Definition of ψ, φ and J

Conservation of	ψ_k	$\vec{J}_k$	φ_k	φ_i
Mass	1	0	0	0
Linear Momentum	$\vec{v}_k$	$-\Pi_k$	$\vec{F}$	0
Energy	$u_k + v_k^2/2$	$\vec{q}_k - \Pi_k \cdot \vec{v}_k$	$\vec{F} \cdot \vec{v}_k$	0
Entropy	s_k	$\vec{q}_k/T_k$	$(\Delta s)_k/\rho_k$	Δs_i

Any attempt at integrating Eqs. (2.2), subject to closure, initial conditions, and also the boundary conditions in Eqs. (2.3), would amount to the computationally impossible task of predicting the flows in detail between continually deforming bubbles or droplets. Therefore, Eqs. (2.2) are averaged over ensembles, time or space, depending on geometry, flow pattern and application. Ensemble or time-averaging produces multidimensional flow equations (Ishii 1975). area-averaged equations are suitable for one-dimensional separated flows (Kocamustafaogullari 1971). Time and area-averaging yields one-dimensional two-phase flow equations for flow patterns with mixed phases (bubbly, churn-turbulent, etc.).

2.2 Averaged Phasic Equations

Since the same mathematical form results from all types of averaging, I present only the time-averaged equations. The reader should refer for further details on time averaging to the book by Ishii (1975).

If t_k is the cummulated time at which phase k occupies a point $\vec{x}$ of interest during the interval t_a of time averaging, then the phase time average of a function f is defined by

$$\bar{f}^k = \frac{1}{t_k} \int_{[t_k]} f\, dt \;, \; [t_k] = t_k \epsilon t_a \;, \tag{2.5}$$

and the local time fraction of phase k is

$$\alpha_k = t_k / t_a \;. \tag{2.6}$$

When k represents the gas phase (k=G), then $\alpha_G = \alpha$ is the void fraction. The time-averaged form of Eqs. (2.2) is the fundamental balance equation:

$$\frac{\partial\{\alpha_k \overline{(\rho_k \psi_k)}^k\}}{\partial t} + \nabla \cdot \{\alpha_k [\overline{(\rho_k \psi_k \vec{v}_k)}] + \overline{\vec{J}_k}^k\} - \alpha_k \overline{(\rho_k \varphi_k)}^k =$$

$$- \sum_{j=1}^{n} (\dot{m}_k \psi_k + \vec{J}_k \cdot \hat{n}_k + \varphi_i)_j / l_i \;, \tag{2.7}$$

where ψ, $\vec{J}$, φ are defined in Table 2.1 above, further

$$l_j = \left| \vec{v}_i \cdot \hat{n}_k \right|_j t_a \tag{2.8}$$

and the summation over j means summation over all n interfaces passing through $\vec{x}$ during the time t_a.

Equations (2.7) reveal already the form of the governing equations for multiphase flow, even though the averages of products need to be replaced by (Ishii 1975)

$$\overline{\rho_k \psi_k}^k = \overline{\rho_k}^k \tilde{\psi}_k \tag{2.9}$$

$$\overline{\rho_k \psi_k \vec{v}_k}^k = \overline{\rho_k}^k \tilde{\psi}_k \vec{v}^k + \vec{J}_k^T \tag{2.10}$$

so that a workable set of partial differential equations is obtained, which contains measurable parameters and standard thermodynamic properties. Here ψ is the mass weighted time average of the flow parameter defined in Table 2.1 and $\vec{J}^T$ is the turbulent part of the convective flow.

There are three equations of the form of Eqs. (2.7) for every phase. As shown by Ishii (1975, p. 188), the two fluid model for two-phase flow, as described by six balance equations in the form of Eqs. (2.7), contains 32 variables, requiring 26 additional equations. These are:

(i) three equations describing interface conditions (see Eqs. 2.3),
(ii) three equations for mechanical, chemical and saturation conditions,
(iii) one axiom of continuity, $\Sigma\alpha_k = 1$,
(iv) four equations of caloric and thermal states,
(v) one surface equation of state, $\sigma(T_i)$,
(vi) eight constitutive equations for molecular and turbulent heat fluxes and stress tensors in each phase,
(vii) two equations for mass transfer and drag,
(viii) one geometrical state equation for the interface,
(ix) two constitutive relations for interface heat flux density and for interface momentum source, and
(x) one balance equation for the evolution of interface area density, $1/l = \sum_i (1/l_i)$ (see Eq. 2.8), namely

$$\frac{\partial}{\partial \tau}\left(\frac{1}{l}\right) + \nabla \cdot (\vec{v}_i/l) = \varphi_L \tag{2.11}$$

The above 26 closure equations are called the *data base* of the two-fluid model. The integration of Eqs. (2.7) requires this data base and also initial and boundary conditions. Thus the mathematical problem formulation consists of the balance equations, closure equations (data base), initial and boundary conditions.

Quite often individual processes are combined for the sake of simplicity: turbulent and molecular conductivities and diffusivities combined into overall conductivities and overall diffusivities, respectively, lump together turbulent and molecular dissipation processes. Spatial averaging replaces some of the gradient terms through algebraic correlations. However, the difficulties remain of specifying, for *n* phases, 2n transfer laws for the transfers of energy and momentum between each phase and the solid walls and 3n(n-1)/2 laws for the transfers of mass, momentum and energy between the phases. This means that n(3n+1)/2 flow pattern-dependent laws are needed as part of problem closure: 2 laws for single-phase flow, 7 laws for two-phase flow, 15 laws for three-phase flow, 26 laws for four-phase flow, etc. These laws are not known, except for a few flow patterns (Ishii and Chawla 1979a and b; Ishii and Mishima 1980; Ishii, Kataoka and Kocamustafaogullari 1981; Kataoka, Ishii and Serizawa 1984), because appropriate two-phase flow instrumentation is not yet available.

Equations (2.7) apply to the open domain of space which is occupied by the multiphase *mixture*. In this domain they are taken to have continuous derivatives* and to represent coexisting continua. If they are applied to a space occupied by both the multiphase mixture and solid structures, as in nuclear reactor core hydraulics codes, then they must be averaged again over the volume V which is small compared with the volume of the system of interest but large relative to the scales of the solid structures and the multiphase mixture. This averaging introduces the theorem on gradients (Slattery 1972; p. 196):

$$1/V \iiint_V \nabla B \, dV = \nabla(1/V \iiint_V B \, dV) + 1/V \iint_{A_W} B \cdot d\vec{A}, \qquad (2.12)$$

Where B represents any second order tensor, vector or scalar and A_W is the solid wall surface inside of the volume V. Only a few two-phase flow models for multidimensional flows account for both terms on the right-hand side of Eq. (2.12) (Fanselau et al. 1980), while most others omit the first one (TRAC-PF1/Mod 1 1986).

Simpler models can be derived from Eqs. (2.7), as was demonstrated by Boure (1975a,b and 1976). A general overview of two-phase flow models and their major characteristics was presented by Wulff (1981). By writing Eqs. (2.7) for each of the phases and by adding the equations, one obtains one mixture balance equation each for mass, momentum and energy

---

*Notice Delhaye, Giot and Riethmuller (1981, p. 175) on double time averaging.

$$\frac{\partial(\rho_m \psi_m)}{\partial t} + \nabla \cdot \{\rho_m \psi_m \vec{v}_m + \vec{J}_m\} - \rho_m f_m = I_m \,, \tag{2.13}$$

in which the exchange terms [right-hand side of Eqs. (2.7)] cancel at least in part on account of Eqs. (2.3). Equation (2.13) is simpler than n phasic equations for n phases, but it has still the same form as Eqs. (2.7). The mixture equations, Eqs. (2.13), can be written in terms of the mixture mass velocity

$$\vec{v}_m = \left(\sum_k \alpha_k \rho_k \vec{v}_k\right) / \sum_k \alpha_k \rho_k \tag{2.14}$$

and the diffusion velocities $\vec{v}_{km} = \vec{v}_k - \vec{v}_m$ to produce the mixture equations of the diffusion model, or they can be written in terms of the mixture volumetric flux

$$\vec{j}_m = \sum_h \alpha_k \vec{v}_k \tag{2.15}$$

and the drift flux velocities $\vec{v}_{jk} = \vec{v}_k - \vec{j}_m$ to yield the mixture equations of the drift flux model. In the former case, $\vec{j}_m$ contains diffusion terms, in the latter case it contains the drift terms (Ishii 1975, p. 94 and 125). Notice that $\vec{j}_m$ is almost always continuous, while $\vec{v}_m$ is often discontinuous.

The forms of Eqs. (2.7) and (2.13) are the same, only the number of equations and the transfer terms change. As shown in Section 2.4 below, the character of Eqs. (2.7), (2.11) and (2.13) depends entirely on their derivative terms. Additional derivative terms may arise from the so-called source terms φ and I. Allowance will be made in the next section for additional first-order derivatives originating from the source or transfer terms.

2.3 Form of Governing Equations

The governing equations, Eqs. (2.7) and (2.11) have been summarized in their principal form in Section 2.2 above. Here I demonstrate the equivalence between these balance equations and a set of quasi-linear first-order partial differential equations in four independent variables. In the next section I present the conditions for, and the importance of, these equations to be hyperbolic, parabolic or elliptic.

Indicial notation will be used in Sections 2.3, 4 and 5 to represent vectors and tensors

$$\vec{v} = \sum_{k=1}^{3} v_k \cdot \hat{e}_k \equiv v_k \hat{e}_k$$

$$A \hat{=} A_{ij}\cdot, \qquad (2.16)$$

where $\hat{e}_k$ are unit vectors. Since the balance equations are valid in any coordinate system, I select rectangular Cartesian coordinates and thereby dispense with the distinction between covariant and contravariant components; also, covariant differentiation becomes simply partial differentiation with respect to fixed spatial coordinates x_j. Repeated indices imply summation (Einstein convection).

The m balance equations [Eqs. (2.7) and (2.11)] define m state variables. Let w_i^* designate the i-th component of the m-dimensional state variables vector $\{\bar{u}^1_1, \bar{u}^2_2, \bar{\rho}^1_1, \bar{\rho}^1_2, (\tilde{v}_1)_1, (\tilde{v}_1)_2, (\tilde{v}_1)_3, (\tilde{v}_2)_1, \ldots, \alpha. \ldots\}^T$, and recognize that the heat flux vector in Table 2.1 is proportional to the temperature gradient

$$q_j = - k_{eff} \frac{\partial T}{\partial x_j} = - k_{eff}\, T_{,j}\ ,\ j = 1, \ldots, 3, \qquad (2.17)$$

where the comma, followed by a subscript j, means partial differentiation with respect to the j-th spatial coordinate x_j, $j = 1,2,3$. If one can express the time-averaged stress tensor as a linear relation of the velocity derivatives (Eringen 1962, p. 238).

$$t_{ij} = -pI_{ij} + \mu_{eff}(v_{i,j} + v_{j,i}) \qquad (2.18)$$

where I_{ij} is the identity tensor with elements δ_{ij} and if the body force F_i is the gradient of a potential field, then the m balance equations, Eqs. (2.7) and (2.11), have this form:

$$A^*_{ij} \frac{\partial w^*_j}{\partial t} = B^*_{ijk}\, w^*_{j,k} + (C^*_{iljk}\, w^*_{k,j})_{,l} + D_i \qquad (2.19)$$

$$i,j = 1,\ldots, m$$
$$k,l = 1,2,3\ .$$

Here, the order of summations is immaterial. A_{ij}^* is an $m \times m$ square matrix, B_{ijk}^* may be viewed as three $m \times m$ square matrices and C_{iljk}^* as nine $m \times m$ square matrices. Equations (2.19) are linear in the highest-order derivatives, namely the second derivatives $w_{k,jl}^*$, but nonlinear in the first-order derivatives because A_{ij}^*, $B_{i\,jk}^*$, D_i and, importantly, C_{iljk}^* indeed depend on t, x_i and w_j^*. Equations (2.19) are most therefore quasi-linear second-order partial differential equations.

Depending on the number of non-zero elements of C_{iljk}^*, we introduce up to 3m new state variable vector components w_s and 3m new state equations

$$C_{ljk}^* \, w_{k,j}^* - \gamma_{il} = 0 \quad , \quad i = 1,\ldots, m \qquad l = 1,2,3 \,. \tag{2.20}$$

The vector components w_s^{**}, $s = 1,\ldots,$ 3m, are the elements of the tensor γ_{il}, $i = 1,\ldots, m$ and $l = 1,2,3$. The index s is computed from the indices i and l by

$$s(i,l) = m(i-1) + l \ , \ i = 1,\ldots, m; \ l = 1,2,3 \,.$$

By substituting Eqs. (2.20) into Eqs. (2.19), one obtains a quasi-linear first-order partial differential equations

$$A_{ij}^* \frac{\partial w_j^*}{\partial t} = B_{ijk}^* \, w_{j,k}^* + \gamma_{il,l} + D_i \ , \ i,j = 1,\ldots, m; \ k,l = 1,3$$

or
$$A_{ij} \frac{\partial w_j}{\partial t} = B_{ijk} \, w_{j,k} + D_i \ , \qquad i,j = 1,\ldots, 4m \qquad k = 1,2,3 \,. \tag{2.21}$$

The system of 4m quasi-linear first-order equations, consisting of m balance equations, Eqs. (2.19), and of 3m supplementary equations, Eqs. (2.20), define the temporal and spatial variations for the 4m vector components of the column vector w_i, $i = 1,\ldots,$ 4m

$$\{w_1, ..., w_{4m}\}^T = \{w_1^*, ..., w_m^*, w_1^{**}, ..., w_s^{**}, ..., w_{3m}^{**}\}^T \qquad (2.22)$$

The system of equations given by Eqs. (2.21) is equivalent to the system of m second-order differential equations given by Eqs. (2.19) (Courant and Hilbert 1962, p. 13). Thus, it is shown that the second-order partial differential equations for multiphase flow can be written in terms of quasi-linear first-order differential equations, provided that the flux vectors $\vec{J}$ can be represented as proportionalities, Eqs. (2.17) and (2.18), where the coefficients k_{eff} and μ_{eff} depend on the independent variables t, x_i, $i = 1...$, 3, and on the dependent variables $w*$, but not on their derivatives $w*_{,j}$.

The analysis of transient multiphase flow leads to a mixed (initial and boundary value) or "propagation" problem.+ With Eqs. (2.21) assumed to be analytic, their specific solutions can be found which satisfy given analytical initial values $w_j^o(t = 0, x_1^o, x_2^o, x_3^o)$ at the initial time $t = 0$ and also satisfy prescribed analytical conditions $w_j(t, x_1^b, x_2^b, x_3^b) = 0$ on open boundaries as discussed in the next section. The integration domain is open so that the far future does not affect the near future, present or past. Instead, all subsequent behavior of the multiphase mixture is to be determined from given initial and boundary data. Conditions under which a unique solution is possible are contained in the Cauchy-Kowalewsky theorem (Garabedian 1964, p. 6; Courant and Hilbert 1962, p. 39; Smirnow 1961, Teil IV, p. 261). In essence, the theorem implies that Eqs. (2.21), the initial and the boundary conditions all must be analytical and the data must be given on such hypersurfaces $\varphi(t, x_1, x_2, x_3) = 0$, that all derivatives, particularly one that is not in a tangent direction to $\varphi = 0$, can be computed uniquely from Eqs. (2.21) and the data. Otherwise the data on φ are restricted.

2.4 Classification of Partial Differential Equations

The methods used for solving partial differential equations depend very strongly on the type of these equations, whether the integration is by analytical or numerical methods. Systems of quasi-linear first-order partial differential equations++ as given by

---

+Sometimes called "marching" problems (Richardson 1925) or simply "initial value" problems.

++Partial differential equations are quasi-linear if they are linear in their highest-order derivatives.

Eqs. (2.21) are either elliptic, parabolic or hyperbolic, depending on A_{ij} and B_{ijk}.

Elliptic equations can represent only pure boundary value problems, that is, steady-state problems. The solution $w_j\ (x_1, x_2, x_3)$ depends everywhere on all the boundary data. This case is of interest only if $A_{ij} = 0$ for all i and j.

Parabolic and hyperbolic equations describe propagation problems, namely transient processes, the future of which is defined entirely by initial conditions and appropriate external effects (boundary conditions).

Equations (2.21) are (within the limitations of this review) the most general three-dimensional multiphase flow equations. Yet A_{ij} and B_{ijk} can take on many different forms, reflecting particular choices of primary or state variables (enthalpy or internal energy, velocity or velocity difference, for example) or of coordinate systems. Multiphase flow analyses are carried out in two- and three-dimensional space, but the vast majority of multiphase flow models are one-dimensional, in close relation to the multiphase instrumentation in test facilities. As shown in Section 2.5 this simplifies the study and the solution of Eqs. (2.21) enormously, as B_{ijk} becomes a single $m \times m$ square matrix.

The reduction of the three-dimensional to a one-dimensional problem entails the first step with important consequences: area-averaging leads to the replacement of dissipative *gradient* terms (such as temperature gradients in lateral directions) by algebraic correlations (convective heat transfer) which results in the removal of some B_{ijk} elements and the addition of some D_i elements. Ultimately, the complete omission of all (perhaps quantitatively unimportant) dissipative flux terms C^*_{ijlk} in Eqs. (2.19), or γ_{il}-terms in Eqs. (2.21), has a profound influence on the solution because the reduced equations are no longer parabolic.

Therefore, the classification of partial differential equations is very important. However, its presentation here is not specific to any particular model. Rather, the method for classification is given here in general terms and should be applied to a particular model on a case-by-case basis, as has been done for the RETRAN code (Eide and Hughes 1981). Such an analysis is an important basis for selecting computational methods.

2.4.1 *Characteristic Surfaces*

We wish to integrate the first-order system of partial differential equations given by Eqs. (2.21). Since it is a first-order system the Cauchy conditions are, that the initial values be specified at $t = 0$ for every point $\{x_1, x_2, x_3\}$

$$w_j\Big|_{t=0} = \psi_j(x_1, x_2, x_3)\ ,\ j = 1,\ldots, 4m\ . \tag{2.23}$$

With these data, one can compute all derivatives at $t = 0$, except $\partial w_j/\partial t$. One can substitute $t = 0$ and Eqs. (2.23) into Eqs. (2.21), and attempt to solve these for $\partial w_j/\partial t$, $j = 1,\ldots,$ 4m. If this is successful, then one has all derivatives in the plane $t = 0$. Otherwise, this surface in $\{t, x_1, x_2, x_3\}$-space would be called a *characteristic* surface. In general, a hyper-surface

$$\omega_1(t, x_1, x_2, x_3) = 0 \tag{2.24}$$

with its initial data w_j is called a characteristic surface if one cannot uniquely compute all first-order partial derivatives on that surface with the given data and the given differential equations, Eqs. (2.21) (Smirnow 1961, Teil IV, p. 412).

To find the conditions under which Eq. (2.24) represents a characteristic surface, we introduce coordinates $\tilde{t}$ and $\tilde{x}_j$, $j = 1,2,3$, by

$$\tilde{t} = \omega_1(t, x_1, x_2, x_3) \tag{2.25}$$

$$\tilde{x}_j = \omega_j(t, x_1, x_2, x_3)\ , \qquad j = 2,3,4 \tag{2.26}$$

such that we can solve Eqs. (2.25) for t, and x_i, $i = 1,2,3$. By expressing the derivatives in Eq. (2.21) in terms of the new coordinates, we write first

$$\left.\begin{aligned} \frac{\partial w_j}{\partial t} &= \frac{\partial w_j}{\partial \tilde{t}}\frac{\partial \omega_1}{\partial t} + \sum_{l=2}^{4} \frac{\partial \omega_j}{\partial x_{l-1}}\ \partial \omega_l \\ \frac{\partial w_j}{\partial x_k} &= w_{j,k} = \frac{\partial w_j}{\partial \tilde{t}}\frac{\partial \omega_1}{\partial x_k} + \sum_{l=2}^{4} \frac{\partial w_j}{\partial \tilde{x}_{l-1}}\frac{\partial \omega_l}{\partial x_k}\ , \end{aligned}\right\} \tag{2.27}$$

then substitute these results into Eqs. (2.21) and finally collect all terms with $\partial w_j/\partial t$. This leads to

$$\left[A_{ij} \frac{\partial \omega_1}{\partial t} - B_{ijk} \frac{\partial \omega_1}{\partial x_k} \right] \frac{\partial w_j}{\partial \tilde{t}} = \ldots , \tag{2.28}$$

where the three periods represent terms which do not contain the unknown derivatives $\partial w_j/\partial \tilde{t}$ and are therefore completely defined by initial data on the surface $\omega_1 = 0$. This surface is characteristic if Eqs. (2.28) cannot be solved to give a unique solution for $\partial \omega_j/\partial \tilde{t}$, $j = 1,\ldots, 4m$, that is, if the determinant

$$\left| \omega_{ij} \right| = 0 , \tag{2.29}$$

where the ω_{ij} are the elements of the square 4m x 4m matrix in the square brackets of Eqs. (2.28), namely

$$\omega_{ij} = A_{ij} \frac{\partial \omega_1}{\partial t} - B_{ijk} \frac{\partial \omega_1}{\partial x_k} . \tag{2.30}$$

Equation (2.29) is a first-order differential equation for $\omega_1(t,x_1, x_2, x_3)$. It is of degree 4m in terms of the four derivatives $\partial \omega_1/\partial t$ and $\partial \omega_1/x_k$, $k = 1,2,3$.

If the determinant on the left-hand side of Eq. (2.29) is not equal to zero on $\omega_1 = 0$ then all derivatives of w_j can be computed and w_j can be expanded in a Fourier series to develop a solution in the neighborhood of $\omega_1 = 0$. The surface $\omega_1 = 0$ is then called a free surface (Courant and Hilbert 1962, p. 173). If, on the other hand, the above determinant is equal to zero, then there are restrictions to be imposed on the initial data on ω_1 (Courant and Hilbert 1962, p. 174).

2.4.2 *Classification*

Equations (2.29) and (2.30) are important for the classification of Eqs. (2.21). Since the coefficients A_{ij} and B_{ijk} for a quasi-linear partial differential equation depend not only on the time t and spatial coordinates x_1, x_2, x_3, but also on the solution itself, the classification of Eqs. (2.21) depends also on its solution. Therefore we consider only some domain D in the neighborhood of the solution w_j^o on some surface ω_1^o in (t, x_1, x_2, x_3)-space.

(a) If Eq. (2.29) has fewer solutions for $\partial\omega_1/\partial t$ and $\partial\omega_1/\partial x_k$, $k = 1,2,3$ than there are equations in the system of Eqs. (2.21), then that system is *parabolic.*

(b) If Eq. (2.29) has no real solutions for $\partial\omega_1/\partial t$ and $\partial\omega_1/\partial x_k$, $k = 1,2,3$, except for $\partial\omega_1/\partial t = \partial\omega_1/\partial x_k = 0$, then the system of Eqs. (2.21) is called *elliptic* (Courant and Hilbert 1962, p. 174),

(c) If Eq. (2.29) possesses (4m) different real solutions for arbitrarily prescribed values of $\omega_2, \ldots, \omega_{4m}$ (or if a corresponding statement is true after a coordinate transformation) then the system of Eqs. (2.21) is totally *hyperbolic* (Courant and Hilbert 1962, p. 174).

Conditions (b) and (c) above imply also the interpretation given by Garabedian (1964) of a Fourier expansion for the solution ω_j to Eqs. (2.21) about a specific point $(t^o, x_1^o, x_2^o, x_3^o)$. If one sets D_i equal to zero then one can show that for constant A_{ij} and B_{ijk}

$$w_j = W_j \, e^{i(\lambda_1 x_1 + \ldots + \lambda_3 x_3)} e^{i\lambda t} \tag{2.31}$$

satisfies Eqs. (2.21), where $\lambda, \lambda_1, \ldots, \lambda_3$ are fixed parameters such that

$$\left| A_{ij}\lambda - B_{ijk}\lambda_k \right| = 0 \tag{2.32}$$

and

$$\lambda A_{ij} U_j - \lambda_k B_{ijk} U_j = 0 \tag{2.33}$$

defines a non-trivial (4m)-dimensional vector. When all λ and $\lambda_1, \ldots, \lambda_3$ are fixed parameters, then Eq. (2.31) is the general term of the Fourier expansion. If every root λ of Eq. (2.32) is real for all choices of real parameters $\lambda_1, \ldots, \lambda_3$ then any oscillatory dependence of Eq. (2.31) on the space variables $x_1, \ldots, x_3$ results in a function w_j that neither grows nor decays exponentially with time t: the solution is stable and depends continuously on initial data, if every root λ is real. This is one of the necessary conditions for a properly posed problem. These observations and the results from two-dimensional quasi-linear problems led Garabedian to say that Eqs. (2.21) are hyperbolic if

(i) A_{ij} is non-singular,

(ii) for every set $(\lambda_1, \lambda_2, \lambda_3)$ the (4m) roots λ of Eq. (2.32) are real, and

(iii) with the (4m) characteristic roots λ are associated 4m linearly independent 4m-dimensional vectors W.

If we now recall that Eqs. (2.21) was derived from the system of m differential equations, Eqs. (2.19), by the addition of 3m equations, Eqs. (2.20), we can see that A_{ij} in Eqs. (2.21) differs from A^*_{ij} by 3m additional rows and columns of zeros, because Eqs. (2.20) contain no time derivatives. This means that Eqs. (2.29) cannot produce 4m solutions (real or complex) for $\partial\omega_i/\partial t$. Therefore, Eqs. (2.19) or their equivalents, Eqs. (2.21) are parabolic.

If one omits the second-order terms $(C^*_{iljk} w^*_{k,j})_l$ in Eqs. (2.19) or $\gamma_{il,l}$ in the first of Eqs. (2.21), then these equations may be elliptic or hyperbolic, depending on A^*_{ij} and B^*_{ijk}.

All that was said here depends on the solution w_j itself. The same model equations may (and actually do) change their classification, depending on the solution.

2.5 Equations for One-Dimensional Multiphase Flow

The vast majority of two-phase flow models used for analyzing such flows in test facilities and industrial systems are transient one-dimensional flow models. Equations (2.7) or (2.13) are area-averaged over the flow cross-sectional area. Some stress and heat flux gradient terms reduce thereby to wall stress and wall heat flux terms, and their associated velocity and temperature gradients are expressed by correlations in terms of mean excess fluid over wall velocity and temperature, respectively. However, dissipations along the streamwise coordinate remain after area averaging.

Equations (2.19) reduce for one-dimensional flow to

$$A_{ij} \frac{\partial w_j}{\partial t} = B_{ij} \frac{\partial w_j}{\partial x} + \frac{\partial}{\partial x} \left(C_{ij} \frac{\partial w_j}{\partial x} \right) + D_i \ , \ i = 1,\dots, m \ , \qquad (2.34)$$

and this system of equations is parabolic (cf. Section 2.4.2).

The second-order terms in Eqs. (2.34) represent dissipation in the flow direction and are frequently found to be quantitatively insignificant. If they are omitted, then Eqs. (2.34) are easier to integrate. In fact, all currently used major systems codes for the analysis of two-phase flow in nuclear reactors impose $C_{ij} = 0$. As pointed out earlier, multiphase flow problems are propagating problems and their governing field equations must be parabolic (as are Eqs. (2.34)) or hyperbolic. Therefore, the dissipative terms can be ignored by setting $C_{ij} = 0$ in Eqs. (2.34) only if

(i) they are quantitatively insignificant *and*

(ii) the determinant $|A_{ij}\lambda - B_{ij}|$ has m distinct and real characterisstic roots λ_k, $k = 1,\ldots, m$, which renders the equations without second-order terms hyperbolic.

Whether or not the dissipative terms can be ignored, depends clearly on the model, i.e., on A_{ij} and B_{ij}. Notice that dissipative processes can be accounted for in D_i, which can contain wall shear, form losses, interphase drag, as well as heat transfer in algebraic relations. To render Eqs. (2.34) in the simple form most frequently used, one needs to be concerned only with dissipation in stream-wise direction and with the coefficients A_{ij} and B_{ij}.

To develop a solution for Eqs. (2.34) with $C_{ij} = 0$; $i,j = 1,\ldots, m$, we pre-multiply Eqs. (2.33) with the inverse A^{-1}_{ik} of A_{ik} to obtain

$$\frac{\partial w_k}{\partial t} = E_{kl}\frac{\partial w_l}{\partial x} + F_k \; ; \; k,l = 1,\ldots, m, \tag{2.35}$$

where $E_{kl} = A^{-1}_{ki}B_{il}$ and $F_k = A^{-1}_{ki}D_i$. Then we transform the state variables w_k into new state variables g_k according to

$$w_k = M_{kl}g_l \,, \tag{2.36}$$

where the elements of M_{kl} depend only on t and x. By substituting Eq. (2.36) into Eq. (2.35) one finds

$$M_{kl}\frac{\partial g_l}{\partial t} = E_{kl}\left[M_{lm}\frac{\partial g_m}{\partial x} + \frac{\partial M_{lm}}{\partial x}g_m\right] + F_k - \frac{\partial M_{kl}}{\partial t}g_l$$

or, by pre-multiplying with the inverse M^{-1}_{ik}, this transformed equation

$$\frac{\partial g_l}{\partial t} = M^{-1}_{li} E_{ij} M_{jk} \frac{\partial g_k}{\partial x} + G_l \,, \tag{2.37}$$

where

$$G_l(t,x,g_k) = M^{-1}_{li} \left\{ \left(E_{ij} \frac{\partial M_{jk}}{\partial x} - \frac{\partial M_{ik}}{\partial t} \right) g_k + F_i \right\} . \tag{2.38}$$

If condition (ii) above is satisfied in some neighborhood of (t^o,x^o) then the determinant $|A_{ij}\lambda - B_{ij}| = 0$ and it can also easily be seen that

$$\left| E_{ij} - \lambda \, I_{ij} \right| = 0 \tag{2.39}$$

has m distinct characteristic roots λ_k. Thus, the matrix M_{jk} in Eqs. (2.37) has the eigenvectors of E_{ij} as columns (Smirnow 1961, Teil III$_1$, p.27) and the product $M^{-1}E_{ij}M_{jk}$ becomes a diagonal matrix

$$M^{-1}_{li} E_{ij} M_{jk} = \lambda_l I_{lk} . \tag{2.40}$$

This in turn reduces Eqs. (2.37) to

$$\frac{\partial g_l}{\partial t} - \lambda_l(t,x) \frac{\partial g_l}{\partial x} - G_l = 0 \,. \tag{2.41}$$

(no summation over l)

Next, we return to the characteristics introduced in Section 2.4.1 by Eqs. (2.25) and (2.26). By comparing Eqs. (2.21) with Eqs. (2.41) above, we obtain $A_{ij} = \delta_{ij}$ and $B_{ijl} = \lambda_l I_{ij}$. Utilizing the definitions given in Eqs. (2.30) and substituting A_{ij} and B_{ijl} into the characteristic equations, Eqs. (2.29), gives

$$\left[\frac{\partial \omega_1}{\partial t} - \lambda_1 \frac{\partial \omega_1}{\partial x} \right] \cdots \left[\frac{\partial \omega_1}{\partial t} - \lambda_m \frac{\partial \omega_1}{\partial x} \right] = 0 \,, \tag{2.42}$$

which can be decomposed into m linear differential equations

$$\frac{\partial \omega_l}{\partial t} - \lambda_l \frac{\partial \omega_l}{\partial x} = 0 \; , \; = 1,\dots, m \; . \tag{2.43}$$

For every solution ω_l, $l = 1,\dots, m$ of this differential equation, there is a family of characteristic curves $\omega_l(t,x) = C$, a constant. Equation (2.43) is equivalent to this ordinary differential equation (Courant and Hilbert 1962, p. 28):

$$\frac{dx}{dt} = - \lambda_l(t,x) \; , \qquad l = 1,\dots, m \; , \tag{2.44}$$

which defines through every point (t,x) of the plane $\{t,x\}$ m characteristic directions. Along any characteristic curve c_l as given by Eq. (2.44) one has, by virtue of Eq. (2.41):

$$dg_l = \left[\frac{\partial g_l}{\partial t} - \lambda_l \frac{\partial g_l}{\partial x}\right] dt = G_l dt \tag{2.45}$$

Thus the solution to Eqs. (2.41) at some point (t_2,x_2) is found by integrating Eqs. (2.44) to find the position x_1 at the starting time t_1 on the curve c_l which passes through the point (t_2,x_2) of interest, and by solving the integral equations along the curve c_l simultaneously for

$$g_l(t_2,x_2) = g_l(t_1,x_1) + \int_{t_1}^{t_2} G_l(g_1,\dots g_m, x(t),t)dt \; . \tag{2.46}$$

Once the g_l are found, Eqs. (2.36) yield the original state variables at the time t_2, provided they are known at the time t_1.

Equations (2.44) and (2.46) are the basis for proving existence and uniqueness of a solution to Equations (2.35) (Smirnow 1961, Teil IV, p. 437), and they are the basis for the solution by the Method of Characteristics (see Section 5 below). The same techniques as shown here can also be applied to systems of n-dimensional quasi-linear first-order equations of hyperbolic type (Courant and Hilbert 1962, p. 173).

Thus, we have shown for the one-dimensional multiphase flow equations: (1) the conditions under which the inherently parabolic second-order quasi-linear equations can be reduced to simpler, hyperbolic first-order equations, (2) the importance of characteristic curves for hyperbolic systems and (3) a formal solution for hyperbolic systems.

It is also clear from Eqs. (2.46), that the solution ω_j for a *hyperbolic* set of equations at some point *(t,x)* can only depend on the history of conditions at points of the {*t,x*}-plane inside the area enclosed by the characteristics passing through *(t,x)* and by the curve, along which initial data were given. This area is called the *interval of dependence* of *(t,x)*. It is equally clear from Eqs. (2.46), that the conditions at some point *(t,x)* can influence only future conditions of points which lie on the characteristics passing through *(t,x)*. This set of points is called the *domain of influence* of *(t,x)*.

In contrast, parabolic sets of equations imply that the conditions at some point *(t,x)* depend on the previous conditions at all locations occupied by the fluid.

2.6 Existence and Uniqueness of Solutions

There is no general mathematical theory to establish existence and uniqueness for multiphase flow field equations, Eqs. (2.7). Even in single-phase flow the theory is lacking (Roache 1972, p. 6), and for multiphase flow the problem is far more difficult.

For quasi-linear hyperbolic equations, existence and uniqueness can be shown (cf. Section 2.4). The Cauchy-Kowalewsky theorem (Courant and Hilbert 1962, p. 39) guarantees existence and uniqueness for systems of quasi-linear homogeneous first-order partial differential equations. This includes certain second-order equations as well (cf. Section 2.3) because of their equivalence with first-order equations.

For multiphase flows, one depends invariably on numerical solutions, and the separate question arises concerning the existence and uniqueness not only of the partial differential equations but also of the numerical scheme. If the numerical scheme produces a unique solution, one needs to ascertain that this solution satisfies the partial differential equations of the mathematical model. This question of consistency will be discussed in Sections 8, 9 and 12.

In the final analysis, all numerical integration schemes fall into one or both of two categories: schemes for integrating ordinary differential equations or schemes for solving systems of nonlinear equations. The question of uniqueness of these schemes will be discussed in Sections 10 and 11, respectively.

2.7 Boundary Conditions

In Section 2.3 was indicated that in the mixed initial and boundary value problem of transient multiphase flow, boundary conditions had to be described on "open" boundaries. The dependencies of elliptic, parabolic and hyperbolic equations on different kinds of boundary conditions was discussed in Section 2.4. Equations (2.46) in Section 2.5 show that for hyperbolic systems of partial differential equations, a set of initial or boundary values is required for every family of characteristic curves. Based on these discussions, we point out here the number and types of parameters which must be prescribed as boundary conditions. When too few boundary conditions are given the solution to the problem will not be unique; no solution exists in general when too many boundary conditions are given (Courant and Hilbert 1962). Initial values, of course, must be specified for all parameters.

One-dimensional multiphase flows in closed loops and circuits accept no boundary conditions, because all conditions are computed from the balance equations and changes are caused by source terms only. Such source terms represent heating, cooling and the effects of pumps.

In purely elliptic problems (steady diffusion problems) state variables must be specified along boundaries which completely enclose the domain of interest. Either variable values or their normal derivatives must be given on each point of any boundary segment.

For hyperbolic problems, the number of boundary (and initial) conditions required at a particular point on the boundary is equal to the number of characteristics pointing into the domain of interest. This number varies between zero* and five for the six-equation, two-fluid model and may change with time.

The parabolic equations have second-order dissipative terms. Here state variables must be prescribed, or their derivatives, at all points of the boundary and boundedness at infinity, if appropriate. A system of partial differential equations for multiphase flow, while parabolic, may contain hyperbolic equations. the mass balance equations, for example, contain no second-order derivatives, and densities as well as void fractions can be imposed only at inflow boundaries, but not at outflow boundaries.

One should always remember: nothing must be imposed as boundary conditions for a mathematical model that cannot be controlled in an associated experiment; conversely, what needs to be

---

*Choked flow at a pipe break.

controlled for a unique evolution of a process in a test facility must also be specified as boundary conditions. In a test facility one controls valves, pump speeds, heating power, pressure regulators, etc., and boundary conditions must ultimately originate from such controls.

3. GENERAL PRINCIPLES OF MODELING AND COMPUTING

In this section are presented six general principles for model selection, model formulation and for the selection of computing algorithms. The purpose of these principles is to achieve the greatest possible modeling fidelity with the minimum of computational efforts. This means, for the present purpose, the smallest possible number of arithmetic and logical operations for computing routinely the evolution of transient multiphase flow (Wulff 1986). The selections of models, computing methods and even of computer architecture are intimately related.

Experience has shown that multiphase flow analysis by digital computers is time consuming and expensive because such flows are complicated. Computing costs have sharply decreased over the years but the imbalance between computing needs and resources has not improved. Those who have hailed the arrival of "zero-cost" computing capabilities have yet to offer the first "zero-cost" computing service. As the size of computing problems grows, even during times of declining computing costs, the challenge to engineering judgment and computational ingenuity will remain. The following six principles are intended to lead toward simple and transparent computing techniques which go hand-in-hand with cost-effectiveness.

3.1 Model Selection

The first principle demands that one should select the least complicated multiphase flow model which accommodates all available experimental information and the particular processes of interest.

No fluid flow model can produce more information than is contained in its data base, regardless of its complexity. Modeling complexity which is not supported by the necessary closure laws (see Section 2.2 for the number of required transfer laws) absorbs computational resources without producing trustworthy results, and hypothetical closure laws, postulated without experimental foundation, lead to results which can hardly be better than postulated answers.

The application of this principle leads to this conclusion:

(i) When the phases of the mixture are well mixed, one should employ mixture equations [cf. Eqs. (2.13)], preferably the drift flux model because it has the strongest experimental support. The more general and more complicated two-fluid model lacks experimental support because needed instrumentation is still under development.

(ii) When the phases are separated and weakly coupled (such as in horizontal stratified flows), one should employ the two-fluid model [cf. Eqs. (2.7)]. Here the interface geometry is relatively regular and transfer laws can be derived from measurement and boundary layer analysis.

3.2 Model Simplifications

The second principle demands that an order-of-magnitude estimate be carried out on all terms in the governing equations (see, for example, Schlichting 1968, p. 118) and that, after comparing the order-of-magnitude of all terms, one eliminates irrelevant, small-order terms.

The elimination of irrelevant terms depends on the flow circumstances and it is subject to the restrictions discussed in Section 2.5. Typical examples of irrelevant terms are the property dependence at high levels of pressure, on local pressure variation, ∇p, further the effects of virtual mass under conditions of moderate acceleration and, finally, the effects of temporal acceleration under conditions of sharp local acceleration near an orifice.

3.3 Analytical Integration

This third principle urges that analytical instead of numerical integrations be used whenever feasible, and that analytical solutions be dynamically evaluated by computer.

Analytical solutions are always derived with conditions (simplifications) which can be quantitatively assessed. Numerical integration is always afflicted with computational errors which are difficult and often impossible to quantify. Numerical integration is associated with inherent tuncation errors, numerical diffusion, damping, dispersion and with issues of convergence, stability and consistency, as will be discussed in Chapters 5, 6, 7 and 8 below. Therefore, it is easier to evaluate by digital computer, dynamically, an analytical solution, than it is to integrate numerically.

3.4 Iterations on Digital Computers

The fourth principle calls for the elimination of iterative procedures from routine calculations to the greatest extent possible by replacing such iterations by interpolation of tabulated results.

Iterations cannot always be avoided. In fact, implicit integration techniques lead to iteration on large state equation systems, the solutions of which cannot be pretabulated. Here, the computational expense of iterating calls for a careful balancing of benefits from implicit integration against the high cost of iteration (see Section 3.6).

The integration of the balance equations, Eq. (2.19), along with the closure equations, when carried out by any means, involves implicit algebraic and transcendental equations. One example would be the computations of wall temperatures which depend on heat transfer coefficients which, in turn, depend on wall temperatures. The associated set of implicit equations must be solved iteratively, either at every time level during the discrete computation of the transient, or *beforehand* for later interpolation during the transient calculations.* The fourth principle here insists on the latter choice, because computer memory for table data storage is inexpensive and interpolation of pretabulated results reduces the number of needed arithmetic operations during the transient calculations by orders of magnitude.

3.5 Function Tabulation

The fifth principle is also based on the reasons just given in support of the fourth principle. It calls for the pretabulation of composite functions for later interpolation during routine calculations.

Thermophysical properties had been pretabulated some years ago with limited success (NRC/BNL 1976) and are now used for large-scale power plant simulators in the United Kingdom (Whitmarsh-Everiss 1985). The fifth principle presented here, however, goes far beyond the pretabulation of elementary property functions and calls instead for pretabulation of, ideally, the coefficient functions A_{ij}, B_{ijk} and D_i in Eqs. (2.21) or of E_{kl} and F_k in Eqs. (2.34). Of course, a compromise must be found by minimizing the number of coefficient functions within the limitations imposed on the

*Implicitly appearing unknowns are often taken from a previous time step. The resulting time lag produces either damping (if all equations involved are analytic) or nonsense (with discontinuous equations).

dimensionality of the tabulated functions by both the available storage memory and the interpolating algorithms.

This and the previous model formulation principle are obviously most important for high-speed simulations and for large-volume simulations. Their application was of crucial importance in the high-speed BWR power plant simulation at Brookhaven National Laboratory (Wulff et al. 1984).

3.6 Implicit and Explicit Integration Method

A fundamental choice has to be made between implicit and explicit methods for numerically integrating Eqs. (2.21) or (2.34). A method for the integration of time-dependent differential equations is explicit, if the dependent variable(s) can be computed for the next following time level explicitly from values at the current and at previous time levels alone, i.e.,

$$y(t + \Delta t) = f\{t, y(t), y(t - \Delta t), \ldots, y(t - n \cdot \Delta t)\} \ , \qquad (3.1)$$

where y and f may be vectors, but of the same dimension, and $t - \Delta t \geq t^o$, the initial time. All other integration methods are implicit. The function f depends on the type of discretization used for replacing the partial differential by difference or by ordinary differential equations.

The sixth principle presented here calls for the rational choice between implicit and explicit integration schemes, based upon five parameters:

(i) the frequency f_ν of prevailing system stimulants (input data) and system responses of interest (expected results),

(ii) the permissible integration step sizes H_i and H_e, associated, respectively, with the implicit and explicit schemes under consideration, and

(iii) the frame times T_i and T_e needed by the computer(s) to advance the implicit and explicit scheme, respectively, from one time level to the next.

The abscissa in Figure 3.1 below represents the relevant frequency f_ν of system stimulation and response, with the steady

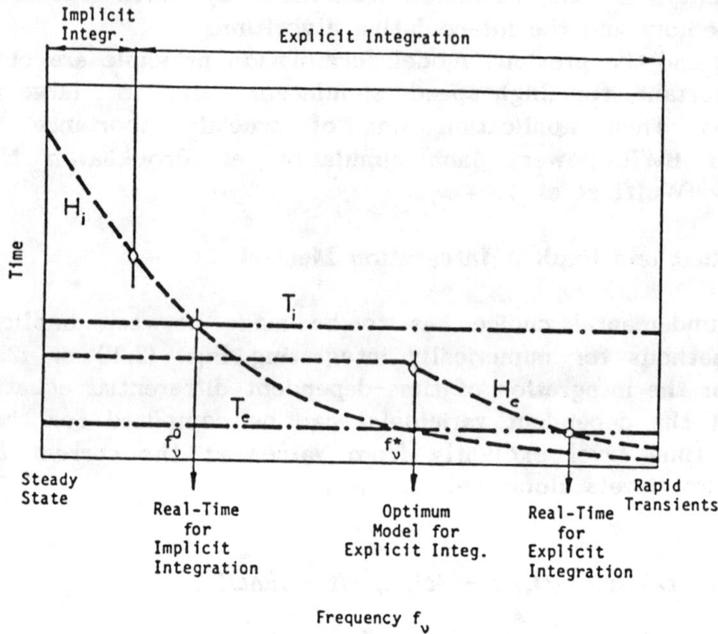

Figure 3.1 Selection of optimum integration method - *H* permissible time step, *T* frame time.

state at the left, and with extremely rapid transients at the right. The frame times T_i and T_e are system independent since they depend only on the number of arithmetic and logical operations carried out in one time step. The vertical time scale is logarithmic and shows that the frame time is much larger for implicit than for explicit algorithms. The permissible time step size H_i for unconditionally stable implicit schemes depends on truncation errors, i.e., on high-order derivatives which increase with f_ν, causing H_i to decrease with increasing f_ν. The permissible time step size H_e of explicit methods is normally larger than H_i because explicit schemes are frequently of higher-order accuracy (i.e., Runge-Kutta methods). However, H_e cannot exceed its stability limit as indicated by the horizontal segment of the H_e-curve in Figure 3.1.

The intersections of *T* and *H* curves give the relevant simulation frequencies for real-time computing speeds *(H/T=1)*. It is clear that explicit integration is superior for fast transients, and implicit for slow, near quasi-steady transients. An optimum model formulation can be achieved by eliminating irrelevant frequency content or numerical stiffness (i.e., sound propagation in liquid) from the model, so that the explicit scheme integrates for most transients of interest with the permissible step size lying just below the stability limit. The ratio H/T is the simulation speed ratio of simulated time over computer run time.

Figure 3.1 above shows that the maximum speed ratio H_e/T_e for explicit integration produces reliably the system response frequencies $f_\nu < f_\nu^*$. The same or better speed ratio can be acheived by implicit integration only for frequencies less than or equal to f_ν^o. This suggests the range of desired response frequencies for the two integration methods as indicated at the top of Figure 3.1.

Figure 3.1 is drawn only qualitatively and it applies only in principle to a particular time step. With appropriate weighting, one can apply the presented principles to variable time steps as well.

4. ANALYTICAL INTEGRATION

It has been pointed out in Section 3.3 that the dynamic computer evaluation of results obtained from analytical integration is superior to numerical integration. Here we review techniques which involve analytical integrations.

4.1 Analytical Solutions

Complete solutions by analytical integration are limited to special initial and boundary conditions (Gonzales-Santalo and Lahey 1971; Wulff 1978; Lahey 1981; Achard, Drew and Lahey 1985; Rizwan-Uddin and Dorning 1986). However, analytical solutions obtained just for particular conditions are indispensable for the verification of numerical methods. It is always possible to execute a general numerical solution under the conditions which are also imposed on a suitable analytical solution, such that the *essential* features of the numerical procedure can be verified (e.g., Wulff 1980, p. 39), including a truncation error assessment. Once the numerical procedure has been verified, one can systematically check out the correctness of the general relations which were excluded from the comparison between analytical and numerical solutions.

In Section 2.5 it was shown how a set of first-order, quasi-linear hyperbolic partial differential equations can be solved by quadrature. To facilitate the analytical integration of, for example, Eqs. (2.34) one needs either to find a closed-form quadrature of the integral in Eq. (2.46), or to simplify that integral until its closed-form quadrature is possible. Finally, one can also evaluate Eq. (2.46) by means of a simple and verified numerical integration technique, such as Simpson's rule, Gaussian quadrature or

Runge-Kutta integration, and thereby verify a newly implemented general integration scheme (finite difference methods, etc.) through a well established and independent method.

4.2 Analytical Results in Support of Numerical Methods

Analytical results can be used to advance a numerical solution in time by imposing certain parameters as constants during a time step. This technique can reduce numerical diffusion associated with spatial discretization (cf. Section 7).

More importantly, analytical methods can be used to simplify the computational task and to alter the classification of the partial differential equation, so as to achieve well-posedness.

For example, if there is no need to trace acoustic waves in the multiphase mixture, one can set the pressure gradient $\nabla p = 0$ and evaluate all thermophysical properties with the volume-averaged pressure and the local enthalpy. This situation is difficult to represent in a finite difference scheme such as that used in TRAC (1986) because the scheme depends on calculating local pressures. Below is shown how analytical integration reduces the one-dimensional mixture mass balance to quadrature and the mixture momentum balance to an ordinary differential equation.

By first adding the two phasic mass balances (cf. Eq. (2.7) and Table 2.1) one can derive the mixture volumetric flux divergence equation, which represents the mixture mass balance.

$$\nabla \vec{j}_m = \frac{\rho_L - \rho_G}{\rho_L \rho_G} \Gamma_v - \frac{\alpha}{\rho_G} \frac{D_G \rho_G}{Dt} - \frac{1-\alpha}{\rho_L} \frac{D_L \rho_L}{Dt} . \tag{4.1}$$

Using the caloric equations of state and the phasic energy balances, along with the condition $\nabla p = 0$, one can integrate Eq. (4.1) along the direction z of the flow (Wulff et al. 1984) to obtain the volumetric flux of the mixture

$$j_m(t,z) = j_m(t,0) + \int_0^z a_1(z) \cdot \Gamma_v(z) dz + \langle \dot{p} \rangle \sum_{k=1}^{2} \int_0^z \alpha_k(z) a_{2k}(z) dz - \sum_{k=1}^{2} \int_0^z \alpha_{3k}(z)\, q_{wk}(z)\, dz \tag{4.2}$$

everywhere and for all times in terms of an arbitrary inlet (or exit) flux $j_m(t,0)$. The second, third and last terms on the right-hand side of Eq. (4.2) account for expansion and contraction due to, respectively, phase change, pressure variation and wall heat transfer. The a-coefficients depend only on global pressure $<p>$ and local enthalpies, and are, therefore, candidates for pretabulation according to Section 3.5:

$$a_1 = \frac{\rho_L - \rho_G}{\rho_L \rho_G} + \frac{h_G - h_g}{\rho_G^2} \left(\frac{\partial \rho_G}{\partial h}\right)_p + \frac{h_G - h_L}{\rho_L^2} \left(\frac{\partial \rho_L}{\partial h}\right)_p \tag{4.3}$$

$$a_{2k} = \frac{\alpha_k}{\rho_k} \left[\left(\frac{\partial \rho_k}{\partial p}\right)_h + \frac{1}{\rho_k} \left(\frac{\partial \rho_k}{\partial h}\right)_p\right], \quad k = G,L \tag{4.4}$$

$$a_{3k} = \frac{1}{\rho_k^2} \left(\frac{\partial \rho_k}{\partial h}\right)_p \frac{\xi}{A}, \qquad k = G,L \tag{4.5}$$

where ξ and A designate heated perimeter and cross-sectional area, respectively. The integrations in Eq. (4.2) are carried out numerically for every time level.

In replacing the mixture mass balance, Eq. (4,1), by Eq. (4.2), we have, at the expense of one unimportant disadvantage, i.e.,

- the loss of ability to track pressure pulses,

gained four important advantages, namely:

- a partial differential equation is reduced to elementary quadratures,

- high accuracy is achievable through simple quadratures (Simpson's rule with fourth-order accuracy* instead of first-order accurate finite difference methods for partial differential equations),

- Eq. (4.2) has infinitely large characteristic speed (cf. Eq. (2.44)) instead of the finite speed for Eq. (4.1), thus there is no stability limit of the kind given by Eq. (8.15), and

---

*See Abramowitz and Stegun (1964) p. 886.

- $j_m(t,z)$ is continuous even across mixture levels* with sharp discontinuities in void fraction, densities, enthalpies and mass velocities (cf. Sections 9 and 10).

Secondly, by integrating the mixture momentum balance for one-dimensional flow, that is, this partial differential equation

$$\frac{\partial G_m}{\partial t} = - \frac{\partial p}{\partial z} - \frac{\partial}{\partial z} (w_G G_G + w_L G_L) - g\rho_m - f_{2\varphi} \frac{G_m|G_m|}{2\rho_L d_h} \qquad (4.6)$$

over the channel length, from $z = 0$ to $z = L$, one obtains an ordinary differential equation for the channel-averaged momentum

$$\frac{dM}{dt} = p(t,0) - p(t,L) + (w_G G_G + w_L G_L)\Big|_{z=0} - (w_G G_G + w_L G_L)\Big|_{z=L}$$

$$- g \int_0^L \rho_m dz - \int_0^L f_{2\varphi} \frac{G_m|G_m|}{2\rho_L d_h} dz \; . \qquad (4.7)$$

For an open channel, with given inlet conditions and given terminal pressures, one can integrate Eq. (4.7) with respect to time by expressing the unknown terminal momentum flux via Eq. (4.2) from the volumetric flux j_m and the drift flux identities for phasic mass flux

$$G_L = \rho_L\left[(1-\alpha C_o)j_m - \alpha v_{gj}\right], \; G_G = \rho_G\left[\alpha(C_o j_m + v_{gj})\right] \qquad (4.8)$$

and for phasic velocities $w_k = G_k/\alpha_k\rho_k$, $k = G,L$. The two quadratures are numerically integrated with local conditions for every time level. For closed loops, one finds that the first four terms in Eq. (4.8) are to be replaced by form losses and possibly by pressure rises across pumps. The solution $M(t)$ can then be solved for $j_m(t,o)$ in Eq. (4.2), since

*If there is no phase change at the surface (Wulff, 1981).

$$M = \int_o^L G_m \, dz$$

$$= \int_o^L \left\{ \left[\rho_L - \Delta\rho\alpha C_o \right] j_m(t,z) - \Delta\rho\alpha v_{gj} \right\} dz. \tag{4.9}$$

where $\Delta\rho = \rho_L - \rho_G$.

As the right-hand side of Eq. (4.9) is linear in $j_m(t,o)$ (cf. Eq. (4.2)) it can be solved to yield the entrance flux in terms of the instantaneous momentum $M(t)$.

The analytical integration of the mixture momentum balance leads therefore to these three advantages:

- a partial differential equation is integrated as an ordinary differential equation (ODE),
- greater accuracy is achieved through available high-order integration procedures for ODEs and integrals,
- problem stiffness (cf. Section 9) has been reduced as one large overall inertia (proportional to L) has replaced many small parts of inertia in each cell of a finite difference scheme.

4.3 Summary on Analytical Techniques

It has been demonstrated in two examples that the use of analytical methods provides:

(i) *computing economy*, since partial differential equations can be integrated as ordinary differential equations and/or through simple quadratures,

(ii) *improved accuracy*, because simplifying assumptions can be quantitatively assessed, because integration procedures can be used which have high-order accuracy and for which error estimates are established, and because there is no numerical diffusion, dispersion or damping.

(iii) *improved stability*, because computational stiffness is reduced and irrelevant processes (pressure pulses for example), along with integration step size limitations (Courant limit) can be eliminated.

5. NUMERICAL METHODS AND FINITE DIFFERENCES

5.1 Predominantly Used Techniques

The computational techniques predominantly used for two- and multiphase flow analyses by digital computers are the

(i) Finite Difference Methods (Ransom et al. 1985, TRAC-PF1 1986, Taylor et al. 1984 and many others),

(ii) Spatial Discretization, Integration and Lumped-Parameter Modeling (Porsching et al. 1971, Holt and Rasmussen 1972, Moore and Rettig 1973, Wulff et al. 1975, Wulff 1978, Wulff et al. 1984).

(iii) Method of Characteristics (Hancox et al. 1975, Kroeger 1976, Hancox and Banerjee 1977, Banerjee and Hancox 1978, Ferch 1978)

(iv) Finite Element Methods or Weighted Residual Methods (Chung 1978, Graf and Romstedt 1983, Graf et al. 1986, Bukhari 1985, Sim 1986), and

(v) Methods of Lines or Space Discretization (Birkhoff and Kimes 1962, Banerjee and Hancox 1977).

It is obvious from the literature that the analyses of two-phase flows in nuclear power plants has stimulated most of the recent interest in numerical methods for multiphase flows. The discussion on the merits and demerits of these methods is deferred until later (cf. Sections 6 through 9).

All computational schemes involve as the first step the transformation of the partial differential equations, Eqs. (2.7) or (2.13),* defining m primary state variables, into either

- *m* ordinary differential equations

---

*and possibly equations of state.

$$\frac{dy_i}{dt} = F_i(t, y_1, \dots, y_m;\ \Delta x_1, \dots \Delta x_3),\quad i = 1, \dots, m \tag{5.1}$$

- or n nonlinear equations

$$G_i(t, y_1, \dots y_m;\ \Delta t, \Delta x_1, \dots, \Delta x_3) = 0,\quad i = 1, \dots, n \tag{5.2}$$

These two distinct categories lead to two entirely different types of, respectively, integration and solution algorithms (cf. Sections 10 and 11). Yet both F_i and G_i in Eqs. (5.1) and (5.2) contain discrete spatial increments Δx_i. These functions G_i and the computer implementation of the dērivatives in Eqs. (5.1) contain also time increments Δt. Both functions F_i and G_i involve the replacement of derivatives by finite difference quotients.

The original issue of finding a unique numerical solution to Eqs. (2.7) or (2.13), with any desirable accuracy, is therefore separable into two distinct problems. The first problem is to determine under what conditions Eqs. (5.1) and (5.2) converge toward the partial differential equations, Eqs. (2.7) or (2.13) as the increments Δt and Δx_j, $j = 1, \dots, 3$ tend toward zero, as suggested by the familiar limit process which leads in calculus from difference quotients to derivatives. The second problem is to establish the conditions under which a chosen and implemented algorithm produces solutions to Eqs. (5.1) or (5.2). The first problem calls for an assessment of the quantitative difference between differentials and finite differences. This is discussed in Section 5.2 below. The second problem is addressed in Sections 10 and 11.

It is far easier to integrate Eqs. (5.1) than it is to solve Eqs. (5.2), primarily because most computing centers have many efficient library subprograms with well-documented characteristics and well-defined criteria for integrating ordinary differential equations, but hardly any programs for solving non-linear equations in general. Most integrating subprograms for ordinary differential equations also have the means for automatically controlling the time discretization error and for accommodating computational stiffness. However, it is more difficult to derive Eqs. (5.1) from the balance equations, Eqs. (2.7) or (2.13), because there are in general subsidiary state variables not defined by balance equations, and ordinary differential equations are therefore supplemented by algebraic equations. These must be solved first to eliminate the subsidiary variables from the balance equations. On the other hand, Eqs. (5.2) can be derived directly by substituting almost mechanically all derivatives by difference quotients.

5.2 Derivatives and Finite Difference Quotients

Introduce for the sake of convenience the differential operator D for a differentiable function $y(x)$ by

$$Dy(x) = dy/dx = y'(x) \; . \tag{5.3}$$

Next, we designate discrete points of the independent variable x by $x_i, x_{i+1},\ldots$ and the associated function values $y(x)$ by $y_i = y(x_i)$. For the sake of simplicity we limit ourselves to equal spacings of x_i, that is $x_{i+1} - x_i = h$, a constant. Then we introduced the forward difference operator by

$$\Delta y(x) = y(x+h) - y(x) \tag{5.4a}$$

or

$$\Delta y_i = y_{i+1} - y_i \tag{5.4b}$$

and the shift operator E by

$$Ey(x) = y(x+h) \tag{5.5a}$$

or

$$Ey_i = y_{i+1} \; , \tag{5.5b}$$

which is related to the forward difference operator Δ by $E = 1+\Delta$, provided 1 is the unity operator: $1 \cdot y_i = y_i$. From the Taylor series expansion of $y(x)$ about x

$$y(x+h) = y(x) = h \left.\frac{dy}{dx}\right|_x + \ldots \frac{h^n}{n!} \frac{d^n y}{dx^n} + \ldots \tag{5.6}$$

one obtains first

$$E = e^{hD} \tag{5.7}$$

and then this expression for the derivative of $y(x)$ in terms of its differences (see also Abramowitz and Stegun 1964, Eq. 25.3.9, p. 883)

$$hD = lnE = ln(1+\Delta)$$

$$= \Delta - \frac{\Delta^2}{2} + \frac{\Delta^3}{3} - \frac{\Delta^4}{4} + -...+(-1)^{n-1}\frac{\Delta^n}{n} + R_n , \tag{5.8}$$

where

$$\Delta y_i = y_{i+1} - y_i$$

$$\Delta^2 y_i = \Delta y_{i+1} - \Delta y_i = y_{i+2} - 2y_{i+1} + y_i$$

$$\Delta^3 y_i = \Delta y^2_{i+1} - \Delta y^2_i = y_{i+3} - 3y_{i+2} + 3y_{i+1} - y_i$$

$$\vdots$$

$$\Delta^n y_i = \Delta y^{n-1}_{i+1} - \Delta y^{n-1}_i = \sum_{k=0}^{n} (-1)^{n-k}\binom{n}{k} y_{i+k} \tag{5.9}$$

and the truncation error with its x-value equal to ξ, $x_i \leq \xi \leq x_{i+1}$:

$$R_n = (hD)^{n+1} y(\xi)/(n+1)! . \tag{5.10}$$

Equation (5.8) gives the first, second and third-order approximations for the first derivative in terms of discrete y-values and the truncation error as follows:

1st order: $y_i' = (y_{i+1} - y_i)/h + (h/2)y''(\xi)_1$ (5.11)

2nd order: $y_i' = (-y_{i+2} + 4y_{i+1} - 3y_i)/(2h) + (h^2/3)y'''(\xi)$ (5.12)

3rd order: $y_i' = (2y_{i+3} - 9y_{i+2} + 18y_{i+1} - 11y_i)/(6h) - (h^3/4)y^{(4)}(\xi)$

$$(5.13)$$

It appears from Eqs. (5.8) and (5.11) through (5.13) as if one could approximate a derivative with any desirable accuracy by forcing the truncation error toward zero, that is by

1) selecting higher-order approximations, or
2) decreasing the size of h.

This would be true if all y_i were exact. All computed y-values have some computational error. If the maximum error of y_i, $i = 1,2...$ is ε_y then the error amplification in the

1st-order approximation is $2\varepsilon_y/h$
2nd-order approximation is $4\varepsilon y/h$
3rd order approximation is $20\varepsilon_y/(3h)$.

Thus the error amplification grows with decreasing h and increasing order of approximations, and one has to minimize the sum of computational noise amplification and truncation error.

Equation (5.8) can be multiplied with itself to produce second and higher-order derivatives

$$(hD)^2 = \Delta^2 - \Delta^3 + \frac{11}{12}\Delta^4 - \frac{5}{6}\Delta^5 + - \ldots \quad (5.14)$$

$$(hD)^3 = \Delta^3 - \frac{3}{2}\Delta^4 + \frac{7}{4}\Delta^5 - \frac{15}{8}\Delta^6 + \ldots \quad \text{etc.} \quad (5.15)$$

Conversely, one can express the difference operator from Eq. (5.7) to get this result:

$$\Delta = E - 1 = e^{hD} - 1 \quad (5.16)$$

or $$\Delta = hD + \frac{h^2}{2!}D^2 + \frac{h^3}{3!}D^3 + \ldots + \frac{h^n}{n!}D^n + \ldots \quad (5.17)$$

$$\Delta^2 = h^2D^2 + h^3D^3 + \frac{7}{12}\, h^4D^4 + \frac{1}{4}\, h^5D^5 + \ldots \tag{5.18}$$

All the relations between derivatives and finite differences were shown here on the basis of forward differences. Similar relation can be given, based on centered differences

$$\delta y_i = y(x_i + h/2) - y(x_i - h/2) \,, \tag{5.19}$$

or on backward differences

$$\nabla y_i = y(x_i) - y(x_{i-1}) \,, \tag{5.20}$$

which are needed for "upwind" differencing under conditions of forward flow (cf. Sections 8 and 9). The choice between forward, centered and backward differences depends not only on the available meshpoints but also on stability (cf. Sections 7,8,9). Centered differences have the smallest truncation error. Derivatives can also be obtained for positions between meshpoints (Abramowitz and Stegun 1964, Eq. 25.3.7, p. 883).

With backward differences, one obtains instead of Eqs. (5.11) through (5.13):

$$\text{1st order:} \quad y_i' = (y_i - y_{i-1})/h + (h/2)y''(\xi) \,, \quad x_{i-1} \le \xi \le x_i \,. \tag{5.21}$$

$$\text{2nd order:} \quad y_i' = (3y_i - 4y_{i-1} + y_{i-2})/(2h) + (h^2/3)y'''(\xi) \tag{5.22}$$

$$\text{3rd order:} \quad y_i' = (11y_i - 18y_{i-1} + 9y_{i-2} - 2y_{i-3})/(6h) + (h^3/4)y^{(4)}(\xi) \,. \tag{5.23}$$

All coefficients in Eqs. (5.11) through (5.13) and (5.21) through (5.23) can also be derived by passing a polynomial through sufficiently many points (x_i, y_i) and then differentiating the polynomial. Moreover, these coefficients and those of higher-order approximations can be found in Table 25.2 on page 914 of Abramowitz and Stegun (1962). In summary, replacing derivatives by differences

In summary, replacing derivatives by differences produces truncation errors and it amplifies computational errors (round-off, etc.). Other consequences will arise later in the context of solving the finite difference equations. It will become apparent in Sections 7, 8 and 9 that the choice between backward, centered and forward spatial differences impacts on computational stability. One might suspect even at this point a relation between the type of suitable differences, the interval of dependence for hyperbolic equations (cf. end of Section 2.5), and the characteristic directions given by Eqs. (2.44).

6. METHOD OF CHARACTERISTICS (MOC)

Rather than starting with the most popular techniques for analyzing multiphase flows, the Finite Difference Method, we start here with the Method of Characteristics because it is more closely related to the analytical solution given by Eq. (2.46). The Method of Characteristics has its name from the fact that it entails integration along characteristics (cf. Sections 2.4.1 and 2.5).

The Method Characteristics (MOC) can only be used if the model formulation is given in terms of m hyperbolic partial differential equations: there must be m real eigenvalues and m eigenvectors for the coefficient matrix E_{ij} in Eq. (2.34).

In multiphase flows, the application of MOC has been limited to one-dimensional flows. MOC is, from a computational point of view, the most expensive method. The reasons for the computational expenses are the need for computing the roots of Eq. (2.40) for every meshpoint and for interpolations in the conversion from characteristic coordinates (curves c_l, Eq. (2.44)) to natural coordinates (t,x), all of this at every time step.

MOC has the potential of producing the most accurate results and is therefore suitable for reference or benchmark calculations. Its high accuracy stems from the fact that MOC is not afflicted by numerical diffusion and it reduces the partial differential equations of the form of Eqs. (2.34) to ordinary differential equations (cf. Eq. (2.45)) or to integral equations of the form of Eqs. (2.46). MOC is the only method which tracks accurately the propagation of discontinuities in the first-order derivatives; the characteristic coordinates are Lagrangian coordinates for such discontinuities.

MOC has been used for transient two-phase flows by the Canadians (Hancox et al. 1975, Hancox and Banerjee 1977, Banerjee and Hancox 1978, Ferch 1978) and also for the study of critical flows

(Liles 1974, Kroeger 1976). In all applications, MOC has led to iterative schemes in the form of Eqs. (5.2).

Hancox, Banerjee and Ferch dealt with homogeneous nonequilibrium two-phase flow; their formulations led to three characteristics, corresponding to the material velocity v and the pressure propagation velocities $v + a$ and $v - a$, where a is the speed of sound (cf. Eqs. (2.44) and (6.3)). Kroeger analyzed nonhomogeneous nonequilibrium flows and dealt with four characteristics (Kroeger 1976, p. 126).

6.1 Transformed Equations

None of the above-referenced authors have used the simplifications discussed in Section 2.5 which consist first of premultiplying Eq. (2.33)[+]

$$A_{ij} \frac{\partial w_j}{\partial t} = B_{ij} \frac{\partial w_j}{\partial x} \mp D_i \ , \quad i = 1,\dots,m \tag{6.1}$$

by A_{ij}^{-1} and then transforming the variables w_j according to Eqs. (2.35). Instead of solving the coupled integral equations, Eqs. (2.46), these authors have chosen to solve a set of implicit first-order differential equations, namely

$$f_{lk} \frac{dw_k}{dt} = e_l \ , \tag{6.2}$$

in terms of the original variables w_k, one equation being valid along each characteristic curve c_l. This approach dispenses with the back-transformation in Eqs. (2.35), but it entails the same computational efforts as Eqs. (2.46).

As an illustrative example, Ferch (1978, p. 267) solved this system of five equations with three distinct characteristics:

(i) *along* c_1: $\dfrac{dx}{dt} = -\lambda_1 = v + a \ , \quad \dfrac{dp}{dt} + \rho a \dfrac{dv}{dt} = e_1$

[+]Diffusion terms are omitted.

$$\left.\begin{array}{lll}
\text{(ii)} & \textit{along } c_2\text{:} & \dfrac{dx}{dt} = -\lambda_2 = v - a\ , \quad \dfrac{dp}{dt} - \rho a \dfrac{dv}{dt} = e_2 \\[2ex]
\text{(iii)} & \textit{along } c_3\text{:} & \dfrac{dx}{dt} = -\lambda_3 = v\ , \quad f_{31} \dfrac{dp}{dt} + \dfrac{d\alpha}{dt} = e_3 \\[2ex]
 & & \qquad\qquad -\dfrac{1}{\rho_G} \dfrac{dp}{dt} + \dfrac{dh_g}{dt} = e_4 \\[2ex]
 & & \qquad\qquad -\dfrac{1}{\rho_L} \dfrac{dp}{dt} + \dfrac{dh_L}{dt} = e_5\ ,
\end{array}\right\} \quad (6.3)$$

where f_{31} and e_l, $l = 1,...,5$ are functions of the state variables pressure p, velocity v, void fraction α, gas enthalpy h_G and liquid enthalpy h_L. All functions are specified in the paper by Ferch. Curves c_1 and c_2 are called the wave lines and c_3 is called the path line, traced in the $\{t,x\}$ plane by fluid particles.

6.2 Discretization

There are two distinct arrangements of discrete meshpoints. The first one is suggested by the analytical solution in Section 2.5 where Eqs. (2.46) serve to advance the computations from one time level t_1 to the next level t_2. These time levels are shown as horizontal lines in Figure 6.1 below. Discrete space coordinates x_{-1}, x_0 and x_1 are shown as vertical lines.

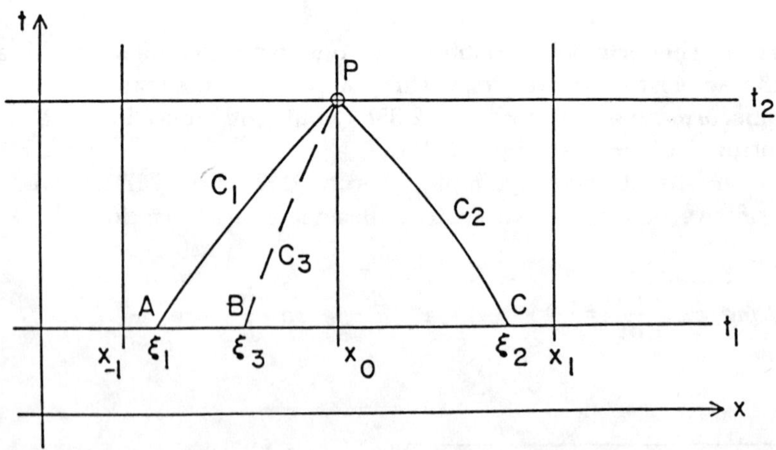

Figure 6.1 Physical plane with characteristics, first arrangement.

Also shown are the three characteristic curves of Eqs. (6.3). Notice that disturbances of pressure and velocity travel from A and C to P, those of void fraction and enthalpies from B to P, while the time advances from t_1 to t_2. The aim is to find new conditions at point $P(t_2,x_0)$, when conditions on the t_1-line are known.

The state variables are given, from initial data or from previous calculations, at the time level t_1 and the locations ..., x_{-1}, x_0, x_1,... A first estimate of the characteristic speeds $-\lambda_i$ is computed from conditions at $\{t_1,t_0\}$, by solving Eqs. (2.40). This yields from Eqs. (2.44) or the first part of Eqs. (6.3) the locations ξ_i of A, B and C:

$$x_0 - \xi_i = \lambda_i (t_2 - t_1),\ i = 1,2,3\ . \tag{6.4}$$

Next all state variables are computed, by interpolation of data at (t_1,x_j), $j=-1,\ 0,\ +\ 1$, for the locations at (t_1,ξ_i), $i = 1,2,3$. Then, the second part of Eqs. (6.3) (Riemann equations) are replaced by finite difference equations (Massan 1899, Sauer 1952)

$$\left.\begin{aligned}
p_P - p_A + \overline{(\rho a)}_1 (v_P - v_A) &= \bar{e}_1 (t_2 - t_1) \\
p_P - p_C - \overline{(\rho a)}_2 (v_P - v_C) &= \bar{e}_2 (t_2 - t_1) \\
\bar{f}_{31} (p_P - p_B) + \alpha_P - \alpha_B &= \bar{e}_3 (t_2 - t_1) \\
-\overline{\frac{1}{\rho_G}} (p_P - p_B) + (h_G)_P - (h_G)_B &= \bar{e}_4 (t_2 - t_1) \\
-\overline{\frac{1}{\rho_L}} (p_P - p_B) + (h_L)_P - (h_L)_B &= \bar{e}_5 (t_2 - t_1)\ .
\end{aligned}\right\} \tag{6.5}$$

Here, the superscript bar denotes evaluation with the arithmetic mean of terminal state variables: for curve C_1 with $(p_A + p_B)/2$, $(v_A + v_B)/2$, etc.

The five equations, Eqs. (6.5) are then solved for the five unknowns p_P, v_P, α_P, $(h_G)_P$ and $(h_L)_P$. These results are used next to calculate $\bar{\lambda}_i$, to compute new locations ξ_i, interpolate again to find new state variables at A, B and C in Figure 6.1 and to solve again

Eqs. (6.5). This process is repeated until convergence is reached for the computed results of point $P(t_2,x_0)$.

The second arrangement of discrete meshpoints is formed by the intersections of wave and path lines as shown in Figure 6.2 (cf. Ferch 1978).

Instead of interpolating data along constant-time curves, the interpolation is carried out here on data along the wave lines (having maxima of $|\lambda_i|$) first at intersection E between the wave line $\overline{AD}$ and the path line from previous data, and then at the intersection B between $\overline{AD}$ and the path line through point D.

The advancement from time level t_1 to the time level t_2 is carried out for the arrangement in Figure 6.2 with the same set of equations, Eqs. (6.5), as before, except that t_1 is replaced by t'_1 in the last three equations.

The second arrangement in Figure 6.2 has the advantage of achieving better starting values for λ_i in Eqs. (6.4), because the characteristics start from points (A and B) with known conditions, rather than from point P in Figure 6.1 of the first arrangement. On the other hand, the locations x'_i *along the* t_2-line fall into unpredictable patterns, which is inconvenient for output editing and for relating results to measurements at fixed locations.

For further details, particularly for the treatments of boundary conditions and of phase change boundaries, the reader is referred to the above references (Kroeger 1976, Ferch 1978). Other computing schemes can be found in Forsythe and Wasow (1960) and in Ames (1969).

6.3 Accuracy and Convergence

The relations between Eqs. (6.1) and (6.3) are exact; the transformation of the partial to ordinary differential equations (cf. Section 5.1 and Eqs. (5.1)) guarantees convergence of the solutions for Eqs. (6.3) to the solutions of Eqs. (6.1). This is the basis for the high accuracy potential of MOC.

The algorithm given by Eqs. (6.4) and (6.3) is the first-order Euler-Cauchy method (cf. Section 10) and converges to the solutions of Eqs. (6.3) as $\Delta t = (t_2 - t_1)$ tends toward zero. The overall truncation error is of order $O\{\Delta t\}$.

7. SPATIAL DISCRETIZATION AND INTEGRATION

In the previous section was presented the first of three methods, the Method of Characteristics, which facilitate the conversion of partial to ordinary differential equations. Later, in

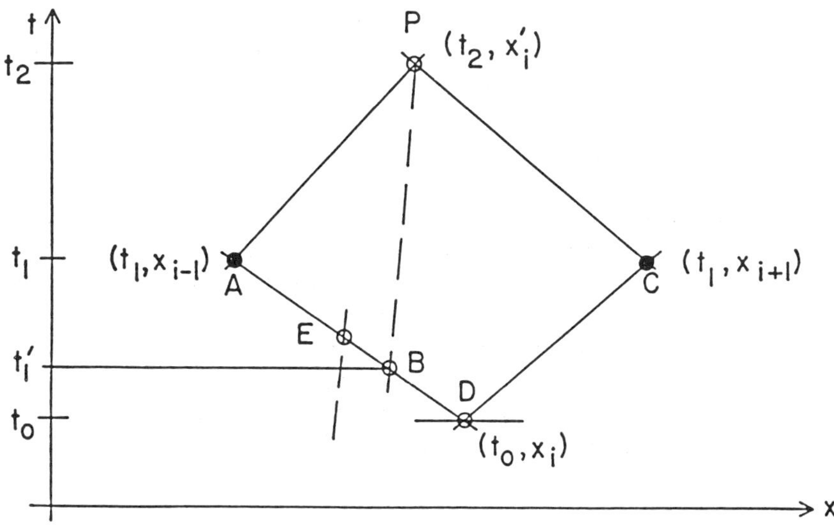

Figure 6.2 Physical plane with characteristics, second arrangement.

Section 9, we present the third method, the Finite Element Method, which can play the dual role of producing both ordinary differential or nonlinear algebraic equations. In this section we present the second category of methods for obtaining ordinary differential equations.

Specifically, we will present the methods of

- Space Discretization
- Space Averaging
- Lumped-Parameter Modeling.

Each of these methods have a variety of names in the literature. The purpose here is to show similarities and distinctions, which appear achievable by using just the above three categories.

The importance of obtaining ordinary from partial differential equations is the ability to utilize the large and well-established pool of integrating procedures for ordinary differential equations. These procedures provide operational criteria for uniqueness and convergence and they offer a rational and quantitative control of the time discretization error. The methods presented here are also important because they work for both hyperbolic and parabolic partial differential equations. On the other hand, the integration of ordinary differential equations becomes most attractive if carried out explicitly (cf. Section 3.6), but much less so if eigenvalues of the Jacobian associated with the ordinary differential equations vary over a wide range of orders of magnitude or if they differ greatly from the response time constants of interest.

7.1 Space Discretization

For space discretization, one replaces only the spatial derivatives in Eq. (2.19) by finite differences as presented in Section 5.2 (cf. Eq. (5.8) for first, Eqs. (5.14) and (5.9) for second derivatives). Depending on the choice between forward, centered or backward (also called upwind) differences, the resulting ordinary differential equations will be unstable or stable. Stability is necessary and achieved with upwind differencing.* A set of ordinary differential equations, Eqs. (5.1) is said to be stable in the present context if all eigenvalues λ_i, $i = 1,\ldots, m$ of the Jacobian

$$J_{ij} = \partial F_i / \partial y_j \tag{7.1}$$

have only negative real parts. Stability in this sense is necessary because the system is expected to approach, under realistic boundary conditions, only states with bounded state variables y_i.

The method of space discretization can be used for multidimensional flows, governed by parabolic or hyperbolic equations. The method is illustrated here for the one-dimensional void propagation equation (Wulff 1981)

$$\frac{\partial \alpha}{\partial t} + C_k \frac{\partial \alpha}{\partial z} = \Omega \,, \qquad 0 \le z \le L \tag{7.2}$$

$$\left.\begin{aligned} \alpha(0,z) = \alpha_0(z) \;;\; \alpha(t,0) = \alpha_{in}(t) \quad & \text{if} \quad C_k(0) > 0 \\ \alpha(t,L) = 1 \qquad & \text{if} \quad C_k(L) < 0 \,, \end{aligned}\right\} \tag{7.3}$$

*Upwind differencing is akin to donor-cell differencing in methods involving computational cells, rather than the discrete mesh discussed in Section 5.2. See also Section 6.

where α, C_k (cf. Eq. (1.2)) and Ω denote void fraction, kinematic wave velocity and characteristic frequency, respectively.

The method of space discretization requires that all second-order space derivatives are continuous and that the scalar functions have continuous derivatives. Discretization, therefore, is most suitable in stream tube geometries; sudden expansions or contractions constitute boundaries. However, discontinuities such as the moving void discontinuity shown in Figure 7.1, can be accommodated with ease.

As shown in Figure 7.1, the channel length is divided into equal segments at $z = z_i$ with $\Delta z = z_{i+1} - z_i$. At $Z(t)$, between z_M and Z_{M+1} is a sharp discontinuity from α below the level to α above the level. After replacing the z-derivative in Eq. 7.2 by first-order backward differences for forward flow, $C_k > 0$, and forward differences for backward flow, $C_k < 0$ (upwind differencing, cf. Section 9) one obtains this first-order approximation for interior points:

$$\frac{d\alpha_i}{dt} = \Omega_i - \frac{1}{2\Delta z}\Big\{(C_{k,i} + |C_{k,i}|)(\alpha_i - \alpha_{i-1})$$

$$+ (C_{k,i} - |C_{k,i}|)(\alpha_{i+1} - \alpha_i)\Big\}$$

$$\alpha_i(0) = \alpha_0(z_i)\ , \quad i = 2,3\ldots,\ M-1,\ M+2,\ldots,\ N-1 \tag{7.4}$$

At the inlet, $z = z_1 = 0$

$$\left.\begin{aligned}\frac{d\alpha_1}{dt} &= \frac{d\alpha_{in}}{dt} \quad \text{for} \quad C_{1,k} \geq 0 \\ &= \Omega_1 - \frac{C_{k,1}}{\Delta z}(\alpha_2 - \alpha_1) \ \text{for} \ C_{1,k} < 0\ .\end{aligned}\right\} \tag{7.5}$$

Equations (7.5) impose the boundary condition at $z = 0$ only when voids propagate into the channel, i.e., in recognition of the physics implied in Eqs. (2.44) and (2.46): C_k is the characteristic velocity, and void fraction information propagates with this velocity. The same considerations apply at the exit, $z = L$.

Below the sharp discontinuity in void fraction, i.e., at $z = z_M < Z(t)$, one uses

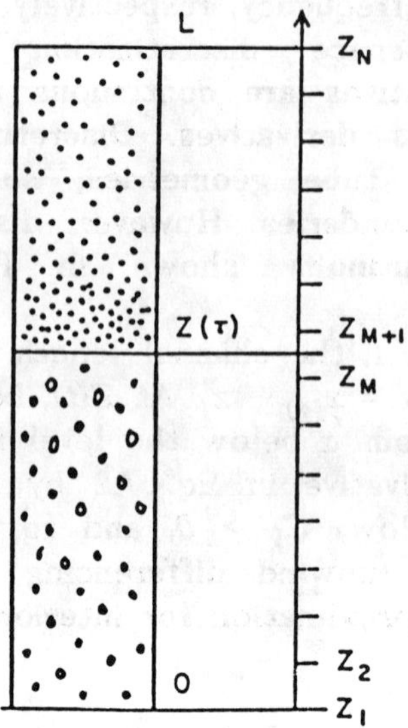

Figure 7.1 Discrete mesh arrangement.

$$\frac{d\alpha_M}{dt} = \Omega_M - \frac{1}{2\Delta z} (C_{k,M} + |C_{k,M}|) (\alpha_M - \alpha_{M-1})$$

$$- \frac{1}{2\Delta z^*} (C_{k,M} - |C_{k,M}|) (\alpha^- - \alpha^*) . \tag{7.6}$$

Above the level, at $z = z_{M+1} > Z(t)$, one uses similarly

$$\frac{d\alpha_{M+1}}{dt} = \Omega_{M+1} - \frac{1}{2\Delta z^{**}} (C_{k,M+1} + |C_{k,M+1}|) (\alpha^{**} - \alpha_{M+1})$$

$$- \frac{1}{2\Delta z} (C_{k,M+1} - |C_{k,M+1}|) (\alpha_{M+2} - \alpha_{M+1}) , \tag{7.7}$$

where $\Delta z^* = \Delta z/2 + Z - z_M$, $\Delta z^{**} = \Delta z + z_{M+1} - Z$, and

$$\alpha(z^*) = \alpha^* = \frac{1}{8} (3 \alpha_M + 6 \alpha_{M-1} - \alpha_{M-2})$$

$$\alpha(z^{**}) = \alpha^{**} = \frac{1}{8} \; (3\, \alpha_{M+1} + 6\, \alpha_{M+2} - \alpha_{M+3}) \tag{7.8}$$

denote the void fractions at the midpoints of the space increments above and below the increment containing the discontinuity. The void fraction just below the discontinuity is computed from parabolic extrapolation

$$\alpha^{-} = \alpha_M \left[1+(\eta/2)(3+\eta)\right] - \alpha_{M-1}\, \eta(2+\eta) + \alpha_{M-2}\, (\eta/2)(1+\eta) \;, \tag{7.9}$$

where $\eta = [Z(t) - z_M]/\Delta z$.

This completes the description of space discretization for Eq. (7.2), even in the presence of a discontinuity.$^+$ The basis principle used is the Taylor series expansion, Eq. (5.6). One could employ higher order differences (cf. Eq. (5.12)) to improve the accuracy, provided the resulting Jacobian (cf. Eq. (7.1)) has eigenvalues with only real negative parts. As all other differencing schemes, the method presented here also introduces its own set of eigenvalues (time constants, frequencies) into the problem, but they are of no practical importance as can be seen from the perfectly smooth time dependence of level position and the smooth spatial variation of void fraction, shown in figures 7.2, 7.3 and 7.4 below. The level position was computed according to the second part of Eq. (1.2), and there is no numerical diffusion at the mixture level.

7.2 Space Averaging

Space averaging for the purpose of transforming partial into ordinary differential equations has been used in systems codes for the analysis of two-phase flows in nuclear power plants for more than 15 years (Porshing 1971; Holt and Rasmussen 1972; Moore and Rettig 1973; Wulff et al. 1975, 1981, 1984). This method can be applied to stream tube geometries as well as to be used also for three-dimensional flow analysis; in fact, spatial differencing is being used as part of three-dimensional finite differencing schemes, as discussed in Section 8.

$^+$The other governing equations are given in the cited reference (Wulff 1981). The logic for switching from Eqs. (7.5) and (7.6) to Eq. (7.4) as the level crosses a cell boundary at z_i depends both on $|Z - z_i|$ and dZ/dt and leads to restart of the integration process.

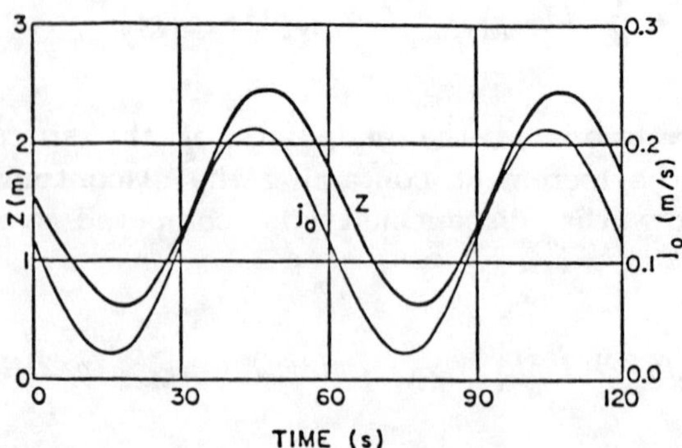

Figure 7.2 Transient level position for low-frequency inlet flow oscillations (0.052 s^{-1}) (Wulff 1981).

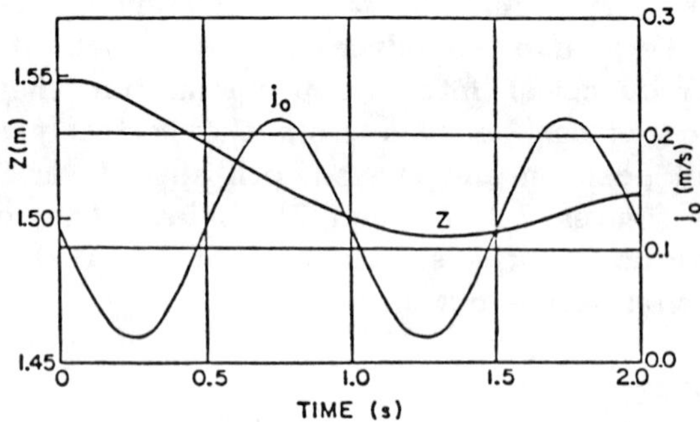

Figure 7.3 Transient Mixture Level position for high-frequency inlet flow oscillation (3.14^{S-1}) (Wulff 1981).

The method is very simple. It relies on two sets of rigid control volumes as shown for one-dimensional flow in Figure 7.5 below. Mass and energy balances are averaged over one set of control volumes with centers at ..., x_{i-1}, x_i, x_{i+1},..., while momentum balances are averaged over the other control volume set, which is displaced or staggered relative to the first set of control volumes and has its centers at ..., $x_{i-1/2}$, $x_{i+1/1}$,... The control volumes are displaced by $\Delta x/2$, where Δx - x_{i+1} - x_i.

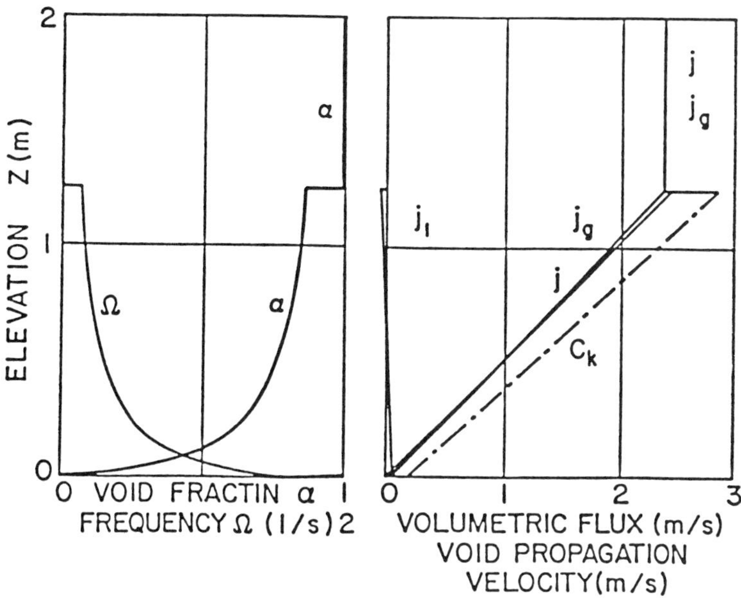

Figure 7.4 Spatial variations of void fraction α, characteristic frequency Ω, volumetric flux j (mixture), j_G (gas) and void propagation velocity C_k (Wulff 1981).

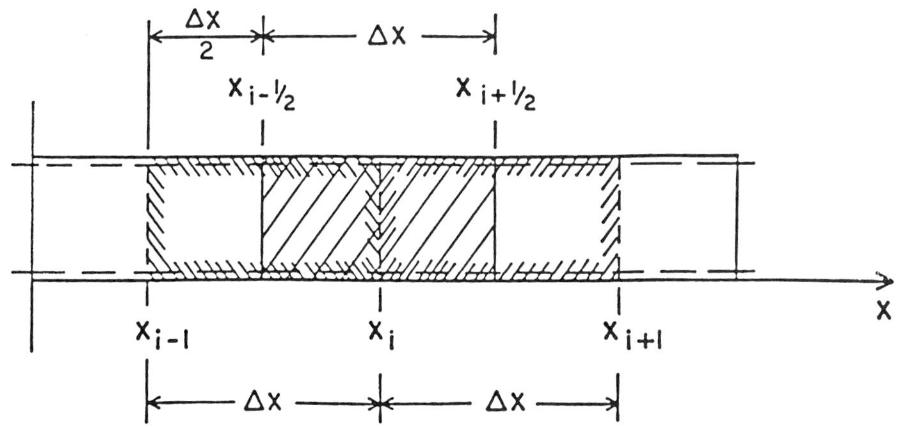

Figure 7.5 Staggered mesh arrangement.

While the staggered arrangement of control volumes, or computational cells, is widely used in several computational schemes,* it is the combination of the following three assumptions which characterize the method of space averaging, as it is used to obtain ordinary from partial differential equations:

(i) Flow parameters assume their volume-averaged values at the locations of the cell centers, namely at $x_{i\pm 1/2}$ for momentum ρv and at x_i for all other parameters.

(ii) At cell boundaries where a fluid phase is leaving the cell whose center is at x_i, the intensive properties of that phase have the same values as at position x_i (cell averaged values).

(iii) Products of volume averages are equal to averages of products.

Equations (2.7) are of the so-called conservatives** form

$$\frac{\partial \varphi}{\partial t} + \nabla \cdot (\vec{v}\varphi + \vec{J}) = S \, , \tag{7.10}$$

where S contains all terms free of derivatives. Integration of Eq. (7.10) over the fixed control volume ΔV_k with center at x_k, $k = i$ or $i + 1/2$, yields at first⁺

$$\iiint_{\Delta V_k} \left\{ \frac{\partial \varphi}{\partial t} + \nabla \cdot (\vec{v}\varphi + \vec{J}) \right\} dV = \iiint_{\Delta V_k} s dV \tag{7.11}$$

or

$$\Delta V_k \frac{d\langle\varphi\rangle}{dt} + \oiint_{A_k} \left\{ \vec{v}\varphi + \vec{J} \right\}_k \cdot d\vec{A} = \Delta V_k \langle S\rangle \, . \tag{7.12}$$

Here, A_k is the *whole* area enclosing ΔV_k, $d\vec{A}$ is the area element pointing outwards from ΔV_k and $\langle\cdot\rangle$ denotes volume-average. By using Assumptions (i) through (iii) in Eq. (7.12) and writing the result for constant cross-sectional area, one finds for the k-th control volume, with $k = i$ or $k = i \pm 1/2$, this ordinary differential equation:

---

*Patankar (1980, p. 115) gives one reason for the widely accepted use of staggered mesh systems.

**Conservative means that internal fluxes cancel correctly after summation to produce exact global balances. Gradient terms have no outer coefficient.

⁺Notice that actually Eq. (7.10) follows from Eq. (7.11).

$$\frac{d\varphi_k}{dt} = S_k + \frac{J_{k-1/2} - J_{k+1/2}}{\Delta z} + \frac{1}{2\Delta z}\Big\{\varphi_{k-1}\left(v_{k-1/2} + \left|v_{k+1/2}\right|\right)$$

$$+ \varphi_k\left(v_{k-1/2} - \left|v_{k-1/2}\right| - v_{k+1/2} - \left|v_{k+1/2}\right|\right) - \varphi_{k+1}\left(v_{k+1/2} - \left|v_{k+1/2}\right|\right)\Big\}$$

(7.13)

Notice that, on the staggered mesh, the integration of Eq. (7.13) yields all intensive fluid parameters at cell centers for the cell with center at x_i and flow velocities at the boundaries of that cell. Moreover, Eq (7.13) has upwind differencing as a result of Assumption (ii) above. Upwind differencing in connection with computational cells is also called donor-cell differencing.

As the previously presented method of spatial discretization, the method of spatial averaging presented here involves also three neighboring φ-values. However, it contains two advective velocities $v_{k+1/2}$, rather than only one. The previously discussed method is centered around the replacement of derivatives by difference quotients (cf. Eq. (5.8)), while integration (cf. Eq. (7.11)) is the fundamental process for deriving Eq. (7.13). The truncation error for Eq. (7.13) can also be derived from Taylor series expansion, and it is found to be of first-order, just as the errors of Eqs. (7.4), (7.6) and (7.7). While the order-of-truncation error can be increased for those equations, it is fixed for Eq. (7.13). The truncation error of Eq. (7.13) can be reduced only by decreasing the space increments Δx. This reduces numerical diffusion and dispersion (cf. Section 8), but it also introduces new eigenvalues, higher order harmonic frequencies and computational stiffness.

The steps from Eq. (7.10) to Eq. (7.12) are quite general. Therefore, this method can be used to represent correctly, geometries of sudden expansions and contractions, particularly with regard to the momentum balance and its account for forces exerted between the fluid and the duct walls (Zuber 1974). Equations (7.12) and (7.13) are exact, yet they produce spatial averages, rather than distributions.

Equation (7.13) implies donor-cell (upwind) differencing, which must be contrasted against the use of centered difference (cf. Section 5.2) for computing convection terms on cell boundaries (Leonard 1973, pp. 66, 69 and 70). The distinction is very important. Spatially centered differences imply linear interpolation in the computation of cell boundary values and produce second-order truncation errors. The disadvantage of centered differences (cf. 5.2) arises from their tendency to suppress convective sensitivity to the cell average. This may produce computational instability, depending on local velocities.

Donor-cell differencing, on the other hand, has convective sensitivity, whence stabilizing feedback, as can be seen from the second term in the curly bracket of Eq. (7.13). Yet, donor-cell differencing amounts to zeroth-order interpolation with first-order truncation errors. It has so much numerical damping that it produces almost always more numerical diffusion than any physical mixing process, unless the discretization is extremely fine (cf. Section 8.4). Leonard (1973, p. 66) points out the fluid dynamicist's uniqueness in accepting, "in a state of false security," large errors from "heavy-handed artificial damping." Numerical damping can be sharply reduced by second-order upwind differencing. Leonard (1973) gives the details for this method (cf. Subsection 8.6.6).

The method presented here differs from the first-order, donor-cell finite difference method (cf. Section 8) only in that it contains time derivatives instead of time differences. It is also related to the less restrictive lumped-parameter modeling discussed next.

7.3 Lumped-Parameter Modeling

This method is based on the integral techniques used in boundary layer analyses (Pohlhausen, v. Karman) and in diffusion problems as summarized by Goodman (1974). It has been used for two-phase flow analysis by Wulff (1978) and by Wulff et al. (1981).

Lumped-parameter modeling differs from the previously presented volume averaging technique primarily because it dispenses with Assumptions (i) and (ii): profile estimates are used for relating fluid properties at cell exits to their respective cell averages and there is no need to identify the location at which the cell average occurs, nor is there a need for a staggered mesh.

In fact, lumped-parameter modeling accommodates moving cell boundaries (Wulff 1978). Moving cell boundaries, or zone boundaries (Theofanous 1986), can be used as boundaries between flow regimes (Wulff 1978). This facilitates, on the one hand, the achievement of high accuracy with a relatively small number of computational cells, and, on the other hand, the use of simple profiles with continuous derivatives within a cell. Yet, the derivation of reliable profile functions requires considerable insight into the two-phase flow processes within each flow regime. The method is particularly attractive for large thermohydraulic systems analyses where individual processes are known but their evolution and their interactions are to be predicted.

The generic form of the balance equation is Eq. (7.11). For one-dimensional flow through volume $V(t)$, one obtains, with the aid of Leibniz' rule for the differentiation of an integral with varying boundaries:

$$\frac{d\langle\varphi\rangle}{dt} = \langle S\rangle + \frac{A}{V}\left\{\left[(v_i - v)\varphi - J\right]_{exit} - \left[(v_i - v)\varphi - J\right]_{inlet}\right\} \quad (7.14)$$

Here v_i is the velocity of the boundary and v is the fluid velocity. The fluid parameters in curly brackets are area-averaged over flow cross sections.

Parameters at inlet and exit cross sections are linked to the volume average by a profile function, such as $f_\varphi(\eta)$:

$$f_\varphi(\eta) = (\varphi - \varphi_{inlet})/(\varphi_{exit} - \varphi_{inlet}) \,, \quad (7.15)$$

where $\eta = (x - x_{inlet})/(x_{exit} - x_{inlet})$ and $\eta = 0$ at the inlet, $\eta = 1$ at the exit. Inlet parameters are given as boundary conditions. Since

$$\langle f_\varphi\rangle = \int_0^1 f_\varphi(\eta)d\eta \quad (7.16)$$

and since $\langle\varphi\rangle$ is computed as the integral of Eq. (7.14), one can compute the unknown exit parameter from

$$\varphi_{exit} = \left\{\langle\varphi\rangle + (\langle f_\varphi\rangle - 1)\varphi_{inlet}\right\}/\langle f_\varphi\rangle \quad (7.17)$$

Equation (7.17) shows that the profile assumptions $f_\varphi(\eta) = 1, 0 < \eta < 1$, and $v_i(t) = 0$ would reduce Eq. (7.14) to Eq. (7.13), once flow reversal is accounted for also in Eq. (7.14). The first assumption implies $\langle f_\varphi\rangle = f_\varphi(1) = 1$ and corresponds to complete mixing within the computational cell.

In fact the profile functions f describe the degree of mixing or, more specifically the balance between advection and diffusion across the length of the cell. There is no universal rule for deriving profile estimates. For "slow" transients, steady-state profile estimates work well (Wulff et al. 1981). For "fast" transients, one can integrate the transport equation $\partial\varphi/\partial t + v\partial\varphi/\partial x - S = 0$ with v and S kept constant during an integration step for Eq. (7.14) (cf. Wulff 1978).

Patankar (1980, pp. 86-96) suggests the steady-state profile from integrating $d/dx(v\varphi - Dd\varphi/dx) = 0$

$$f = \frac{e^{N_{Pe} \cdot \eta} - 1}{e^{N_{Pe}} - 1}, \quad -\infty < N_{Pe} < \infty \tag{7.18}$$

where N_{Pe} is the grid Peclet number $v(z_{exit} - z_{inlet})/D = v\Delta z/D$, and D is the appropriate diffusivity. Other possibilities can be found in the same reference.

It is important to note, that the choice of profile functions affects the eigenvalues of the associated system Jacobian (cf. Eq. (7.1)). Linear profiles, $f = \eta$, for example, introduce imaginary components to the eigenvalues (cf. Wulff 1978). The computational accuracy depends also on the choices of $f(\eta)$, but it is better than or equal to first-order.

In closing, it should be recognized that the lumped-parameter formulation produces directly global balances for powerful engineering analyses and also for checking overall consistency in detailed computations of parameter distributions.

Often, volume-averaged balance equations (also called discrete volume or discrete domain formulations) are contrasted with finite difference equations. The former are said to be "globally exact," while the latter suffer from numerical diffusion. The difficulty with global or lumped-parameter formulations, derived by volume integration or volume averaging, is to find the relations between volume averages and cell exit values. This is as difficult as to cope with numerical diffusion; it is computing economy which may favor lumped-parameter formulations.

8. FINITE DIFFERENCE METHODS

8.1 Main Features and Reference Sources

Mathematical models for multiphase flow are analytical and belong to the continuous mathematics of partial differential equations. In Section 6 and 7 are presented computational techniques based on partially continuous (in time) and partially discrete (in space) formulations which are reflected generically in Eqs. (5.1).

In this section are presented the computational techniques with both time and space discretizations. The partial differential equations are converted completely to nonlinear (and possibly transcendental) difference equations of the form of Eqs. (5.2).

Finite differrence techniques are far more widely used than any other technique. The Finite Difference Method (FDM) is implemented by replacing all derivatives by difference quotients as discussed in Section 5.2. Such implementation appears rather mechanical, yet the outcome is far from unique and not always predictable with regard to computational effectiveness. Finite Difference Methods are implemented faster and produce results sooner than other methods. Even when applied to hyperbolic equations, Finite Difference Methods cannot economically achieve the accuracy of the Method of Characteristics (cf. Section 6) nor that of Finite Element Methods (cf. Section 9).

An extensive survey of Finite Difference Methods (FMD) for fluid dynamics is given by Roache (1972), including a comprehensive bibliography. Current progress is reported in the Journal of Computational Physics and in Mathematics of Computing. Standard texts are those by Forsythe and Wasow (1960), Ames (1969), Mitchell (1969) and Patankar (1980). Stability, convergence and consistency are presented by Lax and Richtmyer (1956), Richtmyer and Morton (1957) and by Hirt (1968). A quantitative assessment on numerical diffusion and dispersion has been presented for a catalog of 14 differencing schemes by Stuhmiller and Ferguson (1979). This assessment is particularly helpful to those who are selecting a finite differencing scheme.

8.2 Important Concepts

We shall discuss finite difference schemes on the basis of model equations

$$\varphi_t + v\varphi_x = \nu\varphi_{xx} + \varepsilon\varphi_{xxx}, \tag{8.1}$$

where subscripts t and x denote partial differentiations.

Plane waves $\quad \varphi = \varphi_k \, e^{-\alpha t + ik(x-ct)}$ (8.2)

will propagate according to Eq. (8.1) with the velocity $c = v + \varepsilon k^2$ and dissipate with $\alpha = \nu k^2$. Here k is the wave number, and $k = 2\pi/l_k$ if l_k is the wavelength. Notice that all waves remain undamped if $\nu = 0$. All waves move with velocity $c = v$ only if the dispersion velocity $\varepsilon k^2 = 0$, i.e., if the dispersion coefficient ε is zero. All higher even-order space derivatives produce fourth or higher order damping, all higher odd-order space derivatives introduce fifth or higher order dispersion, that is wave number-dependent propagation speeds.

As the derivatives in Eq. (8.1) are replaced by difference quotients, there will arise discrete φ_i^n values at locations ..., x_{i-1}, x_i, x_{i+1} ,... and at two or more time levels t^{n-1}, t^n, t^{n+1}. When N time levels appear, the differencing scheme is called an N-level scheme.

A finite differencing scheme is said to be *implicit* or *explicit*, if backward or forward time differences, respectively, were used to replace the time derivatives in its associated differential equation. A time difference $\varphi_i^{n+1} - \varphi_i^n$ is *forward* if all space differences are taken at the n-th and lower time levels; it is *backward* if the space differences are taken at the *(n+1)*-st time level. Explicit finite difference schemes for systems of partial differential equations with only one time-derivative each can be solved explicitly for the new-time values φ_i^{n+1}, while *implicit* schemes must be solved iteratively (cf. Section 11). For the definitions of forward, centered and backward space differences see Section 5.2 above.

We shall judge the accuracy of a differencing scheme by the amount of undesirable numerical dissipation (damping) and dispersion it introduces into the model equation having no physical damping or dispersion

$$L(\varphi) \equiv \varphi_t + v\varphi_x = 0 \ , \tag{8.3}$$

because advection is the primary concern in multiphase flows. Equation (8.3) has the same form as Eqs. (2.34) or (7.2), save for the source term, which is unimportant in the discussion of numerical diffusion dispersion, convergence, stability or consistency. By replacing derivatives in Eq. (8.3) by finite difference quotients, we replace the differential operator L in Eq. (8.3) by a difference operator L_Δ, to produce an approximate numerical solution φ. The operators L and L_Δ may operate on vectors if sets of differential equations are to be solved. The subscript Δ of L_Δ denotes its dependence on differences $\Delta t = t^{n+1} - t^n$, $n = 1, ...;$ $\Delta x = x_{i+1} - x_i$, $i=1,...;$ and possibly Δy and Δz. By saying $\Delta \to 0$ we mean Δt, Δx, Δy, $\Delta z \to 0$. By $\|\cdot\|$ we denote a vector norm such that $\|\vec{v}\| = \{v_1^2 + v_1^2 + ... + v_N^2\}^{1/2}$.

Convergence. The approximate solution Φ converges to the exact solution φ of Eq. (8.3) if

$$\|\Phi - \varphi\| \to 0 \quad \text{as} \quad \Delta \to 0 \ . \tag{8.4}$$

If $\| \Phi - \varphi \| = O(\Delta t^{p1}, \Delta x^{p2}, ...)$, where O denotes "order of," then this quantity is called the error of the approximate solution Φ. The p's are called the rate of convergence.

Consistency. The difference operator L_Δ is called consistent with the differential operator L if

$$\| L_\Delta(\varphi) - L(\varphi) \| \rightarrow 0 \quad as \quad \Delta \rightarrow 0 \tag{8.5}$$

for all smooth functions φ in the domain of L. If, for small values of Δ, $\| L_\Delta(\varphi) - L(\varphi) \| = O\{\Delta t^{q1}, \Delta x^{q2}, ...\}$, then this quantity is called the truncation error and the q's are the order of accuracy of L_Δ. However, there are, in general, restrictions on the rate at which $\Delta t \rightarrow 0$ relative to $\Delta x \rightarrow 0$ (see Eq. (8.15) below, for example).

Stability. A finite difference scheme is said to be stable if the approximate solution produced by this scheme at any fixed time t remains bounded as the number of time steps increases to infinity and $\Delta \rightarrow 0$, i.e.

$$\| \Phi_i^n \| \leq M\, e^{Ct^n}\, \| \Phi_i^o \| \,, \quad \Phi_i^o = \Phi_i\,(t = 0) \tag{8.6}$$

where M and C are independent of Δ and C is less than or equal to the growth rate of the exact solution. This means that computational noise (round-off errors) must decay in time.

Lax has shown (Lax and Richtmyer 1956, or Richtmyer and Morton 1975, p. 45), that for a well-posed (cf. Section 1,.1) linear set of partial differential equations, a consistent and stable difference system will converge. Most stability analyses are based on constant coefficients, they deal therefore with linear approximations to the nonlinear differential equations and, consequently, assess only local stability.

There are three different operational stability criteria, namely the discrete perturbation stability analysis (Thoman and Szewczyk 1966), the von Neuman stability analysis (Richtmyer and Morton 1957, p. 70) and the heuristic stability analysis by Hirt (1968).

For the discrete perturbation stability analysis, one introduces an error ε at location x_i, under the condition that $\Phi_i^n = 0$ for all i at time $t = t^n$. The resulting Φ_i^{n+1} for all i and $t = t^n + \Delta t$ must then be less than the error ε to prevent error growth, and it must have the same sign as ε to prevent "overshooting":

$$0 \le \frac{\Phi_i^1}{\varepsilon} \le 1 \; . \tag{8.7}$$

The von Neuman stability criterion starts with the transformation

$$\Phi^{n+1} = G \, \Phi^n \; , \tag{8.8}$$

which produces the solution vector Φ^{n+1}, containing all unknown state variables at all locations, for the new time level t^{n+1} from the unknown state variables at the current time level t^n. The constant matrix G is called the amplification matrix and depends on the fixed constants of the partial differential equations and on $\Delta = (\Delta t, \Delta x, ...)$. The von Neuman condition requires that the spectral radius, i.e. the maximum magnitude $|r|_{max}$ of all eigenvalues r_i, of the amplification matrix G satisfy

$$|r|_{max} \le 1 + O(\Delta t) \; . \tag{8.9}$$

Equation (8.9) is necessary for stability of the scheme in Eq. (8.8). It is also sufficient only if G is a normal matrix, i.e. a matrix with a complete orthogonal set of eigenvectors. Thus, for two-level difference equations with only one dependent variable, the von Neuman condition is also sufficient for stability.

The stability criterion by Hirt requires that the artificially introduced numerical diffusion coefficient, that is the coefficient of truncation error which is proportional to $\nabla^2\varphi$, be positive (cf. Eq. (8.1)). Else, $\alpha < 0$ in Eq. (8.2) and computational noise grows without bound.

8.3 Explicit Upwind Differencing

Upwind differencing is akin to donor cell differencing (cf. Sections 7.1 and 7.2). It is the most widely used differencing technique and inherently part of most hybrid computing schemes for two-phase flow such as ICE and MAC (Harlow and Amsden 1971), SIMPLE and SIMPLER (Patankar 1980), TRAC (TRAC-PF1 1986, Taylor et al. 1984), RELAP5 (Ransom 1985) and many others. Upwind differencing is presented here with some degree of detail because it is so widely used.

We consider first the *explicit* upwind differencing. For this purpose we apply forward time and upwind space differencing to the model equation, Eq. (8.3), and obtain for $v > 0$

$$(\Delta t) \cdot L_\Delta(\Phi_i) = \Phi_i^{n+1} - \Phi_i^n + \frac{v\,\Delta t}{\Delta x}(\Phi_i^n - \Phi_{i-1}^n) = 0 \,. \tag{8.10}$$

For $v < 0$, one replaces $\Phi_i^n - \Phi_{i-1}^n$ by $\Phi_{i+1}^n - \Phi_i^n$ as in Eqs. (7.4), (7.6) and (7.7). For $v > 0$, x_{i-1} is located upstream or "upwind" from location x_i. Equation (8.10) is a balance for the computational cell with center at x_i and provides information Φ_i at x_i from information Φ_{i-1} at the upwind location x_{i-1}. This is the reason for calling Eq. (8.10) the upwind difference analog to Eq. (8.3).

Equation (8.10) yields the new-time value explicitly

$$\Phi_i^{n+1} = \left(1 - \frac{v\,\Delta t}{\Delta x}\right)\Phi_i^n + \frac{v\,\Delta t}{\Delta x}\,\Phi_{i-1}^n \,. \tag{8.11}$$

8.3.1 *Consistency*

We need to establish that Eqs. (8.10) and (8.11) are consistent with Eq. (8.3) and that they are stable. Then, by Lax's equivalence theorem, we can assure local convergence of the numerical approximation Φ to the analytical solution φ. By substituting for the difference in Eq. (8.10) the appropriate Taylor series expansions, according to Eq. (5.8), one finds for $L_\Delta(\varphi)$

$$L_\Delta(\varphi) = \varphi_t + v\varphi_x + \Delta t\{\varphi_{tt}/2 + \varphi_{ttt}/6\Delta t + \ldots\} + \Delta t\{v[\varphi_{xx}/2 - \varphi_{xxx}/6\Delta x + \ldots]\} \tag{8.12}$$

and whence with Eq. (8.3), and in accordance with Eq. (8.5), the truncation error T.E. of the finite difference scheme of Eq. (8.10)

$$T.E. = \|L_\Delta(\varphi) - L(\varphi)\| = \Delta t\{\varphi_{tt}/2 + \varphi_{ttt}/6\Delta t + \ldots\} + \Delta x\{v[\varphi_{xx}/2 - \varphi_{xxx}/6\Delta x + - \ldots]\} = O\{\Delta t, \Delta x\} \,. \tag{8.13}$$

This establishes consistency, i.e. $T.E. \to 0$ as $\Delta t \to 0$ and $\Delta x \to 0$, and first-order convergence $(p_1 = p_2 = 1)$ for the scheme in Eq. (8.11).

8.3.2 *Stability*

For stability, we first replace in Eq. (8.11) Φ_i^n by $(\Phi_i^n + \varepsilon)$ and set Φ_i^n for all is equal to zero. This discrete perturbation at x_i^i gives at the subsequent time level this consequence

$$\Phi_i^{n+1} = \left(1 - \frac{v\,\Delta t}{\Delta x}\right)\varepsilon \tag{8.14}$$

and, by Eq. (8.7), the requirement that the error should not grow is:

$$0 \leqq \frac{v\,\Delta t}{\Delta x} \leqq 1\ . \tag{8.15}$$

Here $(v\ \Delta t/\Delta x)$ is the Courant number and Eq. (8.15) is the Courant-Friedrichs-Lewy (CFL) condition. It expresses four facts for the scheme of Eq. (8.11):

(i) $v \geq 0$ is required for stability; if $v < 0$ then $\Phi_i^n - \Phi_{i-1}^n$ in Eq. (8.10) must be replaced by $\Phi_{i+1}^n - \Phi_i^n$ (cf. Eq. (7.4)),

(ii) there is, for stability requirements, an *upper* limit on the time increment which depends on the smallest spacing Δx and on the largest velocity anywhere, i.e. $\Delta t \leq v\Delta x$ must be satisfied everywhere,

(iii) space mesh refinement, $\Delta x \to 0$, is possible only if the time increment is also reduced, such that $\Delta x \geq v\Delta t$,

(iv) computations backward in time, with $\Delta t < 0$ are unstable.

Next, to apply the von Neuman stability criterion, we expand the solution Φ into a Fourier series

$$\Phi_j^n = \sum_k \tilde{\Phi}_k^n\ e^{ikj\Delta x}\ , \tag{8.16}$$

substitute this expansion into Eq. (8.11) and then find the relationship between the two subsequent solutions appearing in Eq. (8.8)

$$\tilde{\Phi}_k^{n+1} = \tilde{\Phi}_k^n \left[1 + \frac{v\,\Delta t}{\Delta x}\left(e^{-ik\Delta x} - 1\right)\right] . \tag{8.17}$$

The amplification matrix G of Eq. (8.8) has here rank 1. Its spectral radius r is the term in square brackets on the right-hand side of Eq. (8.17). Since $|r|^2 = r\,\bar{r}$, where $\bar{r}$ is the complex conjugate of r, the von Neuman criterion becomes (cf. Eq. (8.9))

$$|r| = \left\{1 + 2\left(\frac{v\,\Delta t}{\Delta x}\right)\left[1 - \left(\frac{v\,\Delta t}{\Delta x}\right)\right]\left[\cos(k\Delta x) - 1\right]\right\}^{1/2} < 1 + O\{\Delta t\}. \tag{8.18}$$

Clearly, $|r|$ must be less than or equal to unity because Δt occurs only in the ratio $\Delta t/\Delta x$. Then, $|r| \le 1$ in Eq. (8.18) only if

$$0 < \frac{v\,\Delta t}{\Delta x} \le 1 , \tag{8.19}$$

which is the same CFL condition as in Eq. (8.15).

Finally, we investigate the conditions arising from Hirt's stability criterion, requiring that numerically induced dissipation be positive. With the same formalism we assess then also the accuracy of the scheme of Eq. (8.11) in terms of numerical diffusion and dispersion. For this purpose we differentiate $L_\Delta(\varphi) = 0$ in Eq. (8.12) repeatedly to eliminate all time derivatives up to third order. The result is an equation of the form of Eq. (8.1), namely

$$\begin{aligned} \varphi_t + v\,\varphi_x &= (v/2)(\Delta x - v\Delta t)\varphi_{xx} \\ &\quad + (v/6)(\Delta x - v\,\Delta t)(2v\Delta t - \Delta x)\varphi_{xxx} + \ldots \end{aligned} \tag{8.20}$$

More than Eq. (8.12), this equation demonstrates that the scheme in Eq. (8.11) does not integrate the model equation, Eq. (8.3), for which it is intended, but instead it integrates Eq. (8.20). All finite difference methods must imply finite Δt, Δx, $\Delta y, \ldots$, and produce solutions to equations which differ inherently from their object equations. Convergence is achieved only in the limit as $\Delta t \to 0$, $\Delta x \to 0$.

To satisfy Hirt's criterion, Eq. (8.20) requires, by comparison with Eq. (8.1), that for positive dissipation one must have

$$\nu = (v/2)\,(\Delta x - v\,\Delta t) \geq 0\ . \qquad (8.21)$$

This criterion can be met only if the CFL conditions, Eqs. (8.15) or (8.19) are met for $\Delta t > 0$. Thus we have shown consistency and, by three different criteria, the single stability requirements for the scheme in Eq. (8.11). By Lax's equivalence theorem, the scheme is also convergent to the solution of Eq. (8.3), if the CFL conditions are met.

8.3.3 *Accuracy*

As discussed in connection with Eqs. (8.1) and (8.2), the first term on the right-hand side of Eq. (8.20) represents numerical damping, the second term represents numerical dispersion. Denoting the wave length by l_k and the wave number by k, and using the same symbols as in Eqs. (8.1) and (8.2), we get for the damping coefficient

$$\begin{aligned} exp(-\alpha t) &= exp(-k^2 \nu\, l_k/v) \\ &= exp\left[-2\pi^2 (\Delta x/l_k)\,(1 - v\Delta t/\Delta x)\right]\ , \end{aligned} \qquad (8.22)$$

if we select for the time t the time a wave takes to travel through a distance equal to its wave length l_k. The dispersion velocity is, as a multiple of the physical velocity v,

$$k^2\varepsilon/v = (2\pi^2/3)\,(\Delta x/l_k)^2\,(1 - v\Delta t/\Delta x)\,(2v\Delta t/\Delta x - 1)\ . \qquad (8.23)$$

Notice from these two equations that numerical second-order damping and third-order dispersion both vanish whenever the Courant number $v\Delta t/\Delta x$ is unity. Numerical dispersion vanishes also, when $v\Delta t/\Delta x = 1/2$. At the condition of Courant number equal to unity, the upwind differencing scheme for fixed v resembles the Method of Characteristics (cf. Eqs. (2.44) and (6.4)).

However, as already shown by Stuhmiller and Ferguson (1979), the smaller the Courant number is, the larger are numerical diffusion and dispersion when explicit upwind differencing is used. As an example, Eq. (8.22) shows that a sinusoidal wave, having the wave length l_k and moving with velocity v, has its amplitude reduced to 37% of its original magnitude during the time

$$t = l_k/v \tag{8,.24}$$

that the wave travels through one wave length, if Eq. (8.11) is used, if the Courant number equals 1/2 and if there are as many as 10 subdivisions $\Delta x = l_k/10$ within one wave length. It takes 100 spatial subdivisions to maintain 90.6% of the amplitude and 1000 divisions for 99.0%, under the above conditions. An increase in the Courant number from 0.5 to 0.9, however, would maintain an amplitude of 98.0% with "only" a hundred subdivisions. In any case, one has rarely enough subdivisions to maintain 1% accuracy under conditions of time and space-dependent velocity $\bar{v}$. A move from fine to coarse nodalization, implemented often for the purpose of saving computer time, always brings about a significant loss in accuracy. Numerical diffusion is easily much greater than any physical diffusion.

8.4 Implicit Upwind Differencing

With backward time differences instead of the forward difference in Eq. (8.10), the implicit counterpart of this equation is

$$(\Delta t) \cdot L_\Delta (\Phi_i) = \Phi_i^{n+1} - \Phi_i^n + \frac{v\,\Delta t}{\Delta x} (\Phi_i^{n+1} - \Phi_{i-1}^{n+1}) = 0. \tag{8.25}$$

This equation can be solved explicitly for Φ_i^{n+1} in a marching procedure, starting from $i = 1$, if the inlet conditions $\Phi_o(t^{n+1})$ are given and if v is independent of φ.

$$\Phi_i^{n+1} = \left[\Phi_i^n + \frac{v\,\Delta t}{\Delta x} \Phi_{i-1}^{n+1} \right] / \left(1 + \frac{v\,\Delta t}{\Delta x} \right) . \tag{8.26}$$

In general, however, Eq. (8.25) must be solved implicitly as part of a large system of balance equations.

The spectral radius, or amplification factor, of Eq. (8.25) is

$$r = 1/\left[1 + \frac{v\,\Delta t}{\Delta x}\,(1 - e^{-ik\Delta x})\right], \tag{8.27}$$

the magnitude of which is never greater than unity, provided $v \geq 0$. Thus, the implicit upwind differencing scheme is unconditionally stable.

The truncation error of the implicit upwind differencing scheme is

$$T.E. = \Delta t\left\{\frac{1}{2}\,\varphi_{tt} - \frac{\Delta t}{6}\,\varphi_{ttt} + - \ldots\right\} + \Delta x\left\{v\left[\frac{1}{2}\,\varphi_{xx} - \frac{\Delta x}{6}\,\varphi_{xxx} + - \ldots\right]\right\}$$

$$= O\{\Delta t,\, \Delta x\}\,. \tag{8.28}$$

Thus implicit upwind differencing has, as explicit upwind differencing (cf. Eq. (8.13)), first-order convergence. However, the undesirable numerical diffusion coefficient $\nu = (v/2)\,(\Delta x + v\Delta t)$, while desirably positive for all Δx and Δt, is for the implicit scheme *much larger* than for the explicit scheme (compare with Eq. (8.22)):

$$\exp(-\alpha t) = \exp\left[-2\pi^2\,(\Delta x/l_k)\,(1 + v\Delta t/\Delta x)\right]\,. \tag{8.29}$$

Similarly, the relative dispersion velocity for implicit upwind differencing is

$$k^2\varepsilon/v = -\,(2\pi^2/3)(\Delta x/l_k)^2\,(1 + v\Delta t/\Delta x)\,(1 + 2v\Delta t/\Delta x)\,. \tag{8.30}$$

Table 8.1 below serves to compare the accuracies of explicit and implicit upwind differencing schemes. Listed are the results for three and four Courant numbers, respectively, for explicit and implicit schemes and for four degrees of spatial resolution. The ideal relative amplitude would be unity. The desirable relative dispersion velocity is zero. Numerical schemes of transport equations with dispersion velocities less than (-1.0) imply transport in the wrong direction, by backward numerical dirffusion against physical advection.

Table 8.1 Damping and Dispersion for Upwind Differencing

Node Number $l_k/\Delta x$	Courant Number $v\ \Delta t/\Delta x$	Rel. Amplitude,* $e^{-\alpha t}$ Explicit	Rel. Amplitude,* $e^{-\alpha t}$ Implicit	Rel. Dispersion Vel. Explicit	Rel. Dispersion Vel. Implicit
3	0.5	0.037	5×10^{-5}	0	-2.193
10	0.5	0.373	0.052	0	-0.197
100	0.5	0.906	0.744	0	-0.002
1000	0.5	0.990	0.971	0	-2×10^{-5}
3	0.75	0.193	1×10^{-6}	0.091	-3.198
10	0.75	0.610	0.032	0.008	-0.288
100	0.75	0.952	0.708	8×10^{-5}	-0.003
1000	0.75	0.995	0.966	8×10^{-7}	-3×10^{-5}
3	1.00	1.000	2×10^{-6}	0	-4.386
10	1.00	1.000	0.019	0	-0.395
100	1.00	1.000	0.674	0	-0.004
1000	1.00	1.000	0.961	0	-4×10^{-5}
3	2.00	---+	3×10^{-9}	---+	-10.97
10	2.00	---	0.003	---	-0.987
100	2.00	---	0.553	---	-0.010
1000	2.00	---	0.943	---	-1×10^{-4}

*Amplitude, originally equal to unity, after wave has propagated by one length.
+Unstable.

It is well recognized that the unconditional stability (cf. Eq. (8.27)) of the implicit upwind differencing scheme comes at the expense of considerable computational resources, as it requires iterative soutions to nonlinear implicit equation systems. Table 8.1 shows, that the unconditional stability comes also at a painfully large expense of accuracy. Notice that convergence through mesh refinement $(\Delta x \rightarrow 0)$ deteriorates for implicit upwind differencing as the Courant number increases.

It has been claimed that there is no real difference between implicit and explicit schemes because they are both of first-order accuracy. However, first-order accuracy implies only proportionality between the truncation error on the one hand and on the other the time and space differences, taken to the first power; it implies nothing about the magnitude of the proportionality constant. The message from Table 8.1 is that the proportionality constant for implicit integrations is consistently much larger than for explicit integration.

8.5 Applications of Upwind Difference Schemes

Sections 8.3 and 8.4 deal with purely explicit and implicit difference schemes. In practice, the velocity v is not constant and the parameter φ in the model equation, Eq. (8.1), is the product of vapor void fraction, density and velocity or energy (cf. Eq. (2.7)). Clearly, the balance equations of multiphase flow are far more complex than the model equations used in Sections 8.3 and 8.5.

Fully implicit balance equations require not only in the advection term,* i.e. the second term in Eqs. (2.7), (2.13) and (8.1), that all variables be written down at the $(n + 1)$-st time level

$$v\varphi_i^{n+1} = \left\{\alpha_k^{n+1}\ \rho_k^{n+1}\ v_k^{n+1}\ \psi_k^{n+1}\right\}, \tag{8.31}$$

but also in all diffusion and source terms. the source terms in turn imply convective heat transfer and possibly absorption of heat, so that all heat conduction equations for conduit walls and, in the case of two-phase flow in nuclear reactor systems, neutron kinetics equations for fission heat sources, all must either be written in terms of variables at the $(n + 1)$-*st* time level or the corresponding equations must be integrated simultaneously with implicit coupling. Strictly speaking, all ancillary systems which affect the transient multiphase flow, such as control systems, must also be included into a single system of implicit equations.

Fully explicit equations have only one variable each at the new time level t^{n+1}. This may imply the need to invert a matrix A_{ij} as in Eq. (2.33). The explicit equations produce new-time values directly as Eq. (8.11), but with additional terms. This would be akin to first-order Euler-Cauchy integration of Eqs. (7.13).

Upwind differencing is frequently used in spatial mesh arrangements where the momentum balance grid is displaced by a half width relative to the grid of mass and energy balances. This yields the advection velocities between mesh points while relating the mesh point pressures to the mesh point density and energy. The result resembles the method discussed in Section 7.2, if that method is combined with first-order Euler-Cauchy integration.

---

*Also called the convection term.

Fully implicit formulations lead to enormously large systems of equations. The system is largest if the derivatives of products are first fully expanded:

$$\frac{\partial}{\partial x}\{\alpha\rho v\psi\} = \rho v\psi\frac{\partial\alpha}{\partial x} + \alpha v\psi\frac{\partial\rho}{\partial x} + \alpha\rho\psi\frac{\partial v}{\partial x} + \alpha\rho v\frac{\partial\psi}{\partial x} \quad (8.32)$$

and then expressed in terms of variables at the (n + 1)-st time level, possibly even with the density derivative further expanded in terms of pressure and internal energy. No existing computer code* employs this approach, even for thermodhydraulics simulations, much less for complete, large systems simulations.

A first reduction of the required computing resources is achieved by considering the product $\alpha\rho\psi = \varphi$ as a state variable and to compute next-time level variables α, ρ, p, u and v from the products $(\alpha\rho)^{n+1}$, $(\alpha\rho u)^{n+1}$ and from the equations of state.

A far more widely chosen approach is the so-called semi-implicit formulation: all but one variable in a product are written as the known variables at the n-th time level. This selective mix of current and new-time variables was used already in 1971 in the pioneering ICE method (Harlow and Amsden 1971) and is still used in the TRAC code sequence (TRAC-PF1 1986, Taylor et al. 1984). The tendency toward semi-implicit calculations is motivated by the need to reduce the enormous computing effort required for solving the large fully implicit system of equations. The aim is to obtain a smaller set of implicit equations and a second set of explicit equations, thereby achieving an optimum balance between stability and computational efforts.

The numerical diffusion and dispersion for semi-implicit schemes could be expected to lie between explicit and implicit schemes. The exact values of the diffusion coefficient ν and the dispersion coefficient ε (cf. Eq. (8.1)) must be derived by the means discussed in Sections 8.3 and 8.4. Even though the tendency in published and existing schemes is toward implicit pressure field computations (cf. Section 13), nevertheless the choices for selecting implicitly computed variables are in principle so varied that it is outside the scope of this review to include assessments of numerical diffusion and dispersion for semi-implicit schemes. No error estimates have been published with the most widely used two-phase flow thermohydraulics codes

---

*The same applies to implicitly integrated formulations based on global volume balanced as discussed in Section 7.2.

Table 8.1 reveals however a fundamental fallacy underlying the notion that computing speed can be increased without a loss in accuracy by moving from explicit via semi-implicit toward fully implicit integration. By means of Eq. (8.22), one can see, for example, that an accuracy of 1% in maintaining the amplitude of a wave while it travels through one wave length can be achieved with an *explicit* upwind differencing scheme, using a Courant number of 0.9, a spatial resolution of $(l_j/\Delta x) = 200$ and a time step of $\Delta t = 4.5 \times 10^{-3}\ l_k/v$, where v is the propagation velocity. If a fully *implicit* scheme were used for the same circumstances and if the same accuracy were attempted, then, by virtue of Eqs. (8.22) and (8.29) one would have to satisfy first

$$\left(\frac{\Delta x}{l_k}\right)_{impl.} = \left(\frac{\Delta x}{l_k}\right)_{expl.} - \frac{v}{l_k}\left[(\Delta t)_{expl.} + (\Delta t)_{impl.}\right] . \tag{8.33}$$

Clearly, the nodalization would have to be refined. Even in the limit of $(\Delta x) \to 0$ for implicit integration, the maximum permissible time step would be

$$(\Delta t)_{impl.} = \frac{l_k}{v}\left\{\left(\frac{\Delta x}{l_k}\right)_{expl.}\left[1 - \left(\frac{v\,\Delta t}{\Delta x}\right)_{expl.}\right]\right\} , \tag{8.34}$$

which equals $(\Delta t) = 5 \times 10^{-4}\ l_k/v$. Thus the permissible time step, based on accuracy, is nine times smaller for implicit integration than it is for explicit integration, while the computing effort is infinitely larger.

A lot of arguments have been published imploring everyone to use implicit integration. All arguments are based exclusively on stability considerations while accuracy requirements are ignored. Spalding (1982) points out correctly that most reactor safety codes do not have sufficiently fine nodalization to achieve accuracy. Then, however, he claims that only fully implicit schemes can provide affordable computations with sufficiently fine noding. His lengthy and strictly qualitative chain of arguments avoids this key question: Can implicit integration produce the same accuracy as explicit integration with a sufficiently larger permissible time step so as to compensate for the longer frame time (cf. Section 3.6) which is necessitated by iteration? Equation (8.33) shows for the upwind differencing scheme that the time step can only be smaller and the computing effort is always larger unless, of course, the flow fields have only linear spatial variations ($\varphi_{xx} = 0$) or they are stationary ($\varphi_t = 0$).

8.6 Other Finite Differencing Schemes

All finite differencing schemes have numerical dispersion but the error from dispersion is normally less important. The far more important damping, however, varies strongly with the choice of the scheme. Below are summarized a few noticeable schemes. For a more complete catalog of schemes see Stuhmiller and Ferguson (1979) and Birkhoff and Kines (1962). Both references show assessments of amplification factors, phase shift and diffusion.

8.6.1 *Fully Implicit Central Space Differencing*

Considerably less damping than with upwind differencing can be achieved with the unconditionally stable, implicit scheme having *central* space differencing. Referring to the model equation, Eq. (8.3), the difference operator L_Δ yields

$$\Phi_i^{n+1} - \Phi_i^n + \frac{v\,\Delta t}{2\Delta x}\left(\Phi_{i+1}^{n+1} - \Phi_{i-1}^n\right) = 0 \,. \tag{8.35}$$

The amplification factor of this scheme is $r = [1 - i(v\Delta t/\Delta x)\sin(k\Delta x)]^{-1}$ and the truncation error is

$$T.E. = \Delta t\left\{1/2\ \varphi_{tt}\right\} + (\Delta x)^2\left\{v/3\ \varphi_{xxx}\right\} . \tag{8.36}$$

This scheme has the advantage of being independent of the velocity direction.

8.6.2 *Lax-Friedrichs Scheme*

This explicit scheme has also better convergence than the upwind differencing scheme (cf. Section 8.3). L_Δ yields

$$\Phi_i^{n+1} - \frac{1}{2}\left(\Phi_{i+1}^n + \Phi_{i-1}^n\right) + \frac{v\,\Delta t}{2\Delta x}\left(\Phi_{i+1}^n - \varphi_{i-1}^n\right) = 0 \tag{8.37}$$

and its truncation error is

$$T.E. = \Delta t\left\{(1/2)\varphi_{tt}\right\} - \frac{(\Delta x)^2}{\Delta t}\left\{(1/2)\ \varphi_{xx}\right\} , \tag{8.38}$$

while its stability limit is the same as in Eq. (8.15) for the explicit upwind differencing.

8.6.3 *Second-Order Forward Time, Central Space Differencing*

Another explicit scheme with no appreciable diffusion or phase lag for spatial resolutions $(l_k/\Delta k) > 10$ is obtained by expanding first

$$\varphi^{n+1} = \varphi^n + \Delta t \varphi_t + \frac{(\Delta t)^2}{2} \varphi_{tt}$$

$$= \varphi^n + \Delta t(-v\, \varphi_x) + \frac{(\Delta t)2}{2} \left[-v\, \varphi_x \right]_t$$

$$= \varphi^n - v\, \Delta t \varphi_x + \frac{v^2 \Delta t}{2} \varphi_{xx} .$$

The result of replacing their derivatives by difference quotients is:

$$\Phi_i^{n+1} - \Phi_i^n + \frac{v\, \Delta t}{2\Delta x} \left(\Phi_{i+1}^n - \Phi_{i-1}^n \right)$$

$$+ \frac{v^2 \Delta t}{2\Delta x} \left[\frac{\Phi_{i+1}^n - \Phi_{i-1}^n}{\Delta x_{i+1/2}} - \frac{\Phi_i^n - \Phi_{i-1}^n}{\Delta x_{i-1/2}} \right] = 0 , \tag{8.39}$$

where $\Delta x_{i \pm 1/2}$ denotes the left and right-hand space increments, in case Δx is not a constant. The differencing scheme, however, does not depend on flow directions.

8.6.4 *Leapfrog Scheme*

This second-order, fully explicit, three-level scheme has no second-order amplitude damping at all (Courant, Friedrichs and Lewy (1967)

$$\Phi_i^{n+1} - \Phi_i^{n-1} + \frac{v\, \Delta t}{\Delta x} \left(\Phi_{i+1}^n - \Phi_{i-1}^n \right) = 0 \tag{8.40}$$

The amplification factor is

$$r = -i\left(\frac{v\,\Delta t}{\Delta x}\right)\sin(k\,\Delta x) \pm \left[1 - \left(\frac{v\,\Delta t}{\Delta x}\right)^2 \sin^2(k\,\Delta x)\right] \tag{8.41}$$

and indicates the possibility of alternating errors due to the ± sign. The stability requirement is the same as for the explicit upwind differencing scheme and given in Eq. (8.15). The truncation error is, however, *much* smaller:

$$T.E. = (\Delta t)^2\left\{(1/6)\varphi_{ttt}\right\} + (\Delta x)^2\left\{v/6\varphi_{xxx}\right\} . \tag{8.42}$$

There is also a fourth-order leapfrog scheme

$$\Phi_i^{n+1} - \Phi_i^{n-1} + \frac{4}{3}\frac{v\,\Delta t}{\Delta x}\left[\Phi_{i+1}^n - \Phi_{i-1}^n\right] - \frac{v\,\Delta t}{6\Delta x}\left[\Phi_{i+2}^n - \Phi_{i-2}^n\right] = 0 \tag{8.43}$$

which has an only slightly more restrictive stability limit

$$\left|\frac{v\,\Delta t}{\Delta x}\right| \leq 0.7555 . \tag{8.44}$$

It is fully explicit, three-level and requires data storage at two time levels. Its truncation error is low enough to justify coarser spatial nodalization than both schemes above (cf. Eq. (8.36) and (8.39)):

$$T.E. = (\Delta t)^2\left\{(1/6)\varphi_{ttt}\right\} + (\Delta x)^4\left\{const.\varphi_{xxxxx}\right\} . \tag{8.45}$$

8.6.5 *Schemes Without Amplitude Errors*

The finite differencing schemes without amplitude damping deserve some recognition in light of the discussion in Section 8.5. One scheme is described in Subsection 8.6.4 above. Other schemes are the implicit Box Scheme (Birkhoff and Kines 1962, Chan 1978), the implicit Average-time, Central Space Scheme (Roache 1972, p. 84), the implicit Balanced Expansion Technique, Time-Averaged Scheme

(BET-TA) (Chan 1978), the implicit BET-Al Scheme (Chan 1978), the implicit BET-A4 Scheme (Chan 1978) and the implicit Rhomboid Scheme (Chan 1978).

8.6.6 *High-Order Differencing Schemes*

Leonard (1979) discusses the shortcomings of upwind differencing in detail and shows that most arguments offered in the defense and promotion of upwind differencing have no general validity. He addresses schemes which fall, within the structure of this review, in both categories of Finite Differencing Methods and Space Averaging (finite volume, finite domain, etc.) Methods. Both categories lead to the same discrete problem formulation if first-order Euler-Cauchy methods are used for time-integration of space-averaged equations. Leonard's paper is therefore an important contribution to the understanding of numerical dissipation and its relative significance to physical diffusion and advection.

Leonard concentrates on the issue of gradient representations or, equivalently, the interpolation of advected flow properties at locations of computational cell boundaries (cf. Section 7.3 on Lumped-Parameter Modeling). Leonard presents a quadratic upstream interpolation technique as part of an explicit difference scheme. The stability limits are given as functions of grid-Peclet number, physical diffusivity and advection velocity. For further details, the reader is referred to the paper by Leonard (1979). His work applies also to multidimensional flows.

Moreover, Ivanov and Idiyatullina (1978) developed a third-order accuracy scheme for transient flow calculations. It is explicit and first-order in time, but third-order in space. The interested reader will find stability criteria which are less restrictive than Eq. (8.15) and successful applications to single-phase, two-dimensional and compressible flow through converging-diverging nozzles.

8.7 **Closing Remarks**

Finite Differencing Methods are most attractive for computing parameter distributions, that is, differentiable functions in one or multidimensional space. They are less suitable for global computations than lumped-parameter formulations.

Many selected Finite Difference Methods afford control of computational diffusion that is not available for finite volume, finite "domain" and other techniques (cf. Section 7) which have inherently the upwind differencing characteristics discussed in Sections 8.4 and 8.5. For effective damping control, there are only four choices:

Method of Characteristics, selected finite Difference Schemes, some selected Finite Element formulations and numerical quadrature after analytical integration (cf. Section 4.2). Computational accuracy in Finite Difference Methods is achieved by expressing local balance laws in terms of more than one time level and in terms of discrete data at locations of, and also beyond, adjacent mesh points.

9. FINITE ELEMENT METHODS

9.1 Main Features

The computational methods described in Sections 7 and 8 centered on the term-wise approximation to the balance equations. In the Finite Element method (FEM) the entire balance equation is approximated as a whole. Equations in the form of Eq. (2.21) may first be solved for the derivatives $\partial w_i/\partial t$ and then reduced to zero

$$\frac{\partial w_i}{\partial t} - C_{ijk}\, w_{j,k} - D_i = 0 \qquad i = 1,\dots, s \tag{9.1}$$

The Finite Element Method is based on the idea of finding a numerical approximation W_i for w_i, such that the inner products of the residual error ε of the entire Eq. (9.1), evaluated with W_i,

$$\varepsilon_i \equiv \frac{\partial W_i}{\partial t} - C_{ijk}\, W_{j,k} - D_i, \qquad i = 1,\dots, s \tag{9.2}$$

and a weighting function q_l over the domain of interest V is forced to be minimal* or zero.

$$(\varepsilon_i, q_l) = \int_V \varepsilon_i q_l \, dV = 0 \qquad l = 1,\dots, M \tag{9.3}$$

Clearly, the approximation attempted by FEM is in a weighted integral sense global over the domain V and over the balance equation, Eq. (9.1)

FEM was first developed for aircraft structural analysis (Turner et al. 1956). FEM models can be formulated on the basis of

---

*FEM based on variational principle.

variational functional analysis (Rayleigh 1877 and Ritz 1909), or on the basis of weighted residuals, either according to the Galerkin (1915) method, or according to the method of least squares $(q_l = \varepsilon)$ (Chung 1978), or by collocation. For fluid dynamics problems the Galerkin method is more convenient than variational methods, as functions cannot be found in general, whose Euler-Lagrange equations recover the original balance equation. The Galerkin method is also more suitable than the least-square method with its requirement for high-order interpolation functions.

FEM has been applied for two-phase flow by researchers of the Gesellschaft fuer Reaktorsicherheit Garching (Werner 1981; Graf, Romstedt and Werner 1982; Graf et al. (1986) and at the Rensselaer Polytechnic Institute, Troy, NY (Bukhari 1985 and Sim 1986). FEM theory for fluid dynamics applications are found in Chung (1978), Thomasset (1981), in Taylor and Hughes (1981), in Heubner (1975) and in White (1985); selected advanced topics are found in Finite Elements in Fluids, edited by Gallagher et al. (1985).

FEM has two major advantages over the techniques discussed in Sections 7 and 8 above. Firstly, it is the most powerful method to accommodate curved and irregular geometries of boundaries in multidimensional fluid flow fields. Instead of elementary rectangles and rectangular prisms used in rectangular or cylindrical coordinates, FEM covers conveniently triangular, tetrahedral and hexahedral elements. Triangles and tetrahedrons can approximate any arbitrary shape. Secondly, FEM is more economical than all the widely used first-order Finite Difference Methods or Discrete Volume Methods. Werner (1986) has shown that FEM, based on his second-order Asymmetric Weighted Residual Method, achieved convergence with 4 mesh points, where TRAC PF1, Version 12.1, solving the identical benchmark problem, could not achieve convergence with 120 mesh points, consuming 6.3 times the execution time (CPU) needed for the converging FEM calculation. Convergence could not be obtained at all with TRAC because the computing cost escalated prohibitively with grid refinement.

FEM produces discrete formulations in both the forms of Eqs. (5.1) and (5.2), just as all other previously discussed techniques. In fact, for one-dimensional flow, FEM formulations can be identical to FDM formulations.

9.2 Weighted Residual and Collocation Method

Because of space limitations, I confine the presentation here to FEM formulations based on weighted residuals and collocation, and I describe the technique developed by Graf and Romstedt for one-dimensional flow. The somewhat more complicated three-dimensional extension of this method is found in the review by Graf (1986).

We wish to find an approximate solution to Eqs. (9.1) by solving Eqs. (9.3).

The unknown solution vector W_i is first expressed in a series of products of time-dependent coefficient functions $c_j(t)$ and space-dependent test, trial or interpolation functions $\psi_j(x_k)$

$$(W_i)_L^{(N)} = \sum_{j=1}^{N} \left[c_j(t) \cdot \psi_j(x)\right]_i \tag{9.4}$$

The test functions ψ_i are base vectors of the N-dimensional test space $T_N \epsilon L_2(V)$ with continuous second derivatives in the solutions space V. The coefficient functions $c_j(t)$ must be determined such that they satisfy Eqs. (9.3) for every l, $l = 1,\ldots, M$. Equations (9.3) then become ordinary differential equations for $c_j(t)$. When ψ_j and q_j are identical, the method is that of Galerkin.

9.2.1 *Choice of Test Functions*

Test functions $\psi_j(k)$ are chosen to satisfy boundary conditions. They can be linear (Bukhari, p. 229) or any polynomial base vectors. The domain $V = [x_o, x_E]$ of interest between x_o and x_E is subdivided into M segments $V = [x_{l-1/2}, x_{l+1/2}]$, $l = 1,\ldots, M$. The test functions ψ_l for the l-th segment are chosen by Graf (1986) to be polynomials

$$\left.\begin{aligned} \psi_l^{(j)}(x) &= \left[(x - x_l)/(x_{l+1/2} - x_{l-1/2})\right]^j && \text{for } x \text{ in } V_l \\ &= 0 && \text{for } x \text{ not in } V_l \end{aligned}\right\} \tag{9.5}$$

Here x_l is the midpoint of the segment. Clearly, with $j = 0,1,2$ for quadratic test functions, and with M segments, there are 3M test functions. By using the same expansion for the segment V_l as for V, given by Eq. (9.4), we obtain the segment approximation

$$P_l^{(2)}(x,t) = \sum_{j=0}^{2} c_l^{(j)}(t)\, \psi_l^{(j)}(x) \tag{9.6}$$

which is now a second-degree polynomial in x. $P_l^{(2)}$ must equal $(W_i)_L^{(2)}$ at discrete locations $x_{l-1/2}$, x_l and $x_{l+1/2}$. Therefore

$$\left.\begin{aligned} \left[c_l^{(0)}(t)\right]_i &= (W_i)_l \\ \left[c_l^{(1)}(t)\right]_i &= (W_i)_{l+1/2} - (W_i)_{l-1/2} \\ \left[c_l^{(2)}(t)\right]_i &= 2[(W_i)_{l+1/2} - 2(W_i)_l + (W_i)_{l-1/2}] . \end{aligned}\right\} \quad (9.7)$$

The numerical approximation W_i in the total domain V is

$$(W_i)_L^{(2)} = \Big[\sum_{l=1}^{M} \sum_{j=1}^{2} c_l^{(j)} \psi_l^{(j)}\Big]_i = \Big[\sum_{l=1}^{M} P_l^{(2)}(x,t)\Big]_i \quad i = 1,\ldots, s \quad (9.8)$$

Since the segments are connected, and since the W_i's are continuous across the segment joints, there are for every i $2M + 1$ unknowns $(W_i)_{-1/2}$, $(W_i)_1$, $(W_i)_{1+1/2}, \ldots,$ $(W_i)_{M+1/2}$. If M_B boundary values are given as functions of time, then we must have the total of $s(2M + 1)-M_B$ equations for equally many unknown state variables $(W_i)_l$. This number of equations calls for a computational effort approximately equal to that which is required for a finite difference formulation with 2M segments.

Bukhari (1985) selected linear test functions (bilinear for his two-dimensional analysis) for all variables and equations except for pressure and the liquid mass balance, where he chose constant test functions.

9.2.2 *Selection of Weighting Functions*

The choice of the weighting function characterizes the FEM formulation as least-square, collocation or weighted residual method, etc. If the error ε_i in Eq. (9.2) is chosen as weighting function and if

$$\frac{\partial}{\partial c_l} (\varepsilon_i, \varepsilon_i) = 0 \quad (9.9)$$

is solved, then we have the least-square formulation. If the

Diract-delta function is used, then the FEM basis is a collocation technique, and if the test functions are also chosen as weighting functions then we have the Galerkin method.

Graf (1986) selected two sets of weighting functions. The first set is

$$\left.\begin{aligned} q_l^{(1)}(x) &= 1 \quad \text{for } x \text{ in } V_l \\ &= 0 \quad \text{for } x \text{ not in } V_l \,. \end{aligned}\right\} \tag{9.10}$$

The second set is

$$q_l^{(2)}(x) = \delta(x - x_{l-1/2}) \qquad \text{for all } x \text{ in } V \,, \tag{9.11}$$

where δ is the Diract-delta function. It has the property that

$$\int_V f(x) \cdot \delta(x - x_{l-1/2})dx = f(x_{l-1/2}) \tag{9.12}$$

and it implies collocation with upwind asymmetry.

The choice of test and weighting functions strongly affects stability and convergence rate. Chung (1978, p. 206) points out that the inner product in Eq. (9.3) must yield a spatial invariant and concludes that the momentum balance is to be weighted with the velocity test function and the mass balance by the pressure test function if velocity and pressure are computed from mass and continuity equations. For a detailed discussion on interpolating functions see Heubner (1975, pp. 123-191).

9.2.3 *FEM Equations*

The substitution of Eqs. (9.8) and (9.10) into Eqs. (9.3) yields $s \cdot M$ ordinary differential equations with respect to time

$$\left\{ \sum_{j=0}^{2} \frac{dc_l^{(j)}}{dt} \int_{V_l} \psi^{(j)}(x)dx \right\}_i - \int_{V_l} C_{ij} \frac{\partial W_i}{\partial x} \, dx - \int_{V_l} D_i(W_i, \ldots)dx = 0 \,. \tag{9.13}$$

After replacing the time-dependent c_l-coefficients by the linear combinations of W_i given in Eqs. (9.7) and after integrating by parts the second integral in Eq. (9.13), one obtains $s \cdot M$ nonlinear ordinary differential equations with respect to time for $(W_i)_{l-1/2}$, $(W_i)_l$ and $(W_i)_{l+1}$, $i = 1,..., s$; $l = 1,.., M$.

By substituting Eqs. (9.11) into Eqs. (9.3), one obtains $s(M + 1)$ ordinary differential equations in $(W_i)_{l-1/2}$

$$\frac{d(W_i)_{l-1/2}}{dt} - (C_{ij})_{l-1/2} \left.\frac{\partial(W_i)}{\partial x}\right|_{l-1/2} - (D_i)_{l-1/2} = 0 \,. \tag{9.14}$$

The partial derivative with respect to x is evaluated from Eqs. (9.8). Since the space derivatives are discontinuous at $x_{l-1/2}$, where the segments V_l are joined, one must select one, namely the upwind derivatives for fluid flow problems.

Of the $s(M + 1)$ equations are needed only $s(M + 1)-M_B$ equations, since M_B variables are prescribed as boundary conditions. Thus, the total number of ordinary differential equations, given by Eqs. (9.13) and (9.14) is $s(2M + 1)-M_B$. This number is equal to the number of unknowns $(W_i)_l$.

The integrals of $\psi(x)$ in Eq. (9.13) are constants. Equations (9.13) are therefore linear and implicit in the time derivatives, but they can easily be solved to yield the time-derivatives explicitly, thereby conforming to Eqs. (5.1). On the other hand, if the time derivatives in Eqs. (9.13) and (9.14) are replaced by finite difference analogs, then one obtains a set of nonlinear equations of the form of Eqs. (5.2).

9.2.4 *Asymmetric Weighted Residual Formulation*

Graf and Romstedt (1983) achieved stronger spatial coupling and avoided the development of non-physical oscillations, commonly associated with symmetric mesh point arrangements, by using a staggered mesh point system with asymmetry.

The first grid is defined by the set of segments

$$V_l^{(1)} = \left[x_{l-1/2}, x_{l+1/2}\right], ..., \left[x_{M-3/2}, x_{M-1/2}\right], \left[x_{M-1/2}, x_{M+3/4}\right] \tag{9.15}$$

and associated with thermodynamic variables such as pressure, void fraction and internal energy. The second grid is displaced such that

$$V_l^{(2)} = \left[x_{1/2}, x_{7/4}\right], \left[x_{l-1/2} + \Delta x/4, x_{l+1/2} + \Delta x/4\right], l = 2,\ldots, M. \tag{9.16}$$

This staggered grid has three advantages. Firstly, spatial derivatives of variables defined in $V_l^{(1)}$ and needed for variables defined in $V_l^{(2)}$ are always centered differences. Secondly, there is no interpolation required to obtain all the needed parameters for the balance equations. Lastly, spatial derivatives at segment joints are continuous to the extent that they are needed for implementing Eqs. (9.13) and (9.14).

9.3 Accuracy

For transient one-dimensional flow analyses, one can carry out the truncation error and stability analyses as discussed in Section 8 (cf. Chung 1978, p. 57). For transient multidimensional flow, there is presently no reliable error estimate variable, primarily because of nonlinear terms $(v_{i,j} \cdot v_j)$. Chung (1978, p. 227) gives an explicit error estimate, excluding the effects of nonlinear terms, for the steady Navier-Stokes equations.

Chung (1978) has shown that for one-dimensional FEM formulations with k-th-order polynomials as trial functions $\psi_l(x)$, the truncation error is of order

$$T.E. = O\left\{\Delta x^{k+1}\right\}, \tag{9.17}$$

where Δx is the length of segment V_l in Eq. (9.5), while the m-th derivative is of $O\{\Delta x^{k+1-m}\}$ accurate.

A number of approaches to error estimates for multidimensional FEM formulations are given by Chung (1978, p. 147) and in the references cited there.

10. NUMERICAL INTEGRATION OF ORDINARY DIFFERENTIAL EQUATIONS (ODE)

10.1 Importance to Multiphase Flow Calculations

In the preceding Sections 5 through 9 is discussed the first phase in the problem of solving numerically the governing equations of multiphase flow, namely the conversion of the partial differential equations into either partially or completely discrete equations. In this and the next chapter are summarized the methods for integrating and solving these equations.

In Sections 6, 7 and 9 are developed partially discretized equations: space is discretized and time is retained as continuous. The equations are of the form of Eq. (5.1)

$$dy_i/dt = F_i(t, y_1, \ldots y_m, \Delta x_1, \Delta x_2, \Delta x_3)$$

$$y_i(0) = a_i \quad , \quad i = 1, \ldots, m. \tag{10.1}$$

Here it is assumed that any subsidiary relations $z_j(y_1, \ldots) = 0$ are all solved in terms of the state variables y_i and eliminated from Eq. (10.1).

The multiphase flow model formulation in the form of Eqs. (10.1) has been used by Porshing et al. (1971), by Holt and Rasmussen (1972), by Burwell et al. (1983), by Hofer (1976, 1983), by Wulff (1980, 1981), by Wulff et al. (1984) and by many others. Special-purpose computers have been developed for the sole objective of integrating large systems of nonlinear ordinary differential equations at high speed and low cost (Karplus 1977, Fadden 1984). One such computer has been successfully used for integrating nonhomogeneous non-equilibrium two-phase flow equations (Wulff et al. 1984). Most multiphase flow formulations in the form of Eqs. (10.1) are part of complicated systems simulations, such as in full-scope training simulators for operators of nuclear power plants.

There are two major advantages associated with formulating fluid flow problems in the form of Eqs. (10.1). Firstly, most computing centers have libraries of well-documented integration algorithms for ordinary differential equations, with clear definitions of capabilities, limitations and user requirements. These algorithms have been developed, rigorously validated and documented by mathematicians. Those who use them are free to focus their attention on the physics of the problem, rather than on numerics. The use of such well-perfected algorithms has been very successful (Hofer 1983, Wulff et al. 1984).

The second advantage lies in the capabilities of these readily accessible algorithms to converge fast and to provide rigorous control of the time discretization error. Fast convergence is due to commonly available high-order numerical integration, and error control is based on truncation error estimates, as it is supposed to be. It must be recognized that of all the many two-phase flow computer codes developed in recent years for reactor safety, only the IMEX and FEBE series, developed at the Gesellschaft fuer Reaktorsicherheit in Garching, FRG (Hofer 1983), and the HIPA code in the AD10 special-purpose computer of the BNL Plant Analyzer of Brookhaven National Laboratory (Wulff et al 1984) have documented space and time discretization error controls. Spalding (1982) refers to studies on space discretization errors but gives no quantitative results and makes no mention of the time discretization error. Most codes control the integration step size by the magnitude of selected, first-order time derivates (rates of change), which have no relation to truncation errors because first-order time derivatives are retained in even the lowest-order algorithms. Others control the integration step size on the basis of a convergence criterion for satisfying Eqs. (5.2), thereby ignoring altogether the truncation error built into Eqs. (5.2) (cf. Sections 8.6.1 through 8.6.6). Space-discretization errors are quite generally ignored and most time-discretization error controls are meaningless. Therefore, it is very important to have at least rigorous time-step control during a computation.

The greatest advantage of having the multiphase flow formulation in the form of Eqs. (10.1) is for the analysis of fast transients which call for explicit integration (cf. Section 3.6). Explicit integration schemes have only conditional computational stability and are for this reason too often dismissed as unsuitable. Three points are to be made in this respect: (1) In most instances conditionally stable algorithms are rejected on the single criterion of stability and in total disregard for accuracy. The painful consequences of this narrow view are shown in Table 8.1 and are discussed at the end of Section 8.5. (2) If the predominant limit on the time step size is due to stability rather than to truncation error, then the system is stiff, i.e. the eigenvalues of the system Jacobian, associated with Eqs. (10.1),

$$J_{ij} = \partial F_i / \partial y_j \,, \quad i,j = 1,\dots,\ m \tag{10.2}$$

have moduli which vary over several orders of magnitude. In many cases,. this is indicative of improper modeling, in that, for example, irrelevant inertia effects have been modeled. (3) Even when implicit algorithms are called for by unavoidable stiffness, the integration of ordinary differential equations (ODE) still yields a higher rate of convergence in time than most existing first-order schemes in the form of Eqs. (5.2).

The integration of ODEs is not pecular to multiphase flow analyses and will therefore not be presented here in detail. The reader is referred primarily to the nearest computing center for ready-made subroutines, then to standard texts on numerical methods for integrating ODEs (Keller 1968, Gear 1971) or on numerical methods in general (Isaacson and Keller 1966; Carnahan, Luther and Wilkes 1969; Scarborough, Sixth Edition 1966). Good and more reviews can be found in the papers by Benyon (1968) and by Lomax (1967).

10.2 Uniqueness and Well-Posedness

We denote by $\boldsymbol{a}$, $\boldsymbol{y}$ and $\boldsymbol{F}$ all the vector components and by Δ all the space increments in Eq. (10.1) and we use the superscripted dot for time differentiation to write

$$\dot{\boldsymbol{y}} = \boldsymbol{F}\,(t,\boldsymbol{y},\Delta)\ ,\ \boldsymbol{y}(0) = \boldsymbol{a}\ . \tag{10.3}$$

The solution $\boldsymbol{y}(t)$ to Eqs. (10.3) is unique and differentiable, if $\boldsymbol{F}$ is continuous with respect to time in $0 \le t \le t_{max}$ and if $\boldsymbol{F}$ is Lipschitz continuous in $\boldsymbol{y}$ for all $\boldsymbol{y}$ and $\boldsymbol{y}^*$

$$\left|\boldsymbol{F}(t,\boldsymbol{y},\Delta) - \boldsymbol{F}(t,\boldsymbol{y}^*,\Delta)\right| \le L \left|\boldsymbol{y} - \boldsymbol{y}^*\right| , \tag{10.4}$$

where L is the Lipschitz constant.

This condition must be met for integration by most numerical algorithms. Whenever $\boldsymbol{F}$ fails to meet Eq. (10.4), at some time t_d, one integrates normally up to t_d and then restarts the integration with the new $\boldsymbol{F}$ (Wulff 1981, Birta, Oeven and Kettenis 1985). Special techniques have been developed for integrating directly across discontinuities (Halin 1979) but caution is recommended to see that the physics are not lost.

Condition (10.4) is not always met, and the solution need not be unique. The Jacobian of the system (cf. Eq. (10.2)) may be singular at selected points for physical reasons (bifurcation), or $\boldsymbol{F}\ (t,\boldsymbol{y}\ ,\Delta,\boldsymbol{z})$ may be constrained by singular algebraic equations $\boldsymbol{g}\ (t,\boldsymbol{y},\boldsymbol{z}) = 0$ (see Gear 1971, p. 223). In both cases, special techniques must be used for numerical integration (Menzel and Schwetlick 1978).

One expects the solution $\boldsymbol{y}(t)$ to depend continuously on initial conditions $\boldsymbol{a} = \boldsymbol{y}(0)$, and on time dependent disturbances $\boldsymbol{\delta}(t)$, contained in $\boldsymbol{F}$. This property is called well-posedness (Gear 1971, p.

7). Equations (10.3) constitute a well-posed problem if F is continuous in t and Lipschitz continuous in y (cf. Eq. (10.4)).

In practice, one needs Lipschitz continuity only in some finite region of y. If F has continuous derivatives with respect to y, i.e. if the Jacobian given by Eq. (10.2) has continuous elements then L equals the norm of the Jacobian in the region of the solution.

10.3 Numerical Techniques

In selecting algorithms for numerical integration of ordinary differential equations one must consider stability and accuracy. Stability deals with unacceptable growth of computational (round-off) errors (cf. Eq. (8.6)). Accuracy deals with truncation errors and round-off errors. Truncation errors produce damping and phase shift (cf. Section 8.5).

In some recently developed algorithms for numerical analysis of multi-phase flow (based on Eq. (5.1) or (5.2) formulations), some parameters are taken from previous time steps and used at a time level before they are computed for that time level (time lag). This time skewing introduces additional damping and may even render a calculation useless, particularly when F is discontinuous.

10.3.1 *Stability*

One must distinguish between three major types of stability. First, a discrete integration algorithm is stable if there exists an $h_0 > 0$ for each differential equation such that a change in starting values $\mathbf{a}$ by a fixed amount produces only a bounded change in the solution $y(t)$ for all integration steps $0 \le h \le h_0$. This definition is identical to that discussed in Section 8.2. Obviously, numerical stability is to an integrating algorithm what well-posedness is to the differential equations.

As in Eq. (8.3) for finite difference schemes, we define here

$$L(u) \equiv \dot{y} - F(t,y,\Delta) = 0 \tag{10.5}$$

and designate with $L_h(Y)$ the associated integration algorithm. At a particular time t_j we denote the solution Eq. (10.5) by $y(t_j)$ and its numerical approximation by Y_j. The local truncation error is defined consistently with the definition in Section (8.3), but here for a k-step method of ODE integration:

$$(T.E.)_j = L_h\left[y(t_j)\right] - L\left[y(t_j)\right] \,,\; k - 1 \le j \le N \,. \tag{10.6}$$

By k-step method is meant that k time levels are involved in L_h:

$$L_h(Y_j) = L_h\left[\, Y_{j+1}, \ldots, \, Y_{j+1-k}\right] , \; (k - 1) \le j \le N \,. \tag{10.7}$$

The algorithm L_h is *consistent* with the operator L in Eq. (10.5) for $h = t_{max}/(N + 1)$, $t_j = jh$, $j = 1, \ldots, N + 1$, if

$$\lim_{N \to \infty} \left| T.E._j \right| = 0 \,, \quad (k\text{-}1) \le j \le N \,. \tag{10.8}$$

A consistent operator L_h has an order of accuracy p if $T.E._j = 0\{h^p\}$ for $k - 1 \le j \le 1$. The numerical scheme $L_h(Y) = 0$ is *convergent* if

$$\lim_{N \to \infty} \left| Y_j - y(t_j) \right| = 0 \quad 0 \le j \le n + 1 \,. \tag{10.9}$$

Convergence is, as discussed in Section 8.3, guaranteed for a solution Y from the algorithm $L_h(Y_i) = 0$ if the latter is consistent and stable (Keller 1968, p. 23). L_h is stable in the sense discussed above if there exists a constant M, independent of h such that for all solutions X_j and Z_i

$$\left| X_j - Z_j \right| \le M\left\{ \max_{i \le k-1} \left| X_i - Z_i \right| + \max_{i \ge k-1} \left| L_h(X_i - Z_i) \right| \right\} \tag{10.10}$$

for all $k \le j \le N + 1$.

However, this stability criterion promises only convergence for all $h \to 0$. In practice, one must use finite step sizes h, and one must accommodate stiff systems of differential equations, where only eigenvalues of small moduli affect the solution, while the ratio of largest to smallest eigenvalue modulus is much greater than unity. In this case the second concept of stability applies, namely, absolute or A- stability. A numerical integration method is said to be absolutely stable for a given step size h, if the change due to a disturbance of size δ on one value $(Y_i)_j$ at time t_j is no larger than δ for all other values Y at all subsequent times $t_{j+1}, \ldots, t_{\infty}$, and if the method achieves this error decay for*

---

*See Lomax, NASA TN D-4703 for a generalization of this definition.

$$\dot{y}_i = \lambda_i y_i \qquad \text{(no summation over } i\text{)}, \tag{10.11}$$

and for all complex $\lambda_i = \rho_i + i\sigma_i$, $\rho_i < 0$, arctan $|\sigma_i/\rho_i| = \alpha$, which have *large* moduli $|\lambda_i| = (\rho_i^2 + \sigma_i^2)^{1/2}$. For λ's with small moduli (near the origin in the complex plane), it may be necessary to accommodate α in $[0.\pi/2]$. An algorithm which is absolutely stable for all $\rho < 0$ and some α, is said to be $A(\alpha)$ stable. Almost all fully implicit algorithms are absolutely or *unconditionally stable*, while explicit algorithms are only conditionally stable.

Often one is not interested in the accurate approximation of some components of Y, say Y_j with λ_i in region R_1 far to the left from the imaginary axis, where $\rho_j \leq D < 0$, while one needs to have the other components Y_i with λ_i in Region R_2 accurately approximated, for which $D < \rho_i < B$ and $|\sigma_i| < \vartheta$ in R_2. A method is *stiffly stable* if in region R_1 it is absolutely stable and in region R_2 it is accurate. This is the third type of numerical stability.

10.3.2 *Computing Accuracy*

Errors arise and accumulate because of the truncation error defined by Eq. (10.6) and because of the round-off errror. Both errors vary with the integration step size h.

It is shown in Gear (1971, p. 19) that for the widely used one-step, first-order Euler-Cauchy method, the truncation error is, of course, proportional to h and the round-off error is proportional to $1/h$. Their sum grows sharply as $h \to 0$, has a minimum and then approaches from above the truncation error as h increases. As $h \to 0$, one must increase computer precision.

The round-off error is rarely a problem in the integration of ordinary differential equations. Firstly, the round-off error is random and grows with the number N of integration steps as $\sqrt{N/12}$. Secondly, the round-off error r_h is irrelevant in most modern digital computers until $h \to 0$. More precisely, the truncation error becomes important for the Euler-Cauchy method only when $h < \sqrt{2r_h/M}$ where (Isaacson and Keller 1966, p. 376), with F_i defined by eq. (10.1),

$$M = \max_i \left| \frac{\partial F_i}{\partial t} + \frac{\partial F_i}{\partial y_j} F_j \right| , \text{ summation over } j . \tag{10.12}$$

Thus, with $r_h = 10^{-16}$, for example, there is no concern about round-off errors for all $h > 10^{-8}$ second.

The truncation error is controlled by first selecting a large enough order p (cf. Eq. (10.8)) and then controlling the step size h. Dahlquist (1956) has shown that the truncation error order p for stable multistep algorithms is related to the step number k by the limitation $p \leq k + 2$.

One uses truncation error bounds or estimates (Gear 1971, pp. 11-20) to calculate the truncation error. Error bounds are normally too restrictive and derived from upper limits of Lipschitz constants, maximum derivatives, etc. Error estimates are obtained by computing higher-order derivatives of F, by repeating time steps with half size, by comparing the results from two algorithms (Runge-Kutta and Simpson methods; Wulff 1982) or by comparing predictors with correctors. All error estimates provide only "local" error control, i.e. of the error made in one time step. It must be kept so small that the accumulated error is small enough for the results to be acceptable.

In closing, we recognize that the integration step size is governed by stability *and* by accuracy limits. An integration algorithm is efficient for a particular application if it is governed by accuracy requirements just below the stability limit (if any) and if the associated computing effort (frame time) is less than that of alternatives (cf. Section 3.6). See Gear (1971, p. 72) and Benyon (1968) for ideas on how to select an algorithm.

10.4 Integrating Algorithms

Integration methods are categorized in One-Step, Multistep and Extrapolation Methods. One-Step Methods start and restart after a discontinuity, and they can easily be stopped precisely at any condition. Some Multistep Methods have automatic local truncation error estimates and Extrapolation Methods are used because of their computing economy and to treat discontinuities without restart.

10.4.1 *One-Step Methods*

The *Euler-Cauchy* algorithm

$$Y_{i+1} = Y_i + hF(t_i, Y_i, \Delta) \tag{10.13}$$

is conditionally stable, i.e., absolutely stable only for integration step sizes h and for complex eigenvalues λ of the physical system $\dot{y}$ = (t,y), for which Δh falls into the circle with radius one and center (-1,0) in the complex plane

$$\left|1 + \lambda h\right| \leq 1 \tag{10.14}$$

An upper bound for the *accumulated* truncation error is (Gear 1971, p. 13)

$$\left|T.E._N\right| \leq (h/L)\left[K + LZ\right](e^{Lt_N} - 1) + \left|T.E._o\right|e^{Lt_N} \tag{10.15a}$$

which shows convergence as $h \rightarrow 0$ if the error $\left|T.E._o\right|$ of the initial data also tends to zero. Here L and K are Lipschitz constants with respect to y and t, respectively and Z is the maximum of the derivatives dF/dt for $0 \leq t \leq t_N$. If F is differentiable, then the *local* error is approximately $(h^2/2)dF/dt(\xi) = Ch^2/2$, with $t_i \leq \xi \leq t_{i+1}$, and the error bound for eq. (10.13) is*

$$\left|T.E._N\right| \leq hC/(2L)\,(e^{Lt_N} - 1) + \left|T.E._o\right|e^{Lt_N} . \tag{10.15b}$$

Thus the Euler-Cauchy method has first-order convergence even though the local truncation error is of order two. Surprisingly, the method is more widely used than can be justified, based on its low accuracy and limited stability.

Better approximations are based on the Taylor series expansion written here for the sake of simplicity just for one equation:

$$y_{i+1} = y_i + \sum_{j=1}^{f} \frac{h^j}{j!}\,\frac{d^{j-1}F}{dt^{j-1}} , \tag{10.16}$$

where

$$\frac{dF}{dt} = \frac{\partial F}{\partial t} + F\,\frac{\partial F}{\partial y} . \tag{10.17}$$

The numerical approximation is written in terms of an increment function φ which is a weighted average of derivatives in the interval $t_i \leq t \leq t_{i+1}$

---

*As indicated before, this error bound is unduly pessimistic. An error estimate requires information about F.

$$Y_{i+1} = Y_i + h\varphi(t_i, Y_i, h) \ . \tag{10.18}$$

The weighting coefficients are computed so as to satisfy the first r-terms in Eq. (10.16). The algebra required for this match of terms is indicated in Gear (1971, p. 35). The match is not necessarily unique.

Two *fourth-order Runge-Kutta methods* have these increment functions, respectively

$$\varphi = (k_1 + 2k_2 + 2k_3 + k_4)/6 \ , \tag{10.19}$$

where

$$k_1 = F(t_i, Y_i)$$

$$k_2 = f(t_i + h/2, Y_i + hk_1/2)$$

$$k_3 = F(t_i + h/2, Y_i + hk_2/2)$$

$$k_4 = F(t_i + h, Y_i + hk_3);$$

and $\varphi = (k_1 + 3k_2 + 3k_3 + k_4)/9 \ ,$ (10.20)

where

$$k_1 = F(t_i, Y)$$

$$k_2 = F(t_i + h/3, Y_i + hk_1/3)$$

$$k_3 = F(t_i + 2h/3, Y_i - hk_1/3 + hk_2) \ .$$

$$k_4 = F(t_i + h, Y_i + hk_1 - hk_2 + hk_3) \ .$$

The method according to Eqs. (10.19) and (10.20) is absolutely stable if

$$\left|1 + (h\lambda) + (h\lambda)^2/2 + (h\lambda)^3/6 + (h\lambda)^4/24\right| < 1 \tag{10.21}$$

that is if $(h\lambda)$ falls into the shaded region of Figure 10.1. The ellipse

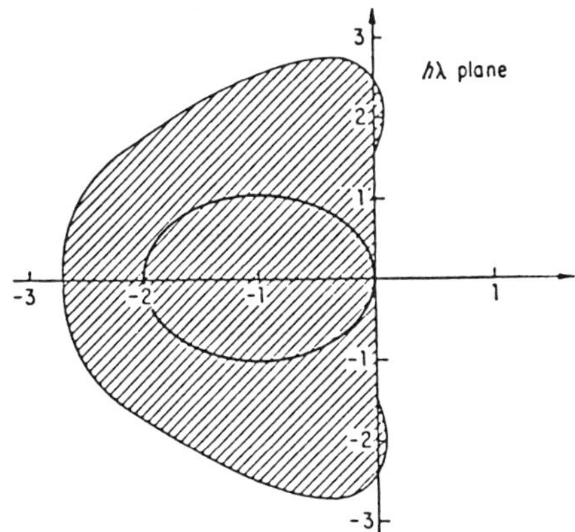

Figure 10.1 Regions of absolute stability for Runge-Kutta and Euler-Cauchy (ellipse) methods.

in that figure shows the stability region for the Euler-Cauchy method, Eq. (10.3). The local truncation error is of order five; the method has a fourth-order convergence rate, but only if F has continuous fifth-order derivatives.

All methods described so far are explicit, in that Y_{i+1} is computed from Y_i and F_i without iteration. For a discussion on highly stable explicit techniques see Lomax (1968). For further details on error estimates for Runge-Kutta methods see Shampine and Watts (1971). The local truncation error for the method given by Eq. (10.20) can be estimated from the difference d between Y_{i+1} computed from this equation and from the standard Simpson rule (Carnahan, Luther and Wilkes 1969)

$$(Y_{i+1})_S = Y_i + (k_1 + 4k_3 + k_4)h/6 \tag{10.19a}$$

A relative error $r = |d/(a + b|Y_{i+1}|)$ is then computed from the difference $d = Y_{i+1} - (Y_{i+1})_S$, from the permissible absolute error a and the permissible relative error b (Wulff 1980). An r is computed for every state variable Y. The maximum value of r is then used to control the truncation error. If $r_{max} \geq 1$, then the result is discarded and the step is repeated with $h/(10\ r_{max})^{1/5}$. If $0.75 \leq r_{max} < 1$, then the calculation is continued with the reduced step size $h/10^{1/5}$. If $0.075 \leq r_{max} < 0.75$, then h is retained and the computation is continued. If $r_m < 0.075$ then the step size h is multiplied by $10^{1/5}$. A fifth-order method produces one additional reliable decimal digit if the step size is reduced by $10^{1/5}$.

It must be pointed out that up to fourth-order, a ν-th order Runge-Kutta method requires ν evaluations of F, for 5-th and 6-th order there are required 6 and 7 evaluations, respectively, and for 7-th and higher orders one needs $\nu + 2$ evaluations. This computing effort is normally more than compensated for by the need for fewer integration steps due to higher accuracy, when compared with first-order Euler-Cauchy. If the evaluation of F is very expensive, then multistep methods are more advantageous. They require more storage memory but fewer F evaluations. One can also solve Eq. (10.13) implicitly by iteration

$$Y_{i+12} = Y_{i+1} + hF(t_{i+1}, Y_{i+1}, \Delta) \tag{10.20}$$

and thereby achieve at the expense of extra computational efforts, absolute stability except inside the circle $|1 - h\lambda| < 1$. Similarly, implicit Runge-Kutta algorithms can be written which have higher order convergence and unconditional stability (Gear 1971, p. 40). However, these implicit Runge-Kutta algorithms have not played any major role since there are better alternatives for computationally stiff systems.

10.4.2 *Multistep Methods*

Multistep Methods are based on the approximation to

$$y_{i+1} = y_{i-p} + \int_{t_{i-p}}^{t_{i+1}} F(t,y)dt \tag{10.21}$$

by either open or closed Newton-Cotes formulae (Abramowitz and Stegun 1964, p. 886). Formulae open at the right (at r_{i+1}) yield explicit expressions, those which are closed at the right yield implicit expressions. It is clear that:

(i) these multistep methods must be started (perhaps by a Runge-Kutta method),

(ii) storage of p Y-values and of n F-values is required, where n is the number of ordinates, F, used for defining an approximating polynomial in the integral of eq. (10.21),

(iii) only *one* new derivative is to be evaluated at every step (as will be shown below, in a predictor-corrector combination, there will be two derivatives needed),

(iv) variable step-size h leads to complications.

With open integration, one can get these frequently used algorithms, both having fifth-order local truncation errors:

$\underline{p = 0}$ and $\underline{n = 4}$

$$Y_{i+1} = Y_i + h\left[55F_i - 59F_{i-1} + 37F_{i-2} - 9F_{i-3}\right]/24 + 0\left\{h^5\right\}, \quad (10.22)$$

$\underline{p = 3}$ and $\underline{n = 3}$

$$Y_{i+1} = Y_{i-3} + 4h\left[2F_i - F_{i-1} + 2F_{i-2}\right]/3 + 0\left[h^5\right]. \quad (10.23)$$

Such algorithms are known as Adams-Bashford Methods (Bashford and Adams 1855). With closed integration, the algorithms become implicit. For example, for $\underline{p = 1}$ and $\underline{n = 3}$

$$Y_{i+1} = Y_{i-1} + h[F_{i+1} + 4F_i + F_{i-1}]/3 + 0\{h^5\} \quad (10.24)$$

is the implicit Simpson Rule and can be solved by successive substitution if $h < 3/(\partial F/\partial y)$ (cf. Section 11). Implicit algorithms should be more accurate than explicit algorithms because the latter are obtained by open integrations which involve the extrapolation of $F(t)$ from t_i to t_{i+1}, while the former involve interpolation.

By combining the formula from an open integration as a predictor with the formula from a closed integration as a corrector, one can achieve quick convergence for the iteration on the implicit corrector and one can get an estimate of the truncation error. The resulting algorithms are known as Milne or Adams-Moulton methods. The fourth-order converging Milne method is:

predictor $$Y^*_{i+1} = Y_{i-3} + 4h(2F_i - F_{i-1} + 2F_{i-2})/3 + 0\{h^5\} \quad (10.25)$$

corrector $Y_{i+1} = Y_{i-1} + h(F_{i+1} + 4F_i + F_{i-1})/3 + O\{h^5\}$ (10.26)

where F_{i+1} is first evaluated with Y^*_{i+1} and then Eq. (10.26) is iteratively solved. Since the predictor has the error of

$$T.E.^* = (14/45)h^5F^{(4)}(\xi) \; , \; t_i \le \xi \le t_{i+1} \tag{10.27}$$

and the corrector has the local error of

$$T.E. = (-1/90)h^5F^{(4)}(\eta) \; , \; t_i \le \eta \le t_{i+1} \; , \tag{10.28}$$

one can first improve the corrector (Carnahan, Luther and Wilkes 1969, p. 386) by

$$(T_{i+1})_{final} = (28Y_{i+1} + Y^*_{i+1})/29 \tag{10.29}$$

and, secondly, estimate the local truncation error

$$T.E. = (Y_{i+1} - Y^*_{i+1})/29 \; . \tag{10.30}$$

A complete list of all multistep methods is given, starting on page 896, in Abramowitz and Stegun (1964).

Stability is guaranteed for a particular multistep algorithm if it satisfies the so-called root condition (Isaacson and Keller 1966, p. 412). Figure 10.2 shows the stability region for Adams-Bashford methods (Gear 1971, p. 131). Absolute stability is obtained with λh inside the contours. Figure 10.3 shows the stability regions for the predictor corrector methods if the corrector is solved iteratively until convergence is obtained. The Adams-Moulton algorithm of order two is stable in the entire negative half plane; the first-order Adams-Moulton method is the implicit Euler method and stable everywhere, except inside the circle $|1 - \lambda h| = 1$.

Figure 10.4 shows the stability regions for all most commonly used algorithms, particularly the Adams-Moulton methods where only one evaluation of the corrector is utilized. Notice that this reduces

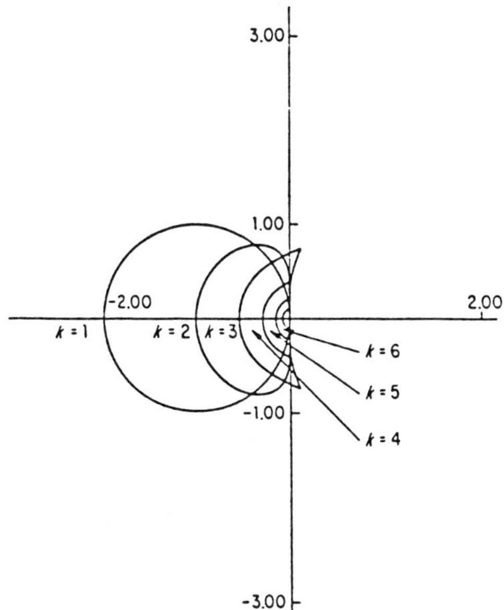

Figure 10.2 Stability regions for Adams-Bashford method (Gear 1971).

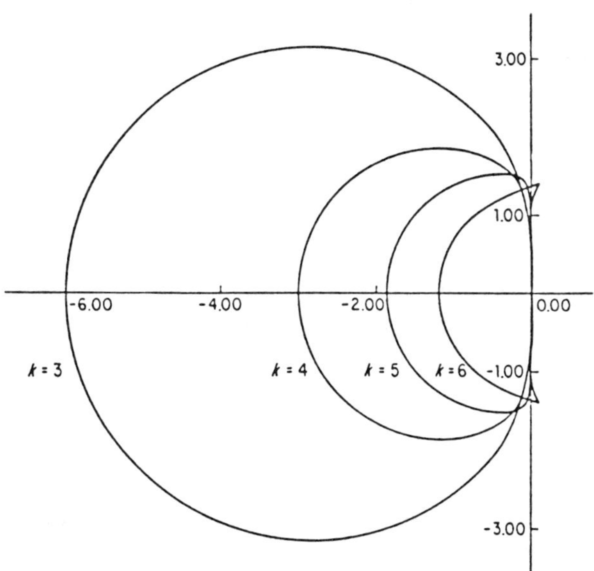

Figure 10.3 Stability regions for Adams-Moulton methods with corrector converged (Gear 1971).

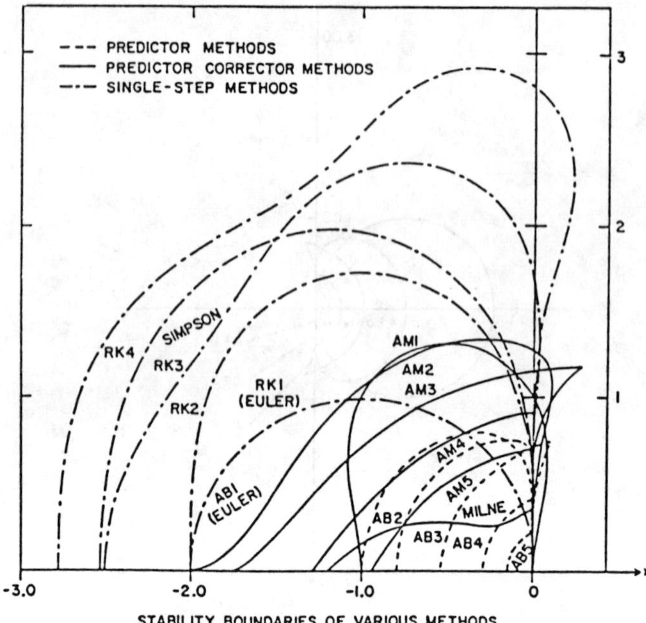

Figure 10.4 Stability regions for Adams-Bashford (AB), Adams-Moulton (AM, no iteration), Runge-Kutta (RK) methods (Benyon 1968).

the stability regions considerably, relative to those shown in Figure 10.3. The figure is from the paper by Benyon (1968).

It should be noticed that round-off errors affect accuracy very strongly for high-order multistep methods, requiring large word lengths (48 bits) in digital computers.

Finally, it should be pointed out that the computing effort is governed primarily by the effort of evaluating F and the number of required F-evaluations. Figure 10.4 may lead one to observe that Runge-Kutta methods allow the longest time step without the need for iterations and then to conclude that Runge-Kutta methods are preferable. This is true only if F is easily evaluated.

10.4.3 *Extrapolation Methods*

Several computer codes for multiphase flow, such as IMEX and FEBE (Hofer 1983) were developed using so-called extrapolation techniques. These methods are used particularly to integrate across discontinuities of F.

The basic idea is to compute the results $Y_{i+1}(h_j)$ at a time level t_{i+1} with several integration step sizes h_j and then to extrapolate a suitable function passing through the ordered pairs $\{h_j, Y_{i+1}, (h_j)\}$ to find $Y_{i+1}(0)$.

One extrapolation technique is based on polynomial approximation and employs the Aitken interpolation method to extrapolate (for Aitken method see Carnaham, Luther and Wilkes 1969, p. 61). The other extrapolation technique is due to Bulirsch and Stoer (1964, 1966) and based on extrapolation by rational polynomials (quotients of polynomials).

Inherently, these methods improve accuracy and have proven themselves to be successful for stiff systems and for local error controls.

10.4.4 *Stiff Systems of Ordinary Differential Equations*

In Section 10.3.1, we characterized locally stiff systems on the basis of their sets of (local) eigenvalues. Many algorithms and computer packages have been developed for integrating stiff systems of differential equations (Certaine 1960, Curtiss and Hirschfelder 1952, Gear 1971, Guderley and Hsu 1972, Jain 1972, Klopfenstein and Davis 1971, Lomax 1968, Porsching et al. 1971, Treanor 1966 and many others).

Porsching's method is implicit and was used in the RELAP2 to RELAP4 series (Moore and Rettig 1968, 1973, Rettig et al. 1970) for nuclear reactor safety analyses. It has been used since that time for more advanced analyses, and often it has been misused. Therefore, I describe it here briefly.

The Taylor series expansion of $y(t)$ about the time t_{i+1} is

$$y_i = y_{i+1} - hF_{i+1} + (h^2/2)dF_{i+1}/dt - + \ldots \qquad (10.31)$$

For an implicit Euler-Cauchy integration, one would have

$$y_{i+1} - y_i = hF(t_{i+1}, y_{i+1}) - (h^2/2)dF/dt(t_{i+1}, y_{i+1}) + - \ldots \qquad (10.32)$$

The t_{i+1} -time evaluation of F is now once more approximated by a Taylor series expansion:

$$F(t_{i+1}, y_{i+1}) = F(t_i, y_i) + h\partial F/\partial t\Big|_{t_i} + \underline{\underline{J}}_i\,(y_{i+1} - y_i) + O\{h^2\}\ , \qquad (10.33)$$

where $\underline{\underline{J}}$ is the Jacobian given in Eq. (10.2) and evaluated at time t_i. Substitution of Eq. (10.33) into Eq. (10.32) yields

$$y_{i+1} - y_i = \left[hF_i + O\{h^2\}\right][\underline{\underline{I}} - h\underline{\underline{J}}]^1 \ . \tag{10.34}$$

After dropping the second-order term from Eq. (10.34) one gets this formula

$$y_{i+1} = y_i + hF[\underline{\underline{I}} - h\underline{\underline{J}}]^{-1} \ , \tag{10.35}$$

where $\underline{\underline{I}}$ is the identity matrix. Porsching et al. (1971) have shown that this algorithm is consistent and locally absolute stable.

Many users of Porsching's method either violate the prerequisite of differentiability or at least of Lipschitz continuity for Eq. (10.1), when flow or heat transfer regime changes produce discontinuities, or they evaluate the Jacobian only partially, excluding heat transfer, wall shear, form losses, mass change rate and interphase drag. The potential users are cautioned to meet all the requirements for consistency and stability of the method (Porsching et al. 1971).

10.5 Closing Remarks

This review on methods for integrating Eqs. (5.1) is certainly not complete or exhaustive, but it is intended to give the reader a glimpse into a very resourceful branch of numerical methods. Particularly in fluid dynamics, this method has been largely ignored.

It is important to remember that the class of methods presented here affords rational control of time-discretization errors. The user of these methods will still have to assess the space-discretization error by repeated calculations with Δx, Δy, $\Delta z \rightarrow 0$, but the task is considerably smaller if one does not have to superimpose another series of calculations with $\Delta t \rightarrow 0$.

All the methods represented in Figure 10.4 have been recognized to have such general utility for a large class of applications that they were selected to be the key algorithms for two special simulation computers, the AD10 and AD100 of Applied Dynamics International (Fadden 1984). One of these computers has already been used successfully to simulate nonhomogeneous nonequilibrium two-phase flow as part of a high-speed simulation facility for a complete nuclear power plant (Wulff et al. 1984).

11. SOLUTION OF NONLINEAR FINITE DIFFERENCE EQUATIONS

In Section 5, I pointed out that there are two classes of numerical methods needed to obtain numerical results from the discretized field equations of multiphase flows. The first class deals with partially or space-discretized equations and is discussed in Section 10. Here I discuss the second class of methods for solving the completely discretized equations, given in the form of Eqs. (5.1). For the sake of simplicity, let us focus our attention on the unknown state and subsidiary variables y at the future time level and write Eq. (5.1) simply as

$$G(y) = 0 , \tag{11.1}$$

where G, and y are n-dimensional column vectors. This set of equations is nonlinear; through correlations for heat transfer, wall shear, interphase transfer and material properties, it is most likely transcendental. It must be solved iteratively. A solution y^* to Eqs. (11.1) is not necessarily an acceptable, much less an accurate, solution to the field equations, Eqs. (2.7) or (2.13). Rather, the solution y^* is a solution to Eqs. (2.7) or (2.13) only if

(i) Eqs. (11.1) are consistent with Eqs. (2.7) or (2.13) and
(ii) $\Delta = \{\Delta x_1, \Delta x_2, \Delta x_3, \Delta t\}$ are sufficiently small for convergence between Eqs. (11.1) and Eqs. (2.7) or (2.13), and
(iii) y^* satisfies Eqs. (11.1) with an acceptable convergence criterion (cf. Eq. (11.25) below).

Many multiphase flow codes, such as RELAP5 (Ransom 1985), TRAC (Taylor 1984, TRAC-PF1 1986), COBRA (Wheeler et al. 1986), RETRAN (McFadden 1980), K-FIX (Rivard and Torrey 1979), PHOENICS (Spalding 1982a), THERMIT (Loomis et al. 1981), URSULA (Singhal and Spalding 1980) are solved, on the basis of complete discretization, by solving Eqs. (11.1) through iteration. Only the RELAP5 code computes without iteration. As there is no document referenced anywhere on a method for solving Eqs. (11.1) without iteration, it must be concluded that RELAP5 calculations need not converge to the solutions of the field equations (cf. Section 13), in RELAP5.

There are far too many methods available (Ortega and Rheinbolt 1970 and references given there) for their complete discussion here. All methods are generally related to the Newton-Raphson method, which leads toward the solution y^* for $G(y) = 0$ through the repeated (iterative) solution of linear sets of equations. However, in reviewing solution methods for two-phase flow analyses, one finds two methods which are very distinct in their executions.

One method entails repeated evaluations of nonlinear expressions and is called successive substitution or functional iteration (Isaacson and Keller 1966, p. 109). This method is used, for example, in the PHOENICS (Spalding 1982b) and SPORTS (Chatoorgoon 1986) codes.

The other most widely used method is the direct Newton-Raphson method, applied simultaneously to all or some of the field equations, and applied either with full quadratic convergence or with computational simplifications and linear convergence.

11.1 Successive Substitution

For this technique, Eq. (11.1) is first rewritten in this form

$$y = g(y) \; , \tag{11.2}$$

where $g = G + y$, *and* y *and* g are both n-dimensional column vectors. A solution y^* to Eqs. (11.2) is sought by starting with some guess y^0 and successively evaluating

$$y^{k+1} = g(y^k) \; , \quad k = 0,1 \ldots \tag{11.3}$$

In practice, this method does not work in general. Those who use it spend a considerable amount of resources to make it work. Whether or not successive substitution works, depends primarily on which of the many ways are selected to pass from G in Eqs. (11.1) to g in Eqs. (11.2); in other words, how the n equations are *ordered* (Spalding 1982b, p. 41 and 1985, p. 11) and which component of y is taken from a particular equation in Eqs. (11.1) to the left side of Eq. (11.2). The iterative scheme will only work if these two minimum requirements are met (Isaacson and Keller 1966, p. 110):

(i) the vector function g must be Lipschitz-continuous with Lipschitz constant L satisfying

$$0 \leq L < 1 \; , \tag{11.4}$$

that is $\| g(x) - g(y) \| \leq L \| x - y \|$ (11.5)

for all x and y within some hypersphere with radius ρ and center y^0, i.e.

$$\| x - y^0 \| \leq \rho \quad \text{and} \quad \| y - y^0 \| \leq \rho, \quad \text{and} \tag{11.6}$$

(ii) the initial guess y^0 must satisfy

$$\| g(y^0) - y^0 \| \leq (1-L)\rho \ . \tag{11.7}$$

If these two conditions are met then all iterates y^k will fall within the hypersphere with radius ρ about the initial guess, the iterates will converge geometrically, as fast as $L^k \to 0$, toward a solution y^*, and this solution y^* is the unique solution within the hypersphere $\|y - y^0\| \leq \rho$. As the norm $\|y\|$ in Eqs. (11.5) through (11.7) can be chosen any norm, for example

$$\|y\|_\infty = \max_i |y_i| \quad \text{or} \quad \|y\|_m = \sqrt[m]{\sum_{i-1}^{n} |y_i|^m}\ , \quad m = 1,2\ . \tag{11.8}$$

The most difficult condition to meet is in Eq. (11.4), since the Lipschitz constant L can vary over a wide range during a transient or as a consequence of model changes. If g is Gateaux-differentiable (Ortega and Rheinbolt, p. 59), i.e., if it is continuous along straight lines, then the Jacobian of g exists

$$g_{ij} = \partial g_i / \partial y_j \tag{11.9}$$

and Eq. (11.4) can be expressed in the more restrictive, but also more familiar form, which illustrates the significance of Eq. (11.4):

$$\max_i \sum_{j=1}^{n} |g_{ij}(y)| < 1, \quad \text{for all} \quad \| y - y^* \| \leq \rho\ . \tag{11.10}$$

This is a very crucial restriction for large systems of equations. In the case of a single equation, one can always force convergence by estimating the slope $dg/dy = g'_e \neq 1$ in the vicinity of the solution of $y = g$, and then by subtracting $g'_e \cdot y$ from both sides of this equation, where g'_e is kept constant. This leads to a new, but equivalent equation

$$y = \left[g(y) - y \cdot g'_e\right]/(1 - g'_e) = \tilde{g}(y) \ . \tag{11.11}$$

It can be seen that $d\tilde{g}/dy = \lambda \approx 0$. The geometric convergence rate λ^k produces therefore very quickly the solution y^*. This idea has been generalized for systems of equations (Isaacson and Keller 1966, p. 120). Convergence is improved by introducing the non-singular diagonal matrix $\underline{\underline{\vartheta}}$ and by writing instead of Eq. (11.3) the equivalent equation, written here with indicial notation

$$y_i^{k+1} = \vartheta_i g_i(y^k) + (1 - \vartheta_i)y_i^k, \quad i = 1,..., n \ . \tag{11.12}$$

If g has continuous derivatives (cf. Eq. (11.9)) in the neighborhood $\|y - y^*\| \leq \rho$ about the solution, then the above algorithm will converge, provided

$$\left|1 - g_{ii}(y)\right| > \sum_{\substack{j=1 \\ j \neq i}}^{n} \left|g_{ij}(y)\right| \ , \quad i = 1,..., n, \tag{11.13}$$

which is less restrictive than the conditions in Eqs. (11.4) or (11.10). Equation (11.13) is often met in practice, i.e., when diagonal terms in the balance equations dominate. This depends, of course, on the ordering of the equations. The ϑ_i's have to lie between certain limits (see Eqs,. 38 on p. 122 of Isaacson and Keller 1966), but the best choice for accelerated convergence is (cf. Eq. (11.11)):

$$\vartheta_i = 1/\left[1 - g_{ii}(y^k)\right] \ . \tag{11.14}$$

Finally, it must be pointed out that uniqueness of a transient solution is only guaranteed by condition (i) and (ii) above, if the equations $G = 0$ are Lipschitz-continuous *and* if the integration step size Δt is small enough for Eq. (11.7) to be satisfied. For two successive, converged solutions, both

$$\left.\begin{aligned} &\|(y^{n+1})^* - (y^n)^*\| \leq \rho \quad \text{and} \\ &\|g^{n+1}\{(y^n)^*\} - (y^n)^*\| \leq (1 - L)\rho \,. \end{aligned}\right\} \tag{11.15}$$

must be satisfied to meet the requirement of Eq. (11.7). This time step limitation to assure uniqueness is important when the radius ρ of the region, where Lipschitz continuity of g prevails, is small.

11.2 Newton-Raphson and Related Methods

The familiar Newton-Raphson Method for single equations can be used for sets of equations also. It has quadratic convergence, which means that each iteration doubles the number of correct significant digits, but it requires the calculation of derivatives and the inversion of a matrix. Derivatives are best evaluated first analytically (by hand or computer) and then numerically; in practice they are often computed from numerical differentiation.

The method is most advantageous for quickly improving the convergence after a good approximation has been found by a less costly but slowly convergent algorithm (for example, if L in Eq. (11.4) is close to unity).

If one writes for g in Eq. (11.2) $y - \underline{\underline{A}}\ G$, one obtains this rather general iterative procedure from Eq. (11.3)

$$y^{k+1} = y^k - \underline{\underline{A}}(y^k) \cdot G(y^k). \tag{11.16}$$

Since the solution of $y - \underline{\underline{A}}\ G = y$ is also a solution to $G = 0$ if $\underline{\underline{A}}$ is non-singular, Eq. (11.16) generates a solution under the conditions discussed in Section 11.1 above, i.e., if the initial guess y^0 is sufficiently close to the solution y^* and if the derivatives of the right-hand side satisfy a condition corresponding to Eq. (11.10). There are, in fact, many choices for the matrix $\underline{\underline{A}}$, but if $\underline{\underline{A}}$ is the inverse of the Jacobian of G and not singular at the solution y^*, i.e., if

$$A_{ij} = (\partial G_i/\partial y_j)^{-1} \neq 0 \quad in \quad \|y - y^*\| \leq \rho \,, \tag{11.17}$$

then Eq. (11.3), written here with indicial notation and with the summation rule

$$y_i^{k+1} = y_i^k - \left[\frac{\partial G_i(y^k)}{\partial y_j}\right]^{-1} G_j(y^k) \,, \quad \begin{matrix} k = 1,... \\ i,j = 1,..., n \end{matrix} \tag{11.18}$$

constitutes Newton-Raphson's method and, if it converges, then it converges quadratically.

It is very important to determine very early in the iteration process, whether or not it will converge. While computing a transient multiphase flow process, one may select a time integration step size too large for convergence of Eq. (11.18), and then one should not have to iterate too many times before recovering with a smaller integration step size. In fact, Isaacson and Keller (1966, p. 115) presented a criterion by which to determine success or failure after the first interaction.

We define for this purpose three constants, a, b and c as follows. After the first iteration is completed, we determine the norms

$$\| (\partial G_i/\partial y_j) \Big|_{y^1}^{-1} \| \leq a \tag{11.19}$$

and $\|y^1 - y^0\| \leq b.$ (11.20)

Next, we establish that for all y in $\|y - y^0\| \leq 2b$

$$n \cdot \sum_{k=1}^{n} \left|\partial^2 G_i(y)/\partial y_j \partial y_k\right| \leq c \tag{11.21}$$

is satisfied, with G having continuous second derivatives. If the three bounds a, b and c satisfy

$$abc \leq 1/2 \tag{11.22}$$

then the iterates from Eq. (11.18) are unique, they lie inside the sphere $[y^0, 2b]$ and tend toward the solution y^* as

$$\| y^k - y^* \| \leq \frac{2b}{2^k} \,. \tag{11.23}$$

Traub (1974) has defined a measure of "computing efficiency" and applied this measure in the comparison of different algorithms for the purpose of selecting an optimum algorithm for solving nonlinear equations. The reader will find criteria in this reference to determine when the Newton-Raphson method is superior to other alternatives.

I noted at the outside of this section, that the matrix $\underline{\underline{A}}$ in Eq. (11.16) did not have to satisfy Eq. (11.17) but only that the right-hand side of Eq. (11.16) has to satisfy Eqs. (11.4) through (11.7). This means that the Jacobian of G does not have to be evaluated for every iterate, perhaps not even for every iteration, i.e., for every integration step of the field equations. It is suggested that the Jacobian be calculated only when the iteration rate decreases below some rate at which this re-evaluation pays off. In principle, it is never crucial for the final result, how many approximations are involved in the iteration process. It is important that the final result y^N satisfies

$$\| y^N - y^{N-1} \| \leq \varepsilon_1 \, \| y^N \| + \varepsilon_2 \tag{11.24}$$

for judiciously selected relative and absolute errors and that the final solution is invariant under further reductions of time and space increments. Instead of the above global error criterion, one can equally well impose individual criteria

$$\left| y_i^N - y_i^{N-1} \right| \leq (\varepsilon_i) \left| y_i^N \right| + (\varepsilon_i)_2 \tag{11.25}$$

which are more meaningful from the fluid dynamicist's point of view. However, the number of approximations do decrease the computing effort per iteration, while increasing the number of necessary iterations. There is an optimum somewhere.

The methods involved for calculating the inverse $\underline{\underline{A}}$ of the Jacobian are not summarized in this review. The reader can easily find suggestions for numerical matrix inversion (Wilkinson 1974), particularly when they are sparse (Evans 1974, Reid 1974 or Pipes and Hovanessian 1969), and others in the standard literature. I would urge multiphase flow analysts to seek ready-made subroutines, for which applications criteria and error limits are clearly defined, and find them most likely in the libraries of computing centers. There are also a number of codes developed for symbolic algebra and calculus which could serve to determine the elements of the Jacobian first analytically. However, the resulting expressions may require far more arithmetic operations than the numerical differentiation, in which case it is not recommended to obtain the Jacobian first in analytical form, then to evaluate each element numerically and invert the resulting matrix. It is easier to obtain each element by numerical differentiation, preferably using centered differences, and then inverting the matrix.

12. STABILITY, ACCURACY AND CONVERGENCE

This section is a summary of sorely needed criteria for accepting computed results, particularly for the complicated analysis of multiphase flows by large complex computer codes. It will be shown in Section 13 for selected but widely used two-phase flow codes for predicting nuclear power plant behavior, that the majority of these codes do not meet rational criteria which guarantee that computed solutions approximate the original partial differential equations.

A number of analysts may seek, more than anything else, a powerful and universal computing technique with which to solve all two-phase flow problems. However, since the applications vary over a wide spectrum, from basic phenomena, studied in a laboratory test facility, to systems studied for immensely complicated industrial installations, the search for a universal computing method is as futile as the search for the universal camera, car or even spouse.

It appears far more important to

(a) assure the validity of computations performed with currently available methods,

(b) optimize modeling techniques, computing methods and computer architectures synergistically to achieve computing speed and accuracy at low cost.

Item (a) is discussed here, Item (b) has been discussed in Section 3.

12.1 The Need for Validation

In Sections 5 through 11, I have divided the process of computational analysis in multiphase flows into two parts: one is to derive discrete equations from the governing continuum equations and the other is to apply an integration or a solution method to the discrete equations.

A solution can therefore be accepted only if *two* criteria are met: if

(i) it satisfies the discrete equations, and
(ii) the discrete equations have converged to the original field equations, boundary and initial conditions.

The *first* requirement implies in the case of partially or space-discretized equations that the truncation error (cf. Eqs. (10.15), (10.30) etc.) is small enough for Eq. (10.8) to be satisfied within some convergence criterion

$$\left|(Y_i)_j - y_i(t_j)\right| \le (\varepsilon_i)_1 \left|y_i(t_j)\right| + (\varepsilon_i)_2 \qquad (12.1)$$

for all vector components with index i and for all times t_j. Here $(\varepsilon_i)_1$ and $(\varepsilon_i)_2$ are the permissible relative and absolute errors, respectively, of the individual state variables. In the case of fully discretized equations, one has to insist on the convergence criterion given by Eq. (11.25).

The *second* requirement implies that the accepted solution has been computed with a set of space and time increments Δx_1, Δx_2, Δx_3 and Δt, which are so small that their further reductions would not affect the leading significant digits of all computed variables, up to the desired number of significant digits. This convergence requirement is expressed in Eq. (8.4) and satisfied either on the basis of truncation error estimates for at least spatial discretization or, in the case of space and time discretization, for both time and space discretization, or it is satisfied by carrying out a sequence of calculations with decreasing space, and if appropriate, time increments (Werner 1986).

Finally, in the absence of discontinuities, one could also differentiate the final results numerically, at every time level, using the formulae given in Eqs. (5.11) through (5.16) or those given on pages 882-885 of Abramowitz and Stegun (1964), but up to one order higher than the order of truncation error incurred in the solution algorithm. By substituting these derivatives into the field equations, one could compute an estimate of the local truncation error and assure a desired overall accuracy.

Detailed acceptance criteria are given in Section 12.3. It is assumed in what follows, that the models are mathematically properly posed (cf. Section 2).

12.2 Accuracy from Current Computer Codes for Multiphase Flows

This subsection does not reflect on the adequacy of currently used multiphase flow modeling, but only on aspects of computing accuracy.

Aside from three exceptions known to the author, the vast majority of computer codes for multiphase flows, which have been described in the literature in recent years, have no rational computing error control. Two of the exceptions are the FEBE code (Werner 1986) and the HIPA code (Wulff et al. 1984), since for both of these both time and space discretization errors have been assessed. The third exception might be considered to be the LAURA code, of which the description (Spalding 1982) contains the statement that spatial truncation errors have been assessed; no results are reported however, and the time discretization error was not addressed at all.

Some codes control convergence (Item (i) in Section 12.1) either by computing a residual error, similar to Eq. (11.24), or by computing a selected discrepancy, such as between the fluid density computed first with the global mass balance and later with the thermal equation of state. The time step size (which is only one part of Item (ii) in Section 12.1) is in most codes controlled to produce a limited change per time step in selected parameters. This criterion is not only meaningless but outright wrong, because it senses the rate of change. The rate of change is fully accounted for, even in algorithms of lowest possible order of truncation error. The time step must be controlled on the basis of local truncation *error*, as discussed, for example, in Section 8.3.3. This truncation error is most likely of second order. It is geometrically represented by curvature and not by slope. Curvature in turn is largest when the slope changes its sign. An algorithm takes off with large time steps if it is controlled by rates of change or by change itself, just when the time-curve shows a zero slope and possibly the greatest curvature. In fact, the time step size should be smallest when the slope of the time plot for the subject parameter is zero at a peak or at a minimum.

Computational stability has become the singular and predominant issue in the numerical analysis of two-phase flow. The broader problem of computational accuracy has been largely ignored.

The preoccupation with stability led to the widespread misconception that implicit integration techniques are not only superior to explicit methods (for example Lekach and Kaufman 1980), but that nothing else has a chance to produce results economically (Spalding 1982). Lekach and Kaufman compared computing times for implicit and explicit methods, analyzing the same transient, and then strongly recommended implicit integration even though it produced completely incorrect results (100% error, Figs. 2, 3 and 4). Spalding (1982) claimed that, because reliable calculations require fine spatial grid spacings, only the implicit integration method affords economical calculations. His line of arguments can easily be recognized to be surprisingly selective: he ignored the time discretization error and failed to show that the integration step size of implicit integration can be sufficiently larger than that for explicit integration so as to compensate for the iteration expense and still produce the desired accuracy. The discussions in connection with Eqs. (8.34) show that this is most likely wrong in general. Computational stability comes always from numerical damping. The more damping an algorithm has, the more time and space mesh points it requires to achieve a selected accuracy. Section 3.6 presents a rational basis for deciding between implicit and explicit integration methods.

The preoccupation with stability has also led several multiphase flow code developers to strive toward exceeding the Courant-Friedrichs-Lewy (CFL) limitation (cf. Eq. (8.15)). Unwittingly or not, those who exceed the CFL limit, based on the magnitude of sound plus material velocities, fail to track sound waves or pressure perturbations. They have no justification for integrating momentum balances locally, i.e., in every computational cell. Their work constitutes a typical mismatch between objectives and model formulation, as they could save considerable computing resources by simply setting the pressure gradient equal to zero, $\nabla p = 0$, in mass and energy balances and by integrating analytically the momentum balance from entrance to exit, or around the closed loop (cf. Eq. (4.7)), as the case requires. Those who take pride in violating the CFL limit based on the magnitude of material or void propagation velocities are not tracking anything, neither thermal nor kinematic waves! Consequently they have no justification to use up computational resources for the integration of partial differential equations. The answers that they can produce accurately while exceeding the CFL limit can also be found from integrating ordinary differential equations.

Finally one needs to recognize that no experimental test facility and no amount of experimental work can prove that a computed result satisfies the original mathematical modeling equations. Agreement between computed results and governing equations can be established only by mathematical means. Experiments are needed to assess the adequacy of the mathematical model formulation. Reliance on a successful comparison between computed and measured results alone reflects a disregard for compensating errors and a disconcern for identifying observed discrepancies as modeling or computing errors.

12.3 Proposed Acceptance Criteria

Multiphase flow calculations are extremely complicated. It is impossible for the recipient of computed results to assess routinely the computational accuracy of the results. Instead, one should expect that graphically presented results are accurate within plotting accuracy, and that tabulated data show only computationally reliable digits. The reader of computed results should not be burdened with a concern about *computational* errors, since he has already enough of an obligation to ascertain the *modeling* fidelity of the analysis.

Any algorithm used for integrating the field and constitutive equations should be documented to show that the results satisfy these equations, particularly for multiflow analyses associated with nuclear reactor safety. Rational error estimates must be derived or appropriate references must be given. The techniques for estimating the computational errors depend on which of the two possible discretization methods are used: spatial or complete discretization.

12.3.1 *Spatial Discretization*

Spatial discretization was shown in Section 7 to lead first to ordinary differential equations. The results of their integration can be accepted if:

(i) the ordinary differential equations are consistent with the governing partial differential equations as defined in Section 8.3.1,

(ii) the spatial mesh size(s) Δx_1 $(\Delta x_2, \Delta x_3)$ are small enough to produce a sufficiently small truncation error, i.e., difference between partial and ordinary differential equations,

(iii) the integration step size Δt is small enough to guarantee convergence of the numerical to the analytical results of the ordinary differential equations.

Item (i) must be shown analytically. In the cases of Space Discretization (cf. Section 7.1) and Space Averaging (cf. Section 7.2), Item (ii) is accomplished either by estimating the damping effect (cf. Section 8.2) or by successively reducing the space increments until the relevant number of leading significant digits of the results remain unaffected by further space increment reductions. Item (iii) is normally quite easy to achieve with automatic time step control available in most O.D.E. integrators.

12.3.2 *Complete Discretization*

Most two-phase flow computer codes fall into this category, which is described in Section 8. Results from this formulation are acceptable if

(i) the nonlinear, algebraic difference equations (cf. Eq. (11.1) have been shown to be consistent with the governing field equations,

(ii) the solution has converged to the difference equations, as indicated by Eqs. (11.24) or (11.25),

(iii) time and space increments, Δt, Δx_1, Δx_2, Δx_3, are small enough such that further reductions of these increments (subject to constraints among the increments) leaves relevant leading significant digits unaltered.

Item (i) must be proven analytically. Item (ii) is satisfied through a side calculation of relative errors. Item (iii) can be met by estimating damping and dispersion coefficients (cf. Eqs. (8.22) and (8.23)) or by a sequence of calculations, carried out with decreasing time and space increments (Werner 1986). Item (iii) in this section is far more expensive than the equivalent Item (ii) in Section 12.3.1 because it involves space *and* time increment reductions.

12.3.3 *Comparison with Exact Solution*

A series of calculations carried out for a range of scalar source terms and constitutive parameters, all kept constant for a particular calculation, and carried out also for the highest expected propagation velocity and the shortest expected wave length, produces results which can be compared with exact solutions for the purpose of error assessment. It is important that all time and space derivatives be contained in the model of the analytical solution. The exact solution may require numerical evaluation by computer, or even computer integration (Wulff 1980); important is only that the exact solution be generated with a known but insignificant computational error.

13. MAJOR APPLICATIONS

13.1 Objective

In this section are discussed the numerical techniques used in some of the major multiphase computer codes. The review will be neither complete nor exhaustive, but it will highlight the most important capabilities, characteristics and limitations of the methods used, and it will serve as a guide for assessing techniques which are not discussed here. The codes are selected here on the basis of widespread use and of exceptional computing features.

The discussions are based upon published articles and on documents graciously provided in 1986 by the code developers to the author upon his request. This generous help is gratefully acknowledged.

It would have been desirable to present for each code included in this discussion such important features as computational speed, accuracy and other merits. Unfortunately, computing speed and accuracy are inseparable and the latter is not documented in general. A full derivation of coefficients of numerical damping and dispersion, while only a relatively small effort when compared with the total effort of developing a multiphase computer code, is most certainly beyond the scope of this review. General guidelines for carrying out such analyses are given in Section 8.

13.2 The TRAC Code

The TRAC code has become the official computer code of the U.S. Nuclear Regulatory Commission for analyzing safety aspects of nuclear power reactor systems. There are two pressurized water reactor code versions, developed at Los Alamos Nationa Laboratory and called TRAC-PD2 (detailed version, NUREG/CR-2054, 1981) and TRAC-PF1 (fast version), the latter one being used almost exclusively (see reference TRAC-PF1/MOD1 1986). There are also two boiling water reactor code versions, developed at Idaho National Engineering Laboratory and called TRAC-BD1 and TRAC-BF1 (Taylor et al. 1984). The TRAC codes have the largest set of descriptions available in any code (but not in the literature) for transfer laws and material properties.

The TRAC code is said to be a two-fluid code with six balance equations for gaseous and liquid water, yet its data base (TRAC-PF1/MOD11 1986, p. 39-41 and 75-93) is to a large extent that of a mixture model (cf. Section 3.1). Noncondensable gases are accounted for by a separate mass balance. TRAC is said to compute three-dimensional two-phase flow in the reactor vessel, yet is lacks the representation of internal fluid shear, i.e., the second term on the right-hand side of Eq. (2.12), with $\mathbb{B}$ representing the shear stress tensor. TRAC's capability of computing multidimensional single or two-phase flow has not been established via experiment, and all spatial grids used so far for three-dimensional flows in the vessel are far too coarse for producing trustworthy velocity distributions.

The numerical methods of the TRAC code are built upon the well-known and widely used Implicit Continuous Eulerian (ICE) method by Harlow and Amsden (1971). The TRAC-PF1 code is a semi-implicit finite difference code and uses upwind differencing (cf. Section 8.5). TRAC uses also a staggered mesh system as described in Figure 7.5: mass and energy balances, producing void fraction, temperatures and importantly, pressures, are written for the cell centers, while the momentum balances are written for "cell edges," for which they produce velocities.

The numerical techniques for TRAC-PF1/MOD1 can be found in Appendix C of TRAC-PF1/MOD1 1986 and in papers by Mahaffy (1979, 1982 and 1983). The distinctive features of the technique lie in

(i) the selection of terms for implicit and explicit treatment,
(ii) the treatment of the momentum balance, and
(iii) the solution sequences.

13.2.1 *Solution Algorithm of TRAC-PF1*

The partial differential equations are time and space discretized to yield a set of equations in the form of Eqs. (11.1). The solution method is called the Stability-Enhancing Two-Step (SETS) method and developed to "eliminate the material Courant stability limit" (p. 2 of Mahaffy and Liles (1982)), which was explained to be a questionable aim in Section 12.2 above). The following *five* steps are used in TRAC's SETS method.

1. The momentum balance is first solved, in the form of the equation of motion, for a first estimate $\tilde{v}^{n+1}$ of the new-time velocities, using
   a) the old-time pressure field, p^n,
   b) two representations for the momentum flux $v\partial v/\partial x$: one for the case of contracting flow ($\beta=0$ on p. 4 of Mahaffy and Liles 1982), namely $v\partial v/\partial x \approx v^n \nabla \tilde{v}^{n+1}$, where ∇ denotes backward differences (cf. Section 5.2), the other for the case of expanding flow ($\beta=1$), namely $v\partial v/\partial x \approx \tilde{v}^{n+1} \nabla \tilde{v}^n + v^n[\nabla(\tilde{v}^{n+1} - \tilde{v}^n)]$, where $\tilde{v}^n$ is unspecified,

c) for form losses and friction, $K|v|v$, the unusual approximation $K^n(2\tilde{v}^{n+1} - v^n)|v^n|$, rather than the more accurate approximation $[K^n(\tilde{v}^{n+1} + v^n)/2|v^n|$.

The first equation is called the "Stabilizer Equation of Motion" or a predictor.

2. In a second step, the equation of motion is written a second time, but now with

d) a first-approximation to the new-time pressure field, $\tilde{p}^{n+1}$,

e) new-time velocities v^{n+1} replacing all first estimates $\tilde{v}^{n+1}$ in the "Stabilizer Equation of Motion," except in the derivative of the momentum flux term.

This second equation is called the "Basic Equation of Motion" and solved (by hand) for the new-time velocities v^{n+1} as functions of old-time densities ρ^n and first estimates of new-time pressures $\tilde{p}^{n+1}$.

3. The result v^{n+1} is substituted into the "Basic Equations of Mass and Energy Conseration," which are written in conservative form and:

f) are explicit in first estimates of density, $\tilde{\rho}^{n+1}$, and in first estimates of volumetric energy density, $\tilde{\rho}^{n+1}\tilde{e}^{n+1}$,

g) contain new-time values v^{n+1} of velocities and first estimates of new-time pressure $\tilde{p}^{n+1}$,

h) contain old-time values T^n_w of wall temperatures, but first estimates $\tilde{T}^{n+1}$ of fluid temperatures.

4. The four phasic mass and energy balances and the phasic equations of state are combined to yield four nonlinear equations in phasic temperatures, void fraction and pressure, which are solved by Newton-Raphson iteration (cf. Section 11.2), while keeping the velocities and wall temperatures fixed.

5. Finally, the mass and energy balances are written again, this time as "Stabilizer Mass and Energy Equations," which

i) are implicit in densities ρ^{n+1} and in volumetric energy densities $\rho^{n+1}e^{n+1}$

j) contain the same velocities v^{n+1} as the previously used "Basic Equations,"

k) contain new-time pressure p^{n+1} and fluid temperatures T^{n+1}, and

l) contain still old-time values of wall temperatures, T^n_w.

These "Stabilizer Mass and Energy Equations" are finally solved for the new-time values of density ρ^{n+1} and energy e^{n+1}. This completes the solution process (Mahaffy and Liles 1982).

Time step control and accuracy criteria can be found on p. 437 of TRAC-PF1/MOD1 (1986). The criteria presented there involve Courant conditions and appear to be in conflict with the claim by Mahaffy and Liles (1982, p. 2), that SETS "eliminates the material

Courant stability limit." Convergence criteria and time step control is based on user-supplied values for

fractional changes of	• pressure
	• vapour void fraction α, if $\alpha > 0.01$,
total mass in reactor vessel, and change of	• liquid temperature,
	• vapor temperature,
	• structural temperatures.

13.2.2 *Discussion of TRAC-PF1 Solution Method*

The TRAC-PF1 is a prominent code and widely used but its numerical solution scheme is incompletely documented. Particularly, the following seven issues are not addressed in its documentation:

1. What is the range of parameters for which the governing equations are well-posed?

2. It is stated that the truncation error of SETS is of order $O\{\Delta t,\Delta x\}$, but the truncation error is not documented. In fact, the truncation of the "Stabilizer Equation of Motion" alone, described in Subsection 13.2.1 above is surprisingly large. For the case of $\beta = 0$, it is

$$T.E. = \frac{2K|v|\Delta t}{1+2K|v|\Delta t}\left\{vv_x + \frac{1}{\rho}p_x + K|v|v\right\}$$

$$- \frac{\left[v_{tt}/2 + v_{tx}\right]\Delta t - \left[v_{xx}/2\right]\Delta x}{1+2K|v|\Delta t} - O\{\Delta t^2, \Delta t\Delta x, \Delta x^2\}\ . \qquad (13.1)$$

Truncation errors proportional to velocities, velocities squared, and to first-order derivatives are indeed unusual. Normally, truncation errors are proportional to second or higher-order derivatives, v_{xx} etc.

3. The equations of motion fail to meet the condition of Lipschitz continuity with respect to $v_{j-1/2}$ and $v_{j+1/2}$. As discussed in Section 11 (see also Ortega and Rheinbolt 1970, p. 299), there is therefore no assurance of a unique solution for the balance equations in TRAC, unless a reference is cited to a proof for uniquenes, which does not require Lipschitz continuity.

4. Velocities are computed in the first two-steps. Thus velocity calculations are excluded from the iterative Newton-Raphson

procedure, and they are not subject to a convergence criterion. Therefore, it must still be documented that *all* the balance equations, which contain velocities, converge in principle to the nonlinear set of difference equations, Eqs. (11.1). As discussed in Subsection 12.3.2, Item (ii), the solution does not satisfy the balance equations, unless the solution converges to their associated difference equations.

5. Damping and dispersion as discussed in Section 8.3.3 must be documented for the SETS scheme. An operational guide for selecting spatial discretization before a calculation, is needed to avoid the need for repetitive computations with decreasing mesh size, until convergence is achieved.

6. Time step size control must be governed by truncation error estimations, and not by the currently used "fractional changes," since these are not related to computing accuracy.

7. Computing error estimates must be specified, or engineering accuracy must be guaranteed automatically for the computed results, relative to the modeling equations.

13.2.3 *Solution Method for TRAC-BD1*

The BWR version of TRAC is documented in the reference by Taylor et al. (1984). TRAC-BD1 is also a semi-implicit finite difference code with upwind differencing on a staggered mesh. In fact, aside from modifications (see Taylor et al. 1984, p. 264) the solution strategy of TRAC-BD1/MOD1 is essentially the same as that of TRAC-PF1/MOD1.

The description of the numerical scheme for TRAC-BD1/MOD1 in Taylor et al. (1984) does not mention the distinction between expanding *(β=1)* and contracting *(β=0)* flows as indicated in Subsection 13.2.1 for TRAC-PF1/MOD1. It appears therefore, as if only Issues 1, 2, 4, 5, 6 and 7 are unresolved for the multiphase flow computations in TRAC-BD1/MOD1.

Finally, it must be said that TRAC and most codes are continually evolving. What is written here applies to the cited documentation and references given there. The author hopes that most of the issues pointed out here will be resolved in the near future.

13.3 The RELAP5 Code

The RELAP5 code has been developed at Idaho National Engineering Laboratory, also for the U.S. Nuclear Regulatory

Commission to analyze the safety of nuclear power plants. As the TRAC codes, the RELAP5 code is also a six-equation two-fluid code for two-phase steam-water mixtures, and it has a separate mass balance for non-condensable cases. RELAP5 has in principle the same two-phase, two-component fluid flow capabilities as the TRAC code and essentially the same modeling capabilities as the TRAC-PF1 code. RELAP5 is preferred by some users, particularly by those with experience in RELAP4 applications, but mostly on the basis of user conveniences. Technically based advantages or disadvantages for using TRAC or RELAP5 for one or another class of PWR analyses have never been documented.

RELAP5 is said to be a one-dimensional, two-component two-phase flow code, but it also relies to a large extent on the data base for mixture models (Ransom et al. 1985, pp. 87-117). RELAP5 has special provisions for simulating horizontal stratified flow. Virtual mass effects are (unnecessarily) computed everywhere.

While based on the same six-equation two-fluid model as the TRAC code, the RELAP5 model formulation involves sums and differences of the balance equations. Detailed documentations of RELAP5/MOD2 can be found in the references by Ransom et al. (1985), Feinauer et al. (1984), by Trapp and Riemke (1985) and by Ransom (1986).

13.3.1 *Numerical Solution Methods in RELAP5/MOD2*

The RELAP5/MOD2 code is a semi-implicit code with a "nearly-implicit" option. The partial differential equations are reformulated in terms of "sum density," "difference density," "sum velocity" and "difference velocity" equations, the first and third representing actually the mixture balances of mass and momentum, the last one representing the acceleration of vapor relative to the liquid, while the second one has no physical meaning.

The system of governing partial differential equations in RELAP5/MOD2 is ill-posed (Ransom et al. 1985, p. 13). Unidentified and unquantified numerical viscosity is used in the momentum balance to render the equations well-posed. It is claimed that computing stability and accuracy have been demonstrated by computer experimentation, but no references are cited to the results of this experimentation.

The modified balance equations are completely discretized, in space by volume averaging (cf. Section 7.2), and in time by semi-implicit backward differencing. A staggered mesh and donor-cell differencing are used. The same formulation could have been obtained by space and time differencing (cf. Section 8). The finite difference equations of RELAP5/MOD2 are nonlinear and of the form of Eqs. (5.2) or (11.1). Standard methods for their solution are discussed in Section 11.

The selection of new-time values in the semi-implicit finite difference formulation is combined with the linearization of the nonlinear difference equations to obtain by algebraci operations, a single system of N linear equations for N unknown pressures in N computational cells. All other variables are then computed by back-substitution.

A "mass error" is computed as the difference between the mixture density obtained from the above procedure and the mixture density as computed from a mixture mass balance in conservative form. This "mass error" reflects in part the difference between the linearized equations used in the general solution process and the original semi-implicit mixture mass balance. It reflects also the difference between a semi-implicit and an implicit mass balance and it contains obviously residual errors since it is evaluated with new-time velocities for which there is no convergence criterion met. The "mass error" is used to control the integration step size, but no relation is documented between the mass error and the truncation error of state variables in general.

The solution time is said to be very fast and to require only 1.3 milliseconds per integration step and per computational cell (3 milliseconds for TRAC). However, the code is not found to run faster than TRAC for reasons given below in Item 4 of Subsection 13.3.2. For more details on the semi-implicit methods, see Ransom (1985) and Trapp and Riemke (1985).

13.3.2 *Discussion of RELAP5/MOD2 Solution Methods*

There are these seven major issues of the solution method in RELAP5/MOD2, which have not been addressed in the available documentation.

1. The finite difference formulation fails to meet Lipschitz continuity whenever a phasic velocity goes through zero. See Eqs. (71) and (72) on p. 39 of Ransom (1985). This discontinuity can easily be avoided (cf. Section 7.1).

2. The physical basis for "velocity averaging" in areas of changes in cross-sectional area is not documented.

3. A large number of serious computational problems are reported along with numerical "fixes" which have no basis in physics or in numerical analysis. Two examples follow:

3.1 The problem of "water packing" (Feinauer et al. 1984, p. 2) results from the misapplication of the two-fluid model (cf. Item (2) in Section 1.3) and is fixed with a "pressure change" multiplication of 10^6 and by raising the interphase drag by *30 orders of magnitude* (Feinauer et al. 1984, p. 3).

3.2 A large heat transfer coefficient of 10^{12} (no units, Ransom 1985, p. 51) is used when "a second phase appears" in a computational cell. There appears to be no reason nor a need for this artifice.

4. Most importantly, there is no evidence that the RELAP5/MOD2 computer solution satisfies the original field equations. A linearized set of balance equations is solved and there is no error estimate for

(i) demonstrating that the solution satisfies the *nonlinear difference equations*, Eq. (11.1), and
(ii) demonstrating that the finite difference equations satisfy the partial differential equations with engineering accuracy.

The problem is, that in RELAP5/MOD2 a nonlinear set of difference equations is solved without iterations. There is no standard theory for *solving nonlinear equations without iteration*. A reference to an open publication, or a derivation, of such a theory is required before RELAP5/MOD2 calculations can be accepted in general.

RELAP5/MOD2 results *may* appear to be correct if space and time increments are *small* enough such that

(i) all parameter changes are in turn so small that the first iterate would satisfy some (unspecified!) convergence criterion (cf. Eq. (11.25), and
(ii) all truncation errors are small enough to assure convergence of the first-order difference to the partial differential equations.

This requirement for small time steps explains the disappointingly slow computing speeds of RELAP5/MOD2.

5. Coefficients of numerical damping and dispersion (cf. Section 8.3.3) must be specified, so that space mesh sizes can be selected before a computation is started and repetitive calculations with decreasing space increments can be avoided.

6. Time step control must be governed by truncation error (i.e., difference between partial differential and finite difference equations). A separate convergence criterion must be imposed to assure that the solution satisfies (for all state variables) the nonlinear finite difference equations.

7. In general, the computational error should be separable from modeling errors.

13.4 The RETRAN Code

The RETRAN code has been developed by Energy Incorporated in Idaho Falls, Idaho, for the Electric Power Research Institute (EPRI) in Palo Alto, CA. The RETRAN code originated from the three-equation RELAP4 code (Moore and Rettig 1973). It has been restructured into a five-equation thermohydraulics code for nonequilibrium nonhomogeneous two-phase flow, and it has become the nuclear power industry's standard for analyzing the nuclear steam supply system under transient and postulated accident conditions. In its evolution, RETRAN has reached and is now maintained as the version called RETRAN-02/MOD002. Detailed documentations of the code can be found in the EPRI report by McFadden et al. (1981), modeling equations are described by Paulsen and Hughes (1983), and numerical methods used in RETRAN-02 are also described by Hughes and Katsma (1983), and by Hughes, Katsma and Paulsen (1985). The history of the RETRAN development was written by Agee (1983, 1985).

The RETRAN-02 code is the only multiphase flow code which passed through a formal Quality Assurance (QA) procedure. This QA program was designed to assure that the code is adequately documented with regard to modeling assumptions, capabilities and limitations, code logic, application rules and requirements for code implementation. The program is also intended to assure that the coding agrees with the documented modeling equations and that the solution satisfies the modeling equations. The QA program includes also the assurance that the code results "agree with relevant test data and similar calculations," and the code meets its stated objective.

RETRAN-02 has these five conservation equations: three equations for mixture mass, momentum and energy conservation, one equation for vapor mass conservation and one equation for the time rate of change $d(v_L - v_G)/dt$ of the liquid velocity relative to the vapor velocity. The last equation is called the "dynamic slip equation." The vapor phase is taken to be at saturation conditions. Only one-dimensional flow is simulated.

13.4.1 *The Solution Method in RETRAN-02*

The partial differential equations governing RETRAN-02 are first spatially discretized by the method described in Section 7.2. The familiar staggered mesh system is used, with the computational cells for momentum balances displaced by half of an increment. Donor cell differencing with perfect mixing assumption is applied. The resulting equations are ordinary differential equations in the form of Eq. (5.1). Paulsen and Hughes (1983, p. 162) state that the RETRAN code has Porsching's solution method (cf. Section 10.4.4). The actual solution method in RETRAN-02 deviates significantly from Porsching's method.

Next, the time derivatives are replaced by first-order backward time differences (implicit Euler-Cauchy method). However, the terms in $\boldsymbol{F}$ on the right-hand side of Eqs. (5.1) are linearized in terms of time and space differences of mixture mass flow rate, mixture mass, mixture energy and phase relative velocity. The linearization involves the Taylor series expansion of the state variable vector $\boldsymbol{Y}$ about some "initial state $\boldsymbol{Y}_0$." In the balance equation, written without source terms,

$$\partial \boldsymbol{Y}/\partial t + \underline{\underline{A}}\partial \boldsymbol{Y}/\partial z = 0 \qquad (13.2)$$

all state variables $\boldsymbol{Y}$ are replaced by $\boldsymbol{Y}_0 + \partial \boldsymbol{Y}$ (Hughes and Katsma 1983, Eq. (18a), p. 171) to obtain

$$\partial \boldsymbol{Y}_0/\partial t + \partial(\delta \boldsymbol{Y})/\partial t + \underline{\underline{A}}_0 \left[\partial \boldsymbol{Y}_0/\partial z + \partial(\delta \boldsymbol{Y})/\partial z \right]$$

$$+ \partial \boldsymbol{Y} \cdot \partial \underline{\underline{A}}/\partial \boldsymbol{Y} \cdot \partial \boldsymbol{Y}_0/\partial z = 0 . \qquad (13.3)$$

Then, without justification, the "initial state $\boldsymbol{Y}_0$ is *assumed* to be a constant, uniform state so that all gradients are zero." By "initial state $\boldsymbol{Y}_0$" is apparently denoted the fluid state at time t_n, as one advances from time t_n to time t_{n+1}. There is no connection made between the assumptions about state $\boldsymbol{Y}_0$ and the corresponding actual state $\boldsymbol{Y}(t_n)$ in the fluid. The assumptions remove all but the terms with $\partial(\delta \boldsymbol{Y})/\partial t$ and $\partial(\delta \boldsymbol{Y})/\partial z$ from Eq. (13.3), and thereby drastically simplify the balance equations.

The linearization leads, after algebraic substitution, to a single system of linear equations in the three differences ΔW, ΔV_{sl} and ΔE

for mixture mass flow rate, slip velocity and mixture energy, respectively, as they are associated with a particular time step (Hughes, Katsma and Paulsen 1985, p. 35). After this linear set of equations is solved by matrix inversion, the results are utilized to obtain mixture mass and vapor mass by back-substitution.

The procedure is called a two-step method, the first step being the above matrix inversion and the second step being the back-substitution. The method is only conditionally stable.

There are options available for computing homogeneous equilibrium and nonhomogeneous equilibrium flows. For these options, stability has been "established" for some conditions analytically and for other conditions by numerical experimentation.

13.4.2 *Discussion of Solution Method in RETRAN-02*

Notwithstanding the formal quality assurance review of RETRAN-02, there are at least these three issues in the cited references:

1. There is no documented evidence that the solution technique used in RETRAN-02 guarantees convergence of the numerical solution to the solution of the governing field equations. The documentation deals only with stability but not with consistency which is necessary for convergence (cf. Section 8.2). Consistency is established for the Porsching method, which is said to be the basis of the RETRAN-02 numerical solution scheme (Paulsen and Hughes 1983, p. 162). However, the linearization, along with the assumption of a "constant and uniform initial state," as discussed in connection with Eq. (13.3), is a part of the RETRAN-02 solution scheme but not a part of Porsching's convergence proof. Therefore, there is a special convergence proof needed for RETRAN-02.

2. Measures of computing *accuracy* are needed, such as the coefficients of numerical damping (cf. Section 8.3.3) and numerical dispersion. The code user should have the means to (a) select space increment sizes to achieve a desired accuracy for a particular transient and (b) distinguish between computational errors and modeling errors.

3. The time discretization error should be controlled via time step size control which is based on truncation error estimates.

13.5 The RAMONA-3B Code

The RAMONA-3B code was developed under the sponsorship of the U.S. Nuclear Regulatory Commission from the RAMONA-III code of Scandpower, Norway, by Brookhaven National Laboratory (Wulff, Cheng, Diamond and Khatib-Rahbar 1984). The RAMONA-3B is a BWR systems code with the capability of computing three-dimensional neutron kinetics and nonhomogeneous nonequilibrium two-phase flows in any number of parallel code channels. RAMONA-3B accounts also for boron transport.

The two-phase flow in the reactor vessel is predicted on the basis of a four-equation slip flow model, consisting of three mixture balance equations for mass, momentum and energy; further, of the vapor mass balance, of a slip correlation and of a model for nonequilibrium vapor generation. The vapor is taken to be at saturation. The code computes liquid subcooling, saturation or superheat temperatures, as the conditions dictate.

13.5.1 *Solution Techniques in RAMONA-3B*

The two-phase flow calculations in RAMONA-3B are carried out by a combination of analytical integrations (cf. Sections 3.3 and 4.2) for mixture mass and mixture momentum balances and of first-order explicit Euler-Cauchy integration for the loop momentum, the vapor mass and the mixture energy equations, all of which have the form of Eq. (5.1). Quadratures arising from analytical integrations are carried out by the trapezoidal rule. Time step control is based on growth rate of selected parameters (see Wulff, Cheng, Diamond and Khatib-Rahbar 1984, p. 289).

The single-phase steam flow in the steam lines is computed by fourth-order Runge-Kutta integration of mass and momentum balances. The space discretization error has been controlled by selecting the space increment size such that its further reduction produces no significant change. The time discretization error is controlled in truncation error estimate and user specification of permissible relative and absolute errors.

13.5.2 *Discussion of Solution Methods in RAMONA-3B*

Aside from the computations of the acoustical effects in the steam line, RAMONA-3B has no effective control for space and time discretization errors. These must be controlled by repeating calculations for a particular case with increasing numbers of channel groups and computational cells, until a converged solution is obtained.

Therefore, it is necessary to have estimates for numerical damping and dispersion coefficients, particularly for fast transients, when pressure waves are reflected in the steam lines and cause void collapse in the core, with resulting positive reactivity insertion, followed by power excursions. Also, the time step size control must be related to estimates of absolute and relative truncation errors.

13.6 The HIPA-BWR4 Code

The HIPA-BWR4 code was developed under the auspices of the U.S. Nuclear regulatory Commission at Brookhaven National Laboratory (Wulff, Cheng, Lekach and Mallen 1984). The HIPA-BWR4 code is part of BNL's BWR Plant Analyzer and simulates the Nuclear Steam Supply System, the Balance of Plant, the containments (Mark I), the control systems and all engineered safety systems. HIPA-BWR4 has a point kinetics model, models for thermal conduction in fuel and structures and a four-equation drift-flux model for nonhomogeneous nonequilibrium two-phase flow, consisting of mass balances for vapor and mixture, of mixture momentum and energy balances, of drift-flux correlations and models for nonequilibrium evaporation and condensation. The vapor is taken to be at saturation, while the nonequilibrium liquid-phase temperature is computed via the nonequilibrium phase change model. HIPA-BWR4 accounts also for boron transport.

HIPA-BWR4 is implemented in the special-purpose peripheral processor AD10 of Applied Dynamics International, Ann Arbor, MI. The AD10 is specifically designed for integrating large systems of nonlinear ordinary differential equations in the form of Eqs. (5.1), at high speed and low cost. The main features of the AD10 are its capability of interpolating multidimensional tables at high speed with task-specified parallel processors and to integrate ordinary differential equations in a special processor with more than 10 different integrating algorithms built in as firmware. The BNL Plant Analyzer has 13 task-specific processors and has been shown to execute a representative two-phase flow model (four-equation slip flow model) for a BWR pressure vessel two orders-of-magnitude faster than the CDC-7600 mainframe computer. The speed is achieved by the combination of selected model formulations, integrating algorithms and computer architecture (cf. Section 3).

The BNL Plant Analyzer is interactively accessible via standard telephone lines from any IBM-PC compatible computer, which serves as terminal and for graphics displays.

13.6.1 *Solution Methods in HIPA-BWR4*

All six modeling principles discussed in Section 3 are implemented in HIPA. Explicit integration was found to be superior to implicit integration, as the system response frequency to be simulated was estimated to be approximately 10 Hz or less. The integration update frequency is 140 Hz. the simulation speed is about seven times faster than real process time for all transients. Acoustic waves are tracked only in the steam lines. Thermophysical properties for the fluid in the pressure vessel are computed with the volume-averaged vessel pressure and the condition of uniform pressure.

The mixture mass balance is integrated as the mixture volumetric flux divergence equation, Eq. (4.2). The mixture momentum balance is integrated around four loops through average, hot and bypass channels in the core, and through the recirculation loop, as shown in Eq. (4.7). All thermohydraulic balance equations are in the form of Eqs. (5.2); for vapor mass and mixture energy they were obtained via space averaging (cf. Section 7.2, Eqs. (7.12) and (7.13)), and for mixture momentum via closed contour integration.

The numerical integration with respect to time of the resulting ordinary differential equations is achieved by combining standard, explicit first-order Euler-Cauchy and third-order Adams-Bashford algorithms. The algorithms are part of the AD10 special-purpose computer's firmware and can be arbitrarily selected and combined from a list of 13 algorithms, through simple interactive keyboard commands. They have been selected empirically and assigned to the various types of differential equations in HIPA-BWR4, so as to achieve optimum accuracy and speed.

Three time step sizes are used in HIPA-BWR4. They are kept in a fixed ratio: system pressure and steam line dynamics are computed with a time step size equal to 1/7, the loop momentum balances are computed with a step size equal to 1/2, of the overall time step size which is at most 60 milliseconds. The extrapolation error within an overall time step, associated with multi-stepping, has been found to be negligible by comparison with reference calculations, obtained with a single time step.

13.6.2 *Discussion of HIPA-BWR4 Solution Methods*

Spatial discretization in HIPA-BWR4 is fixed.* It has been selected by nodalization studies, so as to yield insignificant space

*There are no DO-loops and branches in HIPA-BWR4.

discretization errors for the fastest transients of interest (shortest wave length l_k in Eq. (8.22)). Steady-state conditions have been maintained without computational drift for a twenty-hour long systems simulation.

Time step size is fixed during a transients and limited for all transients by accuracyh requirements, such that integrations with any time step size below the limit produce the same answer within plotting accuracy (approximately four significant digits, since all variables are scaled).

HIPA-BWR4 simulates two-phase flows in a plant-specific fixed geometry and under plant-specific, normal and accidental conditions. This simplifies greatly the control of computing accuracy. Space and time increments are selected for the most pessimistic conditions and then retained for all conditions. The resulting compromise in computing efficiency is compensated for by the savings in computational overhead, which is normally required to afford flexibility in selecting various arrangements of computational cells.

13.7 The PHOENICS Computer Code

The PHOENICS code was developed at the Computational Fluids Dynamics Unit of the Imperial College of Science and Technology in London, U.K. (Spalding 1982).

The PHOENICS code is actually a system of codes that has grown around a central code segment with many satellite segments. Its intent is the prediction not only of multidimensional, multiphase flows and associated conductive heat transfer in channel walls, but also combustion and chemical reactions in a variety of different systems: nuclear reactors, combustion engines, turbo-machines, furnaces, cooling towers and "other equipment items." In essence, PHOENICS is a code package designed to integrate balance equations. The user must supply the models and coding for constitutive laws, transfer laws and special processes.*

The outstanding feature of PHOENICS is its demonstrated capability to integrate the conservation equations under a wide range of conditions. The balance equations are formulated in conservative form. They are spatially discretized by a standard method (cf. Section 7.2), even though the author calls it a "Finite Domain Method," the domain being the volume of a computational cell. The author claims

---

*A limited number of default options appear to be available in PHOENICS, but very little "state-of-the-art" modeling, much less than in TRAC.

credit for having a completely implicit integration algorithm, but he fails to address fully the issues of accuracy and computing time, as has been discussed earlier in Serction 12.2.

The computing method of PHOENICS is based on successive substitution; a method which normally does not work, but has been made to work successfully in PHOENICS (cf. Section 11.1). How difficult it is to succeed with successive substitution can be seen by looking at the PHOENICS development and the 10-year-long process of finding a computing sequence which leads to a converged solution, while building up to two-phase flow computing capabilities. The solution methods in PHOENICS evolved in many stages of improvement from at least four algorithms called SIMPLE (Semi-Implicit Method for Pressure-Linked Equations, Patankar and Spalding 1972), SIMPLER (for SIMPLE-Revised, Patankar 1980), SIMPLEST (for SIMPLE-ShorTened, Spalding 1980) and IPSA (for Interphase Slip Algorithm, Spalding 1981).

13.7.1 *The Solution Method in PHOENICS*

The conservation equations are written in global form (cf. Section 7.2) for open control volumes, enclosed by potentially moving boundaries. A staggered mesh is used, with control volumes for momentum balance displaced relative to the control volumes for mass and energy balances. Donor-cell differencing is employed. Multidimensional flows can be simulated in principle only, since no constitutive descriptions are given for interfacial drag, phasic wall shear or multiphase fluid stress calculations, nor are there any multidimensional, multiphase flow calculations documented with validation (cf. Sec. 12.3.2) of results.

In their discretized form, the balance equations are sums of only flux and source terms. Even storage terms are represented as fluxes through imaginary time planes, all for the purpose of achieving a certain symmetry in the form of the balance equations. Unfortunately, this forced flux description obscures the physics.

The most important feature fo the solution method is (1) that it is fully implicit and (2) that it is consistent. It has also been shown to be stable for a large number of conditions and systems. The success of the method depends entirely on its very important computing sequence (Spalding 1982, p. 53).

The first step consists of computing, from the phasic mass balances, *two* volume fractions, one for each phase, thereby using a "trick" to ensure that they add up to unity. It is not clear why the phasic and mass balances are not written in terms of vapor volume fraction alone, thereby avoiding the need for a trick.

The second step consists of computing velocities from the momentum balances, using initial guesses of pressures.

In the third step is computed the mixture mass error for a computational cell, by using the results from the previous two steps in the mixture mass balance.

The fourth and fifth steps serve to compute pressure corrections from derivatives of the finite difference mass balance with respect to pressure.

In the sixth step, all pressures and velocities are updated and the computational steps are repeated until the mass discrepancy and differences between iterates are less than user-prescribed error bounds.

13.7.2 *Discussion of Computational Methods in PHOENICS*

The solution method is first-order accurate in space and time. The solution process, if converging, guarantees that the solution satisfies the discretized equations within prescribed error bounds. The issue of discretization errors, however, remains unresolved.

The user of PHOENICS needs, therefore, at the time of arranging the distributions of control volumes a criterion for assessing the space discretization error. Spalding (1982) points out the need for unexpectedly fine nodalization, but he presents no specific quantitative results, nor generally applicable guidelines for selecting space increment sizes.

Also, the time step size in PHOENICS must be controlled by truncation error estimates. Currently, the user must repeat every calculation while decreasing both time and space increments, until convergence is achieved.

It is to be expected that the computational effort for PHOENICS in the context of a large systems code for nuclear reactor power plants is considerably larger than expected on the basis of experiences gained so far, particularly because the computational effort of reducing the time discretization error has not yet been assessed (cf. last paragraph in Section 8.5).

The strength of PHOENICS could be in the area of research on separate effects in simple test facilities, provided its overall implementation for a specific application is convenient. In this application PHOENICS serves to integrate the balance equation, and the success depends on the user's ability to formulate and implement mathematical models for the processes of his interest. Spalding likens PHOENICS to a "well-equipped kitchen," the user has to bring his recipes, the ingredients and the skill to compute ("cook").

13.8 The FEBE Algorithm in ALMOD 4

The FEBE algorithm is a general purpose integrator for ordinary differential equations and is used in the multiphase computer codes DRUFAN M2 and ALMOD 4. Of particular interest here is the ALMOD 4/ASWR version for two reasons:

(i) ALMOD 4/ASWR is based on the Finite Element Method (FEM) as discussed in Subsection 9.2.4.
(ii) FEBE in ALMOD 4/ASWR has time step size control which guarantees user-prescribed accuracy relative to time discretization errors.

The FEM formulation affords second-order accuracy for space discretization. Werner (1986) has shown that ALMOD 4/ASWR converges with little computational effort, relative to both time and space discretization, under conditions where TRAC-PF1 (cf. Section 13.2) could not converge at all. TRAC-PF1 failed to converge with 120 mesh cells, where ALMOD 4/ASWR converged with 4 mesh cells requiring 1/6 of the CPU time of TRAC-PF1. Thus, Asymmetric Weighted Residual (ASWR) formulation with the techniques used in FEBE appear to point toward the potentially most economical computing method for two-phase flow. It seems to indicate that high-order space discretization (cf. Section 8.6.6) is required to avoid the costly need for fine nodalization.

One issue remains unresolved, however, with the application of ASWR to discontinuities in multiphase flow. As described by Graf and Romstedt (1983, p. 19), a Lagrangian grid system is used to follow the motion of discontinuities in derivatives of state variables. The velocities with which such discontinuities are computed to move are in general nonphysical.

The velocity of the moving grid point is computed by setting equal the substantial derivatives at both sides of the moving point. The resulting velocity turns out to be the quotient of the difference between up and downstream partial time derivatives over the difference between up and downstream partial space derivatives, all evaluated for a selected parameter.

In the first place, this implies that the velocity is governed by fluid properties. In fact, it may be governed by boundary conditions such as wall heat flux. An important example is the Saha-Zuber correlation for the point of net vapor generation. Secondly, there are many instances where the proposed scheme cannot work. Consider, for example, the point with location $Z(t)$ at which the liquid enthalpy is saturated. This condition is cited by Graf and Romstedt (1983, p.

21) as a candidate application of their method. It is quite clear that the condition $h_L = h_f$ requires at $z = \lim_{\varepsilon \to 0}(Z-\varepsilon),\ 0 < \varepsilon \to 0$, that

$$\frac{\partial h_L}{\partial t} + \dot{Z}\frac{\partial h_L}{\partial z} = \frac{dh_f}{dp}\left[\frac{\partial p}{\partial t} + \dot{Z}\frac{\partial p}{\partial z}\right]. \tag{13.4}$$

and that the method proposed by Graf and Romstedt (1983) does not meet this requirement for the velocity $\dot{Z}$ of the discontinuity. Moreover, it is also clear that Graf and Romstedt's method violated mass conservation as expressed in the second equation of Eqs. (1.2) for computing the velocity U of the two-phase mixture level.

13.9 Multidimensional Multiphase Flow

There are only a few multidimensional flow calculations reported which have successfully predicted phase separation. It was pointed out in Section 13.7 that IPSA in PHOENICS could only in principle be applied to three-dimensional two-phase flows. Carver and Salcudean (1986) have developed the Flow Analysis in Three-Dimensions (FAITH) computer code which is based upon further improvements of IPSA.

FAITH simulates steady-state three-dimensional flow of two incompressible fluids using two mass and six momentum balances, three for each phase in three coordinate directions. The calculations require, for a modest 10 by 10 by 10 mesh point grid, the formidable effort of solving 8,000 nonlinear equations.

The solution is achieved by successive substitution containing an inner iteration on the pressure field by the Newton-Raphson method. The Newton-Raphson iteration involves the differentiation of the mass balance with respect to pressure; the mass flux derivatives are obtained from the finite difference equation of the momentum balance.

It is interesting to note that volumetric fluxes are used instead of mass velocities in the momentum balance (cf. Section 4.2). However, it is peculiar that no justification is given for the extra effort from iterating to find two unknown phasic volume fractions, instead of formulating the problem in terms of just one unknown void fraction.

The problem with the FAITH code description in the reference by Carver and Salcudean (1986) is that the questions of uniqueness and convergence remain unanswered. The equations are linearized; some of their terms are dropped. There is no indication that the original nonlinear equations are satisfied by the numerical solution and if so, with what accuracy. Single-phase flow calculations are used to "demonstrate the correctness of the method," but there is no discussion on the relevance of single-phase calculations for the prediction of phasic velocities and void fractions.

Numerical solutions were obtained by Carver and Salcudean (1986) for steady two-phase flow in elbows and U bends and compared with experimental results. The agreement between computed and measured results is good and also comparable to that achieved earlier by Pratap, Patankar and Spalding (1976) for single-phase flow. However, the calculations demonstrate the uncertainties in the results arising from the uncertainties in identifying flow regimes and corresponding interphase drag laws for two-fluid models. The method used for selecting and adopting these drag laws is not clear from the reference.

Bukhari, Flaherty and Lahey (1986) presented a paper on the successful application of the Finite Element Method (FEM) for the prediction of two-dimensional single and two-phase flow. This work is based upon the previously referenced research by Bukhari (1985) and Sim (1986) on phase distributions in triangular channels and two-dimensional test sections. This work is still in the stage of development, as all calculations were carried out with properties for a liquid that has the density of air and the viscosity of heavy oil at low temperatures. This was done to achieve convergent solutions with limited computer memory. Good agreement has been obtained between analytical results and results computed by the FEMOTH code (Sim 1986). Also, the computed trends of normalized liquid velocity distributions at low and high flows compare well with measured distributions. The FINITE-2D code, developed earlier by Bukhari (1985) has demonstrated the capability of computing two-dimensional flow qualitatively correct, after fluid property tuning, for single-phase flows over a cavity, between parallel plates, past obstacles and through branches.

The important conclusions from this research are, that

(i) large computer memory is required for two-phase flow in multidimensional geometries,

(ii) efficient solution methods for large systems of linear equations are needed, and

(iii) high-order trial functions should be used for economic FEM applications.

(iv) multidimensional, multiphase flow calculations require constitutive laws which are unavailable at this time for most flow regimes.

14. CONCLUSIONS AND NEEDS FOR FUTURE RESEARCH

1. The primary problems of fundamental research in multiphase flow are not in the area of numerical methods, but instead in the development of two-phase flow instrumentation and in the derivation of closure laws. The issues of multiphase flow prediction in large nuclear power reactors are primarily in the areas of scaling and of modeling separate effects in specific but complex geometries.

2. The issues of ill-posed problem formulation are to be resolved analytically and on the basis of physics and not of numerics. A large number of numerical "fixes" are currently used unsuccessfully to remedy incorrect model formulations. A systematic effort is required to eliminate such fixes (cf. Section 1.3).

3. The vast majority of computed results for two-phase flow are presented without assessment of computational errors, composed of round-off, truncation and convergence errors. Many codes have not been shown to converge to their original set of governing equations. The most urgently needed basic research in numerical methods for multiphase flow analyses is for the development of rational rigorous error estimates and for automatic control of space and time discretization errors.

4. Notwithstanding the declining costs for computing, almost all large systems codes with multiphase flow simulation in general-purpose computers are prohibitively expensive to execute, particularly when the computing accuracy is demanded to be equal to or better than measurement accuracies. First-order integration schemes are not economical enough. Research is needed to establish the potential savings in computational cost that can be achieved by higher-order integration methods.

5. Mathematical model formulation, the selection of numerical computing methods and the choice of computer architecture should be optimized as a whole.

ACKNOWLEDGEMENTS

This assessment of computational methods for multiphase flow analysis would not have been possible without the many documents so graciously contributed by Drs. M. Carver of Chalk River Nuclear Laboratories (AEC), E. D. Hughes of Energy Incorporated, Idaho Falls, E. C. Johnson of INEL, R. T. Lahey, Jr. of RPI, Troy, D. Liles of LASL, D. B. Spalding of Imperial College, London, UK, H. Tamm of Whiteshell Nuclear Research Establishment and W. Werner of G.R.s., Garching, FRG. These contributions are gratefully acknowledged.

The difficult task of typing the manuscript was carried out by K. Becker and C. Falkenbach. Their special care and diligence are greatly appreciated.

This work was performed under the auspices of the U.S. Nuclear Regulatory Commission, in the Fall of 1986.

NOMENCLATURE

A, $\vec{A}$	flow area
$\underline{\underline{A}}$, A_{ij}	function matrix
A_{ij}^{-1}	inverse of A_{ij}
a	norm of Jacobian, Eq. (11.19)
B_{ijk}	set of function matrices
$\mathbb{B}$	scalar, vector or second-order tensor
b	see Eq. (11.20)
C_k	void propagation velocity
C_0	Zuber-Findlay distribution parameter
C_{ijkl}	coefficient functions, Eq. (2.19)
c	propagation velocity,
c	see Eq. (11.21)
c_l	coefficient functions, Eq. (9.4)
D	differential operator
D_i	general function vector (P.D.E.)
D/Dt	Lagrangian derivative
D_k/Dt	phasic Lagrangian derivative
d_h	hydraulic diameter

E	shift operator
$\hat{e}_k$	unit vector in k-th coordinate direction
F_i, F	general function vector (O.D.E.)
$\vec{F}$	body force
f	function
f_ν	largest characteristic system frequency
f_φ	shape function
$f_{2\varphi}$	wall friction factor for two-phase flow
G	mass flux
G	amplification factor (matrix)
G_l, G	vector of functions, Eqs. (2.39, 11.1)
g	gravitational constant
g_k	transformed state variables
$g(y)$	vector function
g_{ij}	Jacobian of $g(y)$
H_i,H_e	permissible time steps for implicit, explicit integration, respectively.
h	integration step size (O.D.E.)
h	enthalpy
I	general source term
I_{ij},$\underline{\underline{I}}$	$\sqrt{-1}$ identity tensor
$\vec{J}$,J	general flux

J_{ij}	Jacobian, Eq. (7.1)
$j, \vec{j}$	volumetric flux
k	wave number
k_{eff}	effective thermal conductivity
L	Lipschitz constant
$L(y)$	ordinary differential operator
$L(\varphi)$	partial differential operator
$L_h(y)$	finite difference analog of $L(y)$
$L_\Delta(\varphi)$	finite difference analog of $L(\varphi)$
ℓ_i	reciprocal of interfacial area density
ℓ_k	wavelength
m	number of state variables
$\dot{m}$	mass flux from phase change
M	global momentum
N_{Pe}	grid Peclet number (Eq. 7.18)
n	number of phases
$\hat{n}$	outward normal unit vector
p	pressure
P_ℓ	see Eq. (9.6)
q_ℓ	weighting function
$\vec{q}$	heat flux
r_{max}	spectral radius

s	specific entropy
T	absolute temperature
T_i, T_e	computer frame time for implicit and explicit integration, respectively
$T.E.$	truncation error
t	time
t_k	cumulative time while phase k occupies point of interest
t_{ij}	stress tensor components
U	level velocity
u	specific internal energy
V	volume
$\vec{v}$	velocity
v_{gj}	void fraction weighted, area-averaged drift velocity
W_i	Fourier coefficients
W_j	numerical approximation to w_j
w	axial velocity
w_j	state variables (P.D.E.)
x, x_i	coordinates
$\vec{x}$	position vector
Y	approximation to y
y_i, y	state variables (O.D.E.)
z	axial coordinate

Greek Symbols

α	vapor volume fraction
α	coefficient of numerical diffusion
Γ	vapor generation rate
γ_{ij}	function matrix
Δ	forward difference operator
Δs	entropy source
$\Delta\rho$	$\rho_L - \rho_G$
$\delta \boldsymbol{Y}$	increment of $\boldsymbol{Y}$
δ	centered difference operator
$\delta(x-x_0)$	Dirac-delta function
∇	backward difference operator
∇	gradient, divergence operator
ε	error
ε	dispersion coefficient
η	normalized integration argument
$\underline{\underline{\Theta}}, \vartheta_i$	diagonal matrix, Eq. (11.2)
λ, λ_i	eigenvalues
μ_{eff}	effective dynamic viscosity
ξ	heated perimeter
ξ_i	coordinate along characteristic
Π	stress tensor
ρ	density

ρ radius of hyper-sphere (Section 11)

σ surface tension

Φ numerical approximation to φ

φ general source term (Section 2)

$\varphi(t,x)$ general dependent variable

$\varphi^b(t,x)$ boundary surface

ψ general flow parameter

$\psi_j(\vec{x})$ surface for initial conditions (Section 2.4)

ψ_i test functions (Section 9.2)

Ω characteristic frequency, Eq. (7.2)

ω_i characteristic surface

Subscripts

G gaseous phase

i interface, boundary

k phase k

L liquid phase

m mixture

t partial derivative with respect to time t

v vapor

x partial derivative with respect to coordinate x

. partial derivative

Superscripts

b	boundary
i	interface
$\overline{}k$	averaged over phase k
T	transpose, turbulent
$\sim$	mass weighted, transformed
$\cdot$	ordinary time derivative
0	initial
+,-	above, below level

REFERENCES

Abromowitz, M. and Stegun, I. A., editors (1964), *Handbook of Mathematical Functions*, National Bureau of Standards, Applied Mathematics Series, 55, Superintendent of documents, U.S. Government Printing Office, Washington, D.C. 20402.

Achard, J-L., Drew, D. A. and Lahey, R. T., Jr. (1985), "The Analysis of Non-linear Density-Wave Oscillations in Boiling Channels," *J. Fluid Mech.*, *155*, p. 213.

Agee, L. J. (1983), "RETRAN Overview," *Nuclear Technology*, Vol. 61, No. 2, p. 143.

Agee, L. J. (1985), "RETRAN Overview," *Nuclear Technology*, Vol. 70, No. 1, p. 7.

Ahl, J. P. (1985), "Analytical Description of PWR Pressurizer Transients," EPRI Final Report NP-3850.

Ames, W. F. (1969), *Numerical Methods for Partial Differential Equations*, Barnes and Noble, Inc. NY.

Baek, S. M. and No, H. C. (1986), "A Non-Equilibrium Three-Region Model for Transient Analysis of PWR Pressurizer," *Nuclear Technology*, Vol. 74, p. 260.

Banerjee, S. and Hancox, W. T. (1978), "On the Development of Methods for Analyzing Transient Flow-Boiling," *Int. J. Multiphase Flows*, Vol. 4, p. 437.

Bashford, F. and Adams, J. C. (1855), "An Attempt to Test the Theories of Capillary Attraction," Cambridge Univ. Press, Chapter III.

Beck, J. V., Blackwell, B. and St. Clair, C. R., Jr. (1985), *Inverse Heat Conduction - Ill-Posed Problems*, A. Wiley-Interscience Publication, New York, Chichester, Brisbane, Toronto, Singapore.

Benyon, P. R. (1968), "A Review of Numerical Methods for Digital Simulation," Simulation Councils, Inc., p. 219.

Birkhoff, G. and Kines, T. F. (1962), "CHIC Programs for Thermal Transients," Bettis Atomic Power Laboratory, Pittsburgh, PA, WAPD-TM-245.

Birta, L. G., Ocren, T. I. and Kettenis, D. L. (1985), "A Robust Procedure for Discontinuity Handling in Continuous System Simulation," Trans. Soc. for Computer Simulation, Vol. 2, No. 3, p. 189.

Bouré, J. A. (1975a), "Mathematical Modeling and the Two-Phase Constitutive Equations," European Two-Phase Flow Group Meeting, Haifa, Israel.

Bouré, J. A. (1975b), "A Unified Presentation of the Non-Equilibrium Two-Phase Flow Models", Winter Annual Meeting of ASME, Houston, Texas, *Non-equilibrium Two-Phase Flows*, eds. R. T. Lahey and G. B. Wallis, ASME, United Engineering Center, 345 East 47th St., New York, NY 10017

Bouré, J. A. (1976), "Mathematical Modeling of Two—Phase Flows, its Bases-Problems: A Review," 1st OECD/ISNC Specialists Meeting on Transient Two-Phase Flow, Toronto, Canada.

Bukhari, K. M. (1985), "Phase Separation and Distribution Phenomena in a Two-Dimensional Test Section," Doctoral Dissertation, Rensselaer Polytechnic Institute, Troy, NY.

Bukhari, K. M., Flaherty, J. E. and Lahey, R. T., Jr. (1986), "Finite Element Analysis of Two-Dimensional Single and Two-Phase Flows," 4th Miami Int. Symposium on Multiphase Transport and Particulate Phenomena, Miami Beach, FL.

Bulirsch, R. and Stoer, J. (1964), "Fehlerabschaetzungen und Extrapolationen mit rationalen Functionen bei Verfahren vom Richardson-Typus," *Num. Math.*, Vol. 6, p. 413.

Burwell, M. J., Enix, D., Steinhoff, F. and Wolfert, W. (1983), "DRUFAN-01/MOD2, Volume III: Model Description," Gesellschaft fuer Reaktorsicherheit (GRS), Garching FGR, GRS-A-846.

Carnahan, B., Luther, H. A. and Wilkes, J. O. (1969), *Applied Numerical Methods*, John Wiley & Sons, Inc., New York.

Carver, M. B. and Salcudean, M. (1986), "Three-Dimensional Numerical Modelling of Phase Distribution of Two Fluid Flow in Elbows and Return Bends, Accepted for publication in *Numerical Heat Transfer*, Vol. 10.

Certaine, J. (1966), "The Solution of Ordinary Differential Equations with Large Time Constants," *Math. Methods for Digital Computers*, A. Ralston and H. S. Wilf, eds., Wiley & Sons, Inc.

Chan, R. K.-C. (1978), "A Balanced Expansion Technique for Constructing Accurate Finite Difference Advection Schemes," *Int. J. Numerical Methods in Engineering*, Vol. 12, p. 1131.

Chatoorgoon, V. (1986), "SPORTS, A Simple Nonlinear Thermalhydraulic Stability Code," *Nuclear Engineering and Design, 93*, No. 1, p. 51.

Chung, T. J. (1978), Finite Element Analysis in Fluid Dynamics, Advanced Book Program, McGraw-Hill Book Co.

Courant, R., Friedrichs, K. and Lewy, H. (1967), "On the Partial Difference Equations of Mathematical Physics," *IBM Journal*, (March (1967, pp. 215-234).

Courant, R. and Hilbert, D. (1962), *Methods of Mathematical Physics*, Vol. 2, Partial Differential Equations, Interscience Publishers, John Wiley & Sons, Inc.

Courant, R. and Friedrichs, K. O. (1948), *Supersonic Flow and Shock Waves*, Interscience Publisher, John Wiley & Sons, Inc.

Curtiss, C. F. and Hirschfelder, J. O. (1952), "Integration of Stiff Equations," Proc. Nat. Acad. of Sci., U.S.A., Vol. 38, p. 235.

Delhaye, J. M., Giot, M. and Riethmuller, M. L. (1981), *Thermohydraulics of Two-Phase Flow Systems for Industrial Design and Nuclear Engineering*, McGraw-Hill Book Company.

Delhaye, J. M. and Achard, J. L. (1976), "On the Averaging Operators Introduced in Two-Phase Flow Modeling," Proc. 1st CSN Specialists Meeting on Transient Two-Phase Flow, eds., S. Banerjee and K. R. Weaver, Vol. 1, pp. 5-84 (AECL).

Delhaye, J. M. and Achard, J. L. (1977), "On the Use of Averaging Operators in Two-Phase Flow Modeling," in *Thermal and Hydraulic Aspects of Nuclear Reactor Safety*, Vol. 1: Light Water Reactors, eds., O. C. Jones and S. F. Bankoff, pp. 289-332, ASME, New York.

Delhaye, J. M. (1968), "Equations Fondamentales des Ecoulements Diaphasiques," CEA-R-3429.

Drew, D. A. and Wood, R. T. (1985), "Overview and Taxonomy of Models and Methods for Workshop on Two-Phase Flow Fundamentals, NBS Gaithersburg, MD.

Eide, S. and Hughes, E. D. (1981), "Characteristic and Linear Dispersion Analysis of the RETRAN Two-Velocity Model," Electric Power Research Institute, NP-1917.

Eringen, A. C. (1962), *Nonlinear Theory of Continuous Media*, McGraw-Hill Book Company.

Euler, L. (1768), "Institutiones Calculi Integralis" in *Leonardi Opera Omnia*, Ser. I, Vol. XI, p. 424, Teubner Verlag, Leibzig (1913).

Evans, D. J. (1974), "Iterative Sparse Matrix Algorithms," in *Software for Numerical Mathematics*, D. J. Evans, editor, Academic Press.

Fadden, E. J. (1984), "The System 10 Plus: Broader Horizons," Peripheral Array Processor Conference, Boston, MA, Simulation Councils, Inc.

Fakory, M. R. (1985), "Component Flow-Path Thermal Hydraulic Model for Real-Time Simulation of Natural Circulation Steam Generators," ASME/AIChE Natl. Heat Transf. Conf., Denver, CO, 85-HT-63.

Fauselan, R. W., Thakkar, J. G., Wiesland, J. W. and Casell, D. S. (1980), "CALIPSO Code Report," Electric Power Research Institute NP-1391, Vol. 1.

Feinauer, L. R., Bryce, W. M., Kiser, D. M., Rienke, R. A. and Tsai, C. C. (1984), "Post-Release Improvements for RELAP5/MOD2," Internal Technical Report, SE-CMD-84-011, EG&G, Idaho National Engineering Laboratory.

Ferch, R. L. (1978), "Method of Characteristics Solution for Non-Equilibrium Transient Flow-Boiling," *Int. J. Multiphase Flow,* Vol. 5, p. 265.

Findlay, J. A. (1961), "Pressurizer Analysis and the Pre-Digital Program," KAPL-M-NP-23, UC-32 Mathematics and Computers (TID-4500 16th Ed), General Electric Corp.

Forsythe, G. E. and Wasow, W. R. (1960), *Finite Difference Methods for Partial Differential Equations*, John Wiley & Sons, Inc., NY.

Galerkin, B. G. (1915), "Rods and Plates, Series Occurring in Various Questions Concerning Elastic Equilibrium of Rods and Plates (in Russian), Vestn. Inghenevov, Vol. 19, p. 897.

Gallagher, R. H., Carey, G., Oden, J. T. and Zienkiewicz, O. C., editors, *Finite Elements in Fluids,* Vols. 1-6, Wiley Series in Numerical Methods in Engineering, Wiley Interscience Publication.

Garabedian, P. R. (1964), *"Partial Differential Equation,"* John Wiley & Sons, Inc.

Gear, C. W. (1971), *Numerical Initial Value Problems in Ordinary Differential Equations,* Prentice Hall, Series in Automatic Computation.

Gear, C. W. (1969), "The Automatic Integration of Stiff Ordinary Differential Equations," in Proc. IFIPS Congress, p. 187, North Holland, Amsterdam.

Goldberg, M. (1973), "A Note on the Stability of an Iterative Finite Difference Method for Hyperbolic Systems," *Mathematics of Computing,* Vol. 27, No. 121, p. 41.

Gonzales-Santalo, J. M. and Lahey, R. T. (1971), "An Exact Solution for Flow Transients in Two-Phase Systems by the Method of Characteristics," Trans. ASME, J. of Heat Transfer, Vol. 95, No. 4.

Goodman, T. R. (1964), "Application of Integral Methods in Transient Nonlinear Heat Transfer," *Advances in Heat Transfer,* edited by T. F. Irvine, Jr. and J. P. Harnett, Vol. 1, p. 52, Academic Press.

Graf, U. (1986), "Survey of a Numerical Procedure for the Solution of Hyperbolic Systems of Three-Dimensional Flow," Gesellschaft fuer Reaktorsicherheit mbH, Garching FRG.

Graf, U., Drittler, K., Honsalek, K., Hrubisko, K., Loy, D., Miro, J., Schaefer, A. and Werner, W. (1986), "Zur Absicherung der Ergebnisse technisch-wissenschaft-licher Rechnungen", GRS-A-1274, Gesellschaft fuer Reaktorsicherheit (GRS), Garching FRG.

Graf, U., Romstedt, P. and Werner, W. (1982), "Ein Verfahren der Gewichteten Residuen zur Loesung von Anfangs-Randwertaufgaben der Fluiddynamik mit Unstetigkeiten," Gessellschaft fuer Reaktorsicherheit (GRS) mbH, Garching, FRG; GRS-A-753.

Graf, U., Romstedt, P. and Werner, W. (1986), "Use of Dynamic Grid Adaptation in the Asymmetric Weighted Residual Method," *Nuclear Science and Engineering 92*, p. 66.

Graf, U. and Romstedt, P. (1983), "Probleme der Numerischen Behandlung von Zweiphasenstroemungen, Loesungsmoeglichkeiten mit dem ASWR-Verfahren," KTG-Fachtagung ueber Neuere Berechnungsverfahren in der Fluiddynamik, Kerntechnische Gesellschaft E.V., April 18/19 Munich.

Guderley, K. G. and Hsu, C. C. (1972), "A Predictor-Corrector Method for a Certain Class of Stiff Differential Equations," *Math. Comp.*, Vol. 26, No. 117.

Hancox, W. T. and Banerjee, S. (1977), "Numerical Standards for Flow Boiling Analysis, "*Nuclear Science and Engineering,* 64, p. 106.

Hancox, W. T., Mathers, W. G., and Kawa, D. (1975), "Analysis of Transient Flow Boiling; Application of Method of Characteristics," AIChE Paper No. 42, 15th National Heat Transfer Conference, San Francisco.

Harlow, F. H. and Amsden, A. A. (1971), "A Numerical Fluid Dynamics Calculation Method for All Flow Speeds," *J. Comp. Physics,* Vol. 8, p. 197.

Heubner, K. H. (1975), *The Finite Element Method for Engineers,"* John Wiley & Sons, New York.

Hirt, C. W. (1968), "Heuristic Stability Theory for Finite Difference Equations," *J. Comp. Physics,* Vol. 2, p. 339.

Hofer, E. (1983), "Zeitintegration von Systemen Partieller Differentialgleichungen mit dem ODE-Solver FEBE und seinem Block-Sparse-Matrix-Paket FTRIX," in Neue numerische Verfahren in der Fluiddynamik, Fachtagung der KTG-Fachgruppe Reaktorphysik und Berechnungsmethoden, Muenchen.

Holt, R. and Rasmussen, J. (1972), "RAMONA II, A FORTRAN Code for Transient Analysis of Boiling Water Reactors," Institutt for Atomenergi, Kjeller Research Establishment, Kjeller, Norway.

Hughes, E. D. and Katsma, K. R. (1983), "Numerical Solution Method Improvements for RETRAN," *Nuclear Technology*, Vol. 61, No. 2, p. 167.

Hughes, E. D., Katsma, K. R. (1983), "Numerical Solution Method Improvements for RETRAN Model Equations," *Nuclear Technology*, Vol. 70, No. 1.

Isaacson, E. and Keller, H. B. (1966), *Analysis of Numerical Methods*, John Wiley & Sons, Inc.

Ishii, M. (1975), *Thermo-Fluid Dynamic Theory of Two-Phase Flow*, Eyrolles, Paris.

Ishii, M. Kataoka, I. and Kocamustafaogullari, G. (1981), "Phenomenological Modeling of Two-Phase Flow in Water Reactors at ANL," Proc. Ninth USNRC Water Reactor Safety Research Information Meeting, Gaithersburg, MD.

Ishii, M. (1971), "Thermally Induced Flow Instabilities in Two-Phase Mixtures in Thermal Equilibrium," Ph.D. thesis, School of Mechanical Engineering, Georgia Institute of Technology, Atlanta, GA.

Ishii, M. and Chawla, T. C. (1979b), "Two-Fluid Model and Momentum Integration Between Phases," Proc. Seventh USNRC Water Reactor Safety Research Information Meeting, Gaithersburg, MD.

Ishii, M. and Mishima, K. (1980), "Study of Two-Fluid Model and Interfacial Area," Argonne National Laboratory, Argonne, IL, ANL-80-111, NUREG/CR-1873.

Ishii, M. and Chawla, T. C. (1979a), "Local Drag Laws in Dispersed Two-Phase Flow," Argonne National Laboratory, Argonne, IL, ANL-79-105, NUREG/CR-1230.

Ivanov, M. Ya. and Idiyatullina (1978), "A Method of Third-Order Accuracy for Calculating the Steady Smooth Flows of an Ideal Gas," Zh. vychist. Mat. mat. Fiz., *18*, 4, p. 996.

Jain, R. K. (1972), "Some A-Stable Methods for Stiff Ordinary Differential Equations," Math. Comp., Vol. 26, No. 117, p. 71.

Karplus, W. J. (1977), "Peripheral Processors for High-Speed Simulation," *Simulation*, Society for Computer Simulation, Vol. 29, No. 5, p. 143.

Kataoka, I., Ishii, M. and Serizawa, A. (1984), "Local Formulation of Area Concentration and its Measurements in Two-Phase Flow," Argonne National Laboratory, Argonne, IL, ANL-84-68, NUREG/CR-4029.

Keller, H. B. (1968), *Numerical Methods for Two-Point Boundary-Value Problems*, Blaisdell Publishing Company, Waltham, MA.

Kestin, J. and Podowski, M. Z. (1985), "Report on International Workshop on Two-Phase Flow Fundamentals," NBS Gaithersburg, MD, DOE/ER/13056-1, prepared for USDOE under Contract No. DE-AC02-83ER13056.

Klopfenstein, R. W. and Davis, C. B. (1971), "PECE Algorithms for the Solution of Stiff Systems of Ordinary Differential Equations," *Math. Comp.*, Vol. 25, No. 115, p. 457.

Kocamustafaogullari, G. (1971), "Thermo-Fluid Dynamics of Separated Two-Phase Flow," Ph.D. Thesis, School of Mechanical Engineering, Georgia Institute of Technology, Atlanta, GA.

Kroeger, P. G. (1976), "Application of a Non-Equilibrium Drift Flux Model to Two-Phase Blowdown Experiments," Proc. 1st OECD/DSNI Specialists Meeting on Transient Two-Phase Flows, Toronto, Canada.

Lahey, R. T., Jr. (1981), "Analytical Modules for Nuclear Reactor Simulators," in Proc. Simulation Methods for Nuclear Power Systems," Tucson, Arizona, EPRI WS-81-212.

Lax, P. D. and Richtmyer, R. D. (1956), "Survey of Stability of Linear Finite Difference Equations," Communications on Pure and Applied Mathematics, Vol. IX, p. 267.

Lekach, S. and Kaufman, J. M. (1980), "Implicit Coupling of Two-Phase Flow Calculations," Proc. ANS/ASME Int. Topical Meeting on Nuclear Reactor Thermal Hydraulics, Saratoga Springs, NY, NUREG/CP-0014, Vol. 3, p. 2024.

Leonard, B. P. (1979), "A Stable and Accurate Convection Modelling Procedure Based on Quadratic Upstream Interpolation," *Computer Methods in Applied Mechanics and Engineering,* Vol. 19, p. 59 (North-Holland Publishing Co.).

Liles, D. R. (1974), "Wave Propagation and Choking in Two-Phase Two-Component Flow," Ph.D. Thesis, Georgia Institute of Technology, Atlanta, GA.

Lomax, H. and Bailey, H. E. (1967), "A Critical Analysis of Various Numerical Integration Methods for Computing the Flow of gas in Chemical Non-Equilibrium," NASA TN D-4109.

Lomax, H. (1967), "An Operational Unification of Finite Difference Methods for the Numerical Integration of Ordinary Differential Equations," NASA TR R-262.

Lomax, H. (1968), "On the Construction of Highly Stable Explicit Numerical Methods for Integrating Coupled Ordinary Differential Equations with Parasitic Eigenvalues," NASA TN D-4547.

Loomis, J. et al., (1981), "THERMIT: A Computer Program for Three-Dimensional Thermal-Hydraulic Analysis of Light Water Reactor Cores," Electric Power Research Institute, Palo Alto, CA,. NP-2032.

Mahaffy, J. H. (1983), "The Advantages and Limitations of the Sets Method," CNS Int. Conf. on Numerical Methods in Nuclear Engineering, Montreal, Quebec, LA-UR-83-1860, Los Alamos National Laboratory, Los Alamos, NM 87545.

Mahaffy, J. H. (1979), "A Stability-Enhancing Two-Step Method for One-Dimensional Two-Phase Flow," NUREG/CR-0971, LA-7951-MS, Informal Report, Los Alamos Scientific Laboratory, Los Alamos, NM 87545.

Mahaffy, J. H. (1982), "A Stability-Enhancing Two-Step Method for Fluid Flow Calculations," *J. Comp. Phys. 46*, p. 329.

Mahaffy, J. H. and Liles, D. (1982), " Numerical Solution Package for Transient Two-Phase Flow Equations," Fall 1982 Annual Mtg. Heat Transfer and Energy Conversion Div., AIChE, Los Angeles, CA, LA-UR-82-1483, Los Alamos National Laboratory, Los Alamos, NM 87545.

Massau, J. (1899), "Mèmoire sur l'intègration graphique des èquations aux dèrivèes partielles, F. Meyer-Van Loo, Ghent (monograph).

McFadden, J. H. et al., (1981), "RETRAN-02 A Program for Transient Thermal-Hydraulic Analysis of Complex Fluid Flow Systems," Electric Power Research Institute, Palo Alto, CA, Vol. 1, NP-1850.

Menzel, R. and Schwetlick, H. (1978), "Zur Loesung parameterab haengiger, nichtlinearer Gleichungen mit singulaeren Jacobi-Matrizen," *Numerische Mathematik, 30,* p. 65.

Mitchell, A. R. (1969), *Computational Methods in Partial Differential Equations,* John Wiley & Sons, Inc., NY.

Moore, K. V. and Rettig, W. H. (1973), "A Computer Programe for Transient Thermal-Hydraulic Analysis," Aerojet Nuclear Co. Report ANCR-1127.

Moore, K. V. and Rettig, W. H. (1968), "RELAP2 - A Digital Program for Reactor Blowdown and Power Excursion Analysis," Idaho National Laboratory, IDO-17263.

NRC/BNL (1976), "User's Manual for RELAP 3B-MOD101, A Reactor Systems Code," Office of Nuclear Reactor Regulation, USNRC, NUREG-0004, p. 9.

Oh, S. J. and Sursock, J. P. (1985), "MMS Two-Phase Nonequilibrium Pressurizer Model Qualification," ASME/AIChE Natl. Heat Transf. Conf., Denver, CO, 85-HT-64.

Ortega, J. M. and Rheinbolt, W. C. (1970), *Iterative Solution of Nonlinear Equations in Several Variables,* Computer Science and Applied Mathematics, A Series of Monographs and Texbooks, Academic Press, New York, San Francisco, London.

Patankar, S. V. and Spalding, D. B. (1972), "A Calculation Procedure for Heat, Mass and Momentum Transfer in Three-Dimensional Parabolic Flows," *Int. J. Heat and Mass Transfer,* Vol. 15, p. 1787.

Patankar, S. V. (1980), *Numerical Heat Transfer and Fluid Flow,* Hemisphere Publishing Corporation, McGraw-Hill Book Company.

Paulsen, M. P. and Hughes, E. D. (1983), "Nonequilibrium Model for Operational Transient Analyses," *Nuclear Technology,* Vol. 61, No. 2, p. 153.

Pipes, L. A. and Hovanessian, S. A., *Matrix Computer Methods in Engineering,* John Wiley & Sons, Inc.

Porsching, T. A., Murphy, J. H. and Redfield, J. A. (1971), "Stable Numerical Integration of Conservation Equations for Hydraulic Networks," *Nuclear Science and Engineering, 43,* p. 218.

Pratap, V. S., Patankar, S. V. and Spalding (1976), "Prediction of Turbulent Flow in Curved Pipes," *J. Fluid Mechanics,* p. 539.

Ransom, V. H. et al. (1985), "RELAP5/MOD2 Code Manual," Idaho National Laboratory, NUREG/CR-4312, EGG-2396.

Ransom, V. H. (1986), "RELAP5/MOD2: For PWR Transient Analysis," proposed CNS paper, private communication.

Rayleigh, J. W. S. (1877), *Theory of Sound,* 1st edn., revised, Dover Publications, New York (1945).

Reid, J. K. (1974), "Direct Methods for Sparse Matrices," in *Software for Numerical Mathematics,* D. J. Evans, editor, Academic Press.

Rettig, W. H. et al., (1970), "RELAP3 - A Computer Program for Reactor Blowdown Analysis," Idaho National Laboratory, JN-1321.

Richardson, L. F. (1925), *Mathem. Gaz.,* vol. 12, p. 415.

Richtmyer, R. D. and Morton, K. W. (1957), *Difference Methods for Initial-Value Problems,* 2nd Edition (1967), Interscience Publishers, John Wiley & Sons, Inc., NY.

Ritz, W. (1909), "Ueber eine Neue Methode zur Loesung gewisser Variations-Probleme der Mathematischen Physik," *J. Reine Angewandte Mathematik,* Vol. 135, No. 1, p. 1.

Rivard, W. C. and Torrey, M. D. (1979), "K-FIX: A Computer Program for Transient, Two-Dimensional, Two-Fluid Flow - THREED: An Extension of a K-FIX Code for Three-Dimensional Calculations," Los Alamos Scientific Laboratory, NUREG/CR-0863.

Rizwan-Uddin and Dorning (1986), "Some Nonlinear Dynamics of a Heated Channel," *Nucl. Eng. Design,* 93, p. 1.

Roache, P. (1972), *Computational Fluid Dynamics,* Hermosa Publishers, Albuquerque, NM.

Sauer, R. (1952), *Anfangswertprobleme bei Partiellen Differenfialgleichungen,* Springer Verlag, Berlin.

Scarborough, J. B. (1966), *Numerical Mathematical Analysis,* sixth edition, Johns Hopkins Press.

Shampine, L. F. and Watts, H. A. (1971), "Comparing Error Estimates for Runge-Kutta Methods," *Math. Comp.* 25, No. 1154, p. 445.

Sim, S. K. (1986), "Phase Distribution Phenomena in Triangular Conduits," Doctoral Dissertation, Rensselaer Polytechnic Institute, Troy, NY.

Singhal, A. K. and Spalding, D. B. (1980), "The URSULA2 Computer Program," Electric Power Research Institute, Palo Alto, CA, NP-1315, Vol. 1 and 2.

Slattery, J. C. (1972), *Momentum, Energy and Mass Transfer in Continua,* McGraw-Hill, Inc.

Smirnow, W. I. (1961), Lehrgang der Hoeheren Mathematik, VEB Deutscher Verlag der Wissenschaften, Berlin.

Snidow, N. L. and Wilson, J. L. (1984), "U-Tube Steam Generator for the Modular Modeling System," (B&W), Int. Conf. on Power Plant Simulation, Cuernavaca, Mexico.

Spalding, D. B. (1980), "Mathematical Modeling of Fluid Mechanics, Heat Transfer and Chemical Reaction Processes," A Lecture Course, CFDU No. HTS/80/1. Imperial College of Science and Technology, Exhibition Road, London SW7, UK.

Spalding, D. B. (1982), "A Note on Numerical Solution Procedures Used in the LAURA Computer Code and Other Codes for Nuclear Reactor Thermal Hydraulics," CFDU, Imperial College of Science and Technology, Exhibition Road, London SW7, UK, CFD/82/11.

Spalding, D. B. (1982a), "Four Lectures on the PHOENICS Computer Code," Computational Fluid Dynamics Unit, Imperial College of Science and Technology, Exhibition Road, London SW7, U.K., CFD/82/5.

Spalding, D. B. (1985), "The Numerical Computation of Multiphase Flows," Industrial Fluid Flow Computational Conference, Teesside Polytechnic; CFDU, Imperial College of Science and Technology, Exhibition Road, London SW7, U.K., CFD/85/7.

Stewart, H. B. and Wendroff, B. (1984), "Two-Phase Flow: Models and Methods," Review Article, *J. of Computational Physics* 56, p. 363.

Stuhmiller, J. H. and Ferguson, R. E. (1979), "Comparison of Numerical Methods for Fluid Flow," Electric Power Research Institute, Interim Report NP-1236.

Taylor, C. and Hughes, T. G. (1981), Finite Element Programming of the Navier-Stokes Equation, Pineridge Press Ltd., Swansea, U.K.

Taylor, D. D. et al. (1984), "TRAC-BD1: An Advanced Best Estimate Computer Program for Boiling Water Reactor Transient analysis," Vol. 1: Model Description, Idaho National Laboratory, NUREG/CR-3633.

Thoman, D. C. and Szewczyk, A. A. (1966), "Numerical Solution of Time-Dependent Two-Dimensional Flow of a Viscous, Incompressible Fluid over Stationary and Rotating Cylinders,"

Techn. Report No. 66-14, Dept. of Mech. Engineering, U. of Notre Dame, Notre Dame, Ind.

Thomasset, F. (1981), "Implementation of Finite Element Methods for Navier-Stokes Equations," Springer Series in Computational Physics, Springer Verlag, NY.

TRAC-PF1/MOD1: An Advanced Best Estimate Computer Program for Pressurized Water Reactor Thermal-Hydraulic Analysis, (1986), Safety Code Development Group, Energy Division, Los Alamos National Laboratory, Los Alamos, NM, NUREG/CR-3858, LA-10157-MS.

Trapp, J. A. and Riemke, R. A. (1985), "Completion of Nearly-Implicit Hydrodynamic Numerical Scheme for RELAP5/MOD2," Internal Technical Report, SE-RSTD-85-006, EG&G, Idaho National Engineering Laboratory.

Traub, J. F. (1974), "Theory of Optimal Algorithms," in *Software for Numerical Mathematics,* D. J. Evans, editor, Academic Press.

Treanor, C. E. (1966), "A Method for the Numerical Integration of Coupled First-Order Differential Equations with Greatly Different Time Constants," *Math. Comp.,* Vol. 20, No. 93, p. 39.

Turner, M. J., Clough, R. W., Martin, H. C. and Tropp, L. P. (1956), "Stiffness and Deflection Analysis of Complex Structures," *J. Aeronautical Science,* Vol. 23, No. 9, p. 805.

Vernier, Ph. and Delhaye, J. M. (1968), "General Two-Phase Flow Equations Applied to the Thermohydrodynamics of Boiling Water Nuclear Reactors," *Energie Primaire,* Vol. 4, No. 1-2, pp. 5-46.

Werner, W. (1981), "Weighted Residual Methods for the Solution of Fluid Dynamics Problems," Topical Mtg. on Advances in Mathematical Methods for the Solution of Nuclear Engineering Problems, Munich, Germany.

Wheeler, C. L., Thurgood, M. J., Guidotti, T. E. and DeBellis, D. E. (1986), "COBRA-NC: A Thermal-Hydraulic Code for Transient Analysis of Nuclear Reactor Components," Pacific Northwest Laboratory, Richland, WA, NUREG/CR-3262, PNL-4710, Vol. 1.

White, R. E. (1985), *"An Introduction to the Finite Element Method with Applications to Nonlinear Problems,"* John Wiley & Sons, New York.

Whitmarsh-Everiss, M. J. (1985), "Fluid Properties and Look-up Tables," Plant Design Analyzer Symposium, 2nd Meeting - Nuclear Plant Analyzers, CEGB, Gloucester, U.K.

Wilkinson, J. H., (1974), "Linear Algebra Algorithms," in *Software for Numerical Mathematics,* D. J. Evans, editor, Academic Press.

Wulff, W. (1986), "Modeling Principles and Computing Technology for High-Speed Simulation of LWR Transients," Proc. ANS topical Meeting on Reactor Physics and Safety, Saratoga Springs, NY, NUREG/CP-0080, Vol. 2, p. 944.

Wulff, W., Cheng, H. S., Leckach, S. V. and Mallen, A. N. (1984), "The BWR Plant Analyzer," Final Report, Brookhaven National Laboratory, NUREG/CR-3943, BNL-NUREG-51812.

Wulff, W. (1981), "The Kinematics of Moving Flow Regime Interfaces in Two-Phase Flow," Proc. 3rd OECD/ISNC Specialists Meeting on Transient Two-Phase Flow, Pasadena, CA.

Wulff, W. (1978), "Lumped-Parameter Modeling of One-Dimensional Two-Phase Flow," Proc. 2nd CSNI Specialists Meeting on Transient Two-Phase Flow, M. Reocreux and G. Katz, CEA, Editors (Paris), Vol. 1, p. 191.

Wulff, W. (1975), "Development of a Computer Code for Thermal Hydraulics of Reactors (THOR)," First Quarterly Progress Report, Thermal Reactor Safety Div., Brookhaven National Laboratory, BNL 199978-RP-1038.

Wulff, W. (1980), "Steam Line Dynamics - A Computer Program," Topical Report, Brookhaven National Laboratory, NUREG/CR-1438, BNL-NUREG-51186.

Wulff, W. et al., (1981), "THOR-1 (PWR): A Computer Code for Predicting the Thermohydraulic Behavior of Nuclear Reactor Systems," Informational Report BNL-NUREG-29781, Brookhaven National Laboratory, Upton, NY.

Wulff, W., Cheng, H. S., Diamond, D. J. and Khatib-Rahbar, M. (1984), "A Description and Assessment of RAMONA-3B/MOD.0 - Cycle 4: A Computer Code with Three-Dimensional Neutron Kinetics for BWR Systems Transients," Brookhaven National Laboratory, Upton, NY 11973, NUREG/CR-3664, BNL-NUREG-51746.

Wulff, W. (1981), "Major Systems Codes, Capabilities and Limitations," in Simulation Methods for Nuclear Power Systems", eds. D. L. Hetrick and P. G. Bailey, Tuscon, Arizona, EPRI WS-81-212.

Zemansky, M. W. (1957) *Heat and Thermodynamics,* 5th ed. (1968), McGraw-Hill.

Zuber, N. (1974), "Volume-Averaged Conservation Equations," U.S.N.R.C. Memorandum, N. Zuber to L. S. Tong, RSR, dated June 15, 1974.

CHAPTER 4

INERTIAL COUPLING IN TWO-PHASE FLOW: MACROSCOPIC PROPERTIES OF SUSPENSIONS IN AN INVISCID FLUID

Graham B. Wallis

Professor, Thayer School of Engineering, Dartmouth College, Hanover, New Hampshire 03755, USA

1. INTRODUCTION

A major theoretical development of the last few decades has been the *two-fluid* or *separated* theory of two-phase flow. The phases are treated as interwoven continua, with effective average properties, and each obeys the usual conservations laws in some *macroscopic form* (Delhaye, (1968), Drew (1971), Bourè (1975), Harlow and Amsden (1975), Ishii, (1975), Yadigaroglu and Lahey (1976), Delhaye and Achard (1978), Nigmatulin (1979), Banerjee and Chan (1980), Delhaye (1981a), Delhaye (1981b), Drew (1983), Ishii and Mishima (1984), Stewart and Wendroff (1984), Drew and Wood (1985), Bourè (1987), Ishii (1987).

Although there is little question that the fundamental equations of fluid mechanics apply at the *microscopic* level, in other words for motions on the scale of the smallest particles or entities, there are at least two major considerations in formulating the *macroscopic conservation laws* for subregions of the flow that are much larger than the particles:

- o The relationship of *higher order averages,* such as occur in the momentum flux, $\langle \rho \mathbf{uu} \rangle$, to averages of individual properties, such as velocity, $\langle \mathbf{u} \rangle$.

- o *Interactions* between the phases, manifest as terms that *couple* the conservation equations, introducing expressions describing the motion of one phase, or the relative motion, in the equations pertinent to the other.

The former consideration is well-known in single-phase flow in the case of turbulence and has spawned several analytical techniques and nomenclature, such as *Reynolds stresses.* The additional complication that occurs in two-phase flow is that averaging must be done with appropriate rigor, considering the effects of discontinuities in the flow field across the phase boundaries. In particular, this affects the operations of vector calculus, the average gradient of pressure in one phase not equalling the gradient of average pressure, for example.

The latter consideration is an inherent property of two-phase flows. The *interaction terms* may be defined as integrals over interfaces, but they can rarely be exactly related to the *macroscopic properties,* such as average velocity, that one would like to use as the *solution variables* for which the resulting equations are to be solved.

Many authors (e.g. Drew (1971), Ishii (1975), Delhaye and Achard (1978), Nigmatulin (1979), Banerjee and Chan (1980), Delhaye (1981b), Drew (1983), Stewart and Wendroff (1984), Drew and Wood (1985)) have addressed the mathematical problem of averaging microscopic conservation laws in order to derive macroscopic ones. However, the result is always a set of equations in which *extra terms* involving integrals over the microscopic domains remain. While various hypotheses may be made about *closure laws* expressing these *extra terms* as functions of the *solution variables,* it is not clear that such laws always exist, what form they should take and what approximations may be implied in their use. It would therefore be very useful to have developed a *set of exact solutions* to well-defined situations that can be used to illustrate general principles that may be implied in the closure laws and to provide special *reference cases* against which any theory claiming general validity can be tested. The present article takes this approach, seeking to build toward a general theory by taking successive steps each of which involves a simple physical situation. Should doubt about the rigor of some of the overall results arise, it is hoped that this will stimulate more sophisticated derivations that may elucidate more comprehensively what approximations are implied and what possible restrictions should be imposed.

Although the motion of any real fluids usually contains complexities that must be handled to some extent empirically, it is useful to start with simpler cases that offer hope of complete solution. Therefore, in this article, viscosity is ignored and solutions are sought to examples of *ideal* fluid flow. This makes it possible to use the classical techniques described, for example, by Lamb (1932) and Milne-Thomson (1949).

Even such a restricted approach to two-phase flow does not appear to have been developed in much detail before, although, if the *two-fluid theory* is to have a rigorous base, it should be possible to develop it along classical lines, deriving effective macroscopic equations of motion, "Bernoulli" equations, potential flow formulations and solutions for the *two independent phase potentials* as well as theorems involving kinetic energy, vorticity and so on. It is only by developing such a self-consistent body of analysis, and applying it to numerous specific examples, that a mature theory can be created and used as a basic for further development.

This article takes the first few steps along such a long road. Classical solutions for the motion of a single sphere are reviewed and extended to give new results. A similar approach is adopted for the macroscopic properties of a suspension of spheres. Specific results are derived for self-consistent *models* that invoke simplifying assumptions. The results are compared with some of the major expositions of continuum theory that have appeared in the literature. It is not the intention, at this stage, to make premature assertions about what the *general macroscopic equations* should be, if they exist.

A major result of this work is the recognition that Maxwell's solution (or some improved alternative) for the electrical conductivity of a suspension of spheres can be used to derive the *added mass* of the same array as well as the added kinetic energy due to the relation motion. From this it is possible to deduce the difference between the average pressures in each phase and some other macroscopic properties such as the force exerted on the stationary dispersed phase due to gradients in void fraction in any direction. The averaged stresses in both phases, as well as the momentum flux tensor of the continuous phase, are also deduced. It is shown how a number of different definitions of *added mass coefficients* are compatible with each other and can be incorporated into the separate equations of motion of each phase in simple cases.

In order to illustrate how the general approach relates to specific flow regimes or constitutive assumptions, three examples are presented in Sections 4, 5 and 6. They serve both to reinforce some of the simpler conclusions already developed and to raise questions that need to be answered by future work.

These developments have not yet led to a completely general formulation of the conservation laws in the two-fluid model. Indeed, comparisons with equations from the literature show many compatibilities but also some disagreements with the present formulation even in very simple situations, indicating the need for a more fundamental approach of the type attempted here. The overall picture that emerges is a tantalizing one, in which many pieces of the puzzle appear to fit together neatly while some incompatibilities remain in achieving a mating of the remaining parts. It is hoped that this article will both consolidate the established pattern and provide clues to the successful assembly of the remainder.

2. A SINGLE SPHERE IN A POTENTIAL FLOW

2.1 Fluid at Rest at Infinity

The potential for the motion of a sphere of radius "*a*" moving at speed, *w*, in fluid at rest at infinity, is, in spherical coordinates,

$$\varphi = wa^3 \cos\vartheta/2r^2 \tag{2.1}$$

On the sphere, $r = a$ and the potential is

$$\varphi = wa \cos\vartheta/2 \tag{2.2}$$

Using this result in (A.8), the kinetic energy of the fluid is

$$K = \frac{\rho w^2 a}{4} \int_0^{\pi/2} 2\cos\vartheta 2\pi a^2 \sin\vartheta \cos\vartheta d\vartheta \tag{2.3}$$

$$= \rho w^2 \pi a^3/3 = \frac{1}{2}\rho w^2 \left(\frac{1}{2}\right)\left(\frac{4}{3}\pi a^3\right) \tag{2.4}$$

Therefore, *C*, the coefficient of added mass, as defined in (A.15) is equal to 1/2.

If the sphere is accelerated by a force, *F*, in the direction of motion, we have, from energy conservation

$$Fw = F\frac{dx}{dt} = \frac{dK}{dt} = \rho w \frac{dw}{dt}\frac{1}{2}\frac{4}{3}\pi a^3 \tag{2.5}$$

or

$$F = \rho C V_0 \frac{dw}{dt} \tag{2.6}$$

which is just what would be obtained by using (A.14) and the principle of momentum conservation.

2.2 Flow Past a Stationary Sphere

Imposing a velocity $-w$ on the motion in the previous section we obtain the velocity potential for flow past a stationary sphere as

$$\vartheta = wr\cos\vartheta\,(1 + a^3/2r^3) \tag{2.7}$$

The radial velocity is zero on the surface of the sphere and the tangential velocity is

$$u_\vartheta = -\frac{1}{r}\left(\frac{\partial\varphi}{\partial\vartheta}\right)_{r=a} = w\sin\vartheta\left(1 + \frac{a^3}{2r^3}\right)_{r=a} = \frac{3}{2}\,w\sin\vartheta \tag{2.8}$$

Therefore, the pressure distribution around the sphere is, from Bernoulli's equation

$$p = p_0 - \frac{9}{8}\,\rho w^2\sin^2\vartheta \tag{2.9}$$

where p_0 is the stagnation pressure. The average pressure around the sphere is

$$\langle p_i\rangle = \frac{1}{4\pi a^2}\int_0^\pi p2\pi a^2\sin\vartheta\, d\vartheta \tag{2.10}$$

Using (2.9) in (2.10) we get

$$\langle p_i\rangle = p_0 - \frac{3}{4}\,\rho w^2 = p_0 - \frac{1}{2}\,\rho w^2\,(1 + C) \tag{2.11}$$

which differs from the static pressure of the fluid by the amount $-1/2\rho w^2 C$.

We may illustrate the methods of Section A.3.1 by deriving (2.11) in a general way for the flow past any object placed in a uniform flow.

Firstly, consider the object moving with speed w in the x-direction in stationary fluid. The fluid kinetic energy will be

$$K_1 = \frac{1}{2}\rho C V_0 w^2 = \frac{1}{2}\rho \int (u_x^2 + u_y^2 + u_z^2)\, dV \tag{2.12}$$

where the integral is taken over all the fluid.

The fluid momentum in the x-direction will be

$$M_x = \rho C V_0 w = \rho \int u_x dV \tag{2.13}$$

Now, impose a velocity $-w\hat{x}$ on the motion. The fluid kinetic energy is

$$K = \frac{1}{2}\rho \int \left\{(u_x - w)^2 + u_y^2 + u_z^2\right\} dV \tag{2.14}$$

$$= K_1 - wM_x + \frac{1}{2}\rho w^2 \int dV \tag{2.15}$$

$$= \frac{1}{2}\rho C V_0 w^2 - \rho C V_0 w^2 + \frac{1}{2}\rho w^2 \left[(V_0 + V) - V_0\right] \tag{2.16}$$

The term involving the total volume, $(V_0 + V)$, in (2.16) represents the total fluid kinetic energy, K_0, in the fluid in the absence of the object; therefore, (2.16) can be expressed as

$$K = K_0 - \frac{1}{2}\rho w^2 (1 + C) V_0 \tag{2.17}$$

If the object changes its volume in a way that keeps C constant, (i.e. it preserves its shape) we may use (2.17) in (A.29) to show that

$$\langle p_i \rangle = p_0 - \frac{1}{2}\rho w^2 (1 + C) \tag{2.18}$$

which is a general form of (2.11) and is valid whether it is the object or the fluid which is moving, as long as w is interpreted as the relative velocity and p_0 is the pressure at a point where the fluid has the same velocity as the object.

2.3 A Sphere in a Non-uniform Steady Flow

Kelvin, as quoted by Taylor (1928a), derived a more general result than (2.17) for the combined kinetic energy of a small sphere (small compared with distances over which velocities change) and the surrounding fluid in *cyclic* motion.

$$K = K_0 + \frac{1}{2} v_s^2 V_0 (\rho_s + C\rho_f) - \frac{1}{2} \rho_f v_f^2 V_0 (1 + C) \tag{2.19}$$

The subscripts s and f refer to the sphere and the fluid respectively. v_f is the average velocity of the fluid over a region much bigger than the sphere.

Several results may be derived from (2.19).

Case A

If the sphere is at rest, $v_s = 0$. If we consider a perturbation in which the sphere is moved an amount dx in the direction of fluid motion, energy conservation requires that an external force F_e be exerted on the sphere and do work:

$$F_e dx = \frac{dK}{dx} dx = - \rho_f v_f \frac{dv_f}{dx} V_0 (1 + C) dx \tag{2.20}$$

Now, the pressure gradient in the fluid in the absence of the sphere is

$$\left(\frac{dp}{dx}\right)_f = -\rho_f v_f \frac{dv_f}{dx} \tag{2.21}$$

Therefore, (2.20) is equivalent to

$$F_e = V_0 (1 + C) \left(\frac{dp}{dx}\right)_f \tag{2.22}$$

for the external force on the body. The force on the sphere from the fluid is the negative of this:

$$F = - V_0 (1 + C) \left(\frac{dp}{dx}\right)_f = +\rho_f V_0 (1 + C) v_f \frac{dv_f}{dx} \tag{2.23}$$

F_e exceeds the product of the sphere's volume and the *pressure gradient* by the factor *(1 + C)*, as shown by Taylor (1928a). However, if we use the average pressure over the surface of the sphere from (2.11), the force is

$$F_e = V_0 \frac{d <p_i>}{dx} \tag{2.24}$$

where the derivative indicates the rate of change of $<p_i>$ with motion of the sphere in the x-direction (or p_i could be defined as a continuous function for a highly dispersed suspension of spheres).

Equations (2.22) and (2.24) indicate that care must be taken with appropriate definitions of *pressure*, as discussed, for example, by Prosperetti and Jones (1984).

Case B

If the sphere is at rest but is allowed to grow uniformly in all directions (which keeps *C* constant), the work done by the *interfacial* pressure less the static pressure at infinity is equal to the change in overall kinetic energy; therefore

$$(<p_i> - p_0)\, dV_0 = - \frac{1}{2} \rho_f v_f^2 (1 + C)\, dV_0 \tag{2.25}$$

from which

$$<p_i> = p_0 - \frac{1}{2} \rho_f v_f^2 (1 + C) \tag{2.26}$$

as in (2.11) for a more restricted example.

Case C

If the sphere as well as the fluid is moving in the x-direction, the equivalent of (2.20) is

$$F_e = \frac{dK}{dx} = v_s \frac{dv_s}{dx} V_0 (\rho_s + C\rho_f) - \rho_f v_f \frac{dv_f}{dx} V_0 (1 + C) \tag{2.27}$$

or

$$F_e = v_s \frac{dv_s}{dx} \rho_s V_0 - v_f \frac{dv_f}{dx} \rho_f V_0 + \left(v_s \frac{dv_s}{dx} - v_f \frac{dv_f}{dx} \right) \rho_f V_0 C \tag{2.28}$$

The three terms in (2.28) could be regarded as components due to the acceleration of the sphere times its own mass, the volume times the pressure gradient that would exist without the sphere and a final term proportional to *relative acceleration* and to the *added mass.*

The case in which the sphere is both moving and changing size cannot be solved in this way because it is impossible to separate the work done by the force and the work done by the interfacial pressure, unless further information is introduced.

2.4 A Sphere in the Neighborhood of a Source

Case A in the previous section may be checked by examining the exact solution for a sphere in the neighborhood of a source, as given by Milne-Thomson (1949). The force due to a source (or sink) of strength "*m*" distant "*c*" from a sphere of radius "*a*" is directed toward the source (or sink) and of magnitude

$$F = 4\pi\rho \frac{m^2 a^3}{c(c^2 - a^2)^2} \tag{2.29}$$

Now, the velocity and velocity gradient at the center of the sphere due to the same source alone are

$$v_f = \frac{m}{c^2} \tag{2.30}$$

$$\frac{dv_f}{dx} = - \frac{2m}{c^3} \tag{2.31}$$

So, the *approximate* force on the sphere in the x-direction (away from the source) from (2.23) is

$$F_1 = \rho v_f \frac{dv_f}{dx} V_o (1 + C) = - \rho \frac{2m^2}{c^5} \frac{4}{3} \pi a^3 \frac{3}{2} = - 4\pi\rho \frac{m^2 a^3}{c^5} \qquad (2.32)$$

which agrees with (2.29) when $c >> a$ but contains an error of 10% when $c = 4a$, 27% when $c = 3a$, etc.

The error may alternatively be expressed as a function of

$$\frac{a}{c} = \frac{a}{2|v_f|} \left| \frac{dv_f}{dx} \right| \qquad (2.33)$$

which is in a useful form because, in evaluating the force on a particle moving in a fluid, one is interested in the relative importance of added mass and drag forces, i.e.

$$\frac{F_a}{F_d} \cong \frac{\rho v \frac{dv}{dx} C \frac{4}{3} \pi a^3}{\frac{1}{2} \rho v^2 C_D \pi a^2} = \frac{8C}{3C_D} \frac{a}{v} \frac{dv}{dx} \qquad (2.34)$$

If $C = 0.5$ and $C_D = 0.44$, this is equal to about six times the expression in (2.33), so that the error in the added mass force is 10% when it is one and one half times the drag force, 27% when it is twice the drag force and almost 80% when it is three times the drag force. Of course, potential flow theory is not compatible with the turbulent wake flow that causes the conventional drag, but the important point is that the simple expression that predicts a force proportional to relative acceleration is increasingly inaccurate the more it tends to dominate the motion.

2.5 Use of Enclosed Singularities to Evaluate Forces on Objects

When the steady potential flow around a submerged body can be presented by a velocity field $\mathbf{w}$, without the body, together with a set of sources and sinks, m_i, in the body, the resultant force on the body is

$$- \sum_i 4\pi\rho m_i \mathbf{w} \qquad (2.35)$$

If some of the sources are dipoles of strength $\mathbf{d}_i$, their contribution takes the form

$$-\sum_i 4\pi\rho \mathbf{d}_i \cdot \nabla \mathbf{w} \tag{2.36}$$

and so on for higher order singularities (Taylor (1928b), Lagally (1922), Yih (1977), Landweber and Miloh (1980)).

Using (2.35) and (2.36) is an alternative to computing the net force from the pressure distribution,

$$\mathbf{F} = -\int p d\mathbf{s} = \int \left(\frac{1}{2}\rho u^2 - p_0\right) d\mathbf{s} = \frac{1}{2}\rho \int u^2 d\mathbf{s} \tag{2.37}$$

where p_0 is the stagnation pressure and $d\mathbf{s}$ is an element of area with a normal pointing outward from the body.

To illustrate the use of (2.35) we will derive (2.29) from it. The image system for a sphere in the neighborhood of a source (Milne-Thomson (1949)) consists of a source of strength ma/c at the inverse point P', distant a^2/c from the center of the sphere, together with a uniform line sink of strength $(-m/a)$ stretched along an axis from P' to the center of the sphere (Figure 2.1). In this case the velocity $\mathbf{w}$ is due entirely to the external source and the resultant force, F, on the body towards the source is given by

$$\frac{F}{4\pi\rho} = \frac{m(ma/c)}{(c - a^2/c)^2} + \int_0^{a^2/c} \frac{m(-m/a)dz}{(c - z)^2} \tag{2.38}$$

where z is measured in the negative x-direction from the center of the sphere. Therefore,

$$\frac{F}{4\pi\rho m^2} = \frac{ac}{(c^2 - \alpha^2)^2} - \frac{1}{a}\left[\frac{a}{c - z}\right]^{a^2/c} = \frac{a^3}{c(c^2 - a^2)^2} \tag{2.39}$$

in agreement with (2.29).

We note that the approximate answer (2.32) becomes valid in the limit when the source is far from the sphere and the image system collapses to a dipole at the center of the sphere.

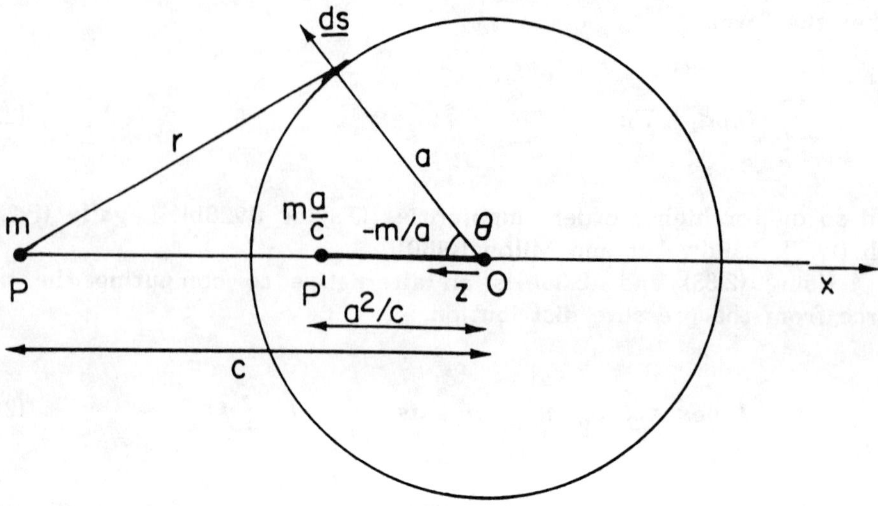

Figure 2.1 A sphere in the vicinity of a source.

2.6 A Stationary Sphere in the Combined Field of a Source and a Uniform Flow

Superimpose on the situation illustrated in Figure 2.1 a uniform flow with speed $(-\ v_S)$ in the x-direction and place a dipole of strength $d\ =\ a^3v_S/2$ at the center of the sphere, pointing in the positive x-direction. The dipole and the source remain fixed. Since the sphere is a streamline surface for the added flow field, it will remain a streamline surface.

The force on the sphere is a result of the combined velocity fields from the source and the uniform flow acting on the source's images plus the dipole.

The uniform flow produces no force since the net sum of the sources in the body is zero.

The only extra force beyond that calculated in Section 2.5 is from the interaction between the dipole and the external source. Its value is

$$4\pi\rho\ \frac{a^3v_S}{2}\ \frac{2m}{c^3}\ =\ 4\rho\pi m\ \frac{a^3}{c^3}\ v_S \tag{2.40}$$

which could also be written as

$$F_2 = -\rho V_s\ (1 + C) v_s \frac{dv_f}{dx} \tag{2.41}$$

If we denote by F_1 the approximate force on the sphere in (2.32), the approximate net force is

$$F_1 + F_2 = \rho V_s (1 + C)(v_f - v_s) \frac{dv_f}{dx} = \rho V_s (1 + C)\, w \frac{dw}{dx} \tag{2.42}$$

where w is the combined velocity, relative to the sphere, from the two external fields.

This result could clearly be generalized by distributing any number of sources and sinks along an axis passing through the sphere.

It would have been possible to express the force on the sphere using (2.37), in the form

$$F = \frac{1}{2} \rho \int (\boldsymbol{u}_m + \boldsymbol{u}_d - \boldsymbol{v}_s)^2\, ds \tag{2.43}$$

where $\boldsymbol{v}_s$ is a vector of magnitude v_s pointing in the positive x-direction, $\boldsymbol{u}_d$ is the velocity field generated by the dipole and $\boldsymbol{u}_m$ is the velocity field arising from the external source and its images. Expanding (2.43) it can be written as

$$F = F_1 + F_2 = \frac{1}{2} \rho \int (\boldsymbol{u}_m^2 + \boldsymbol{u}_d^2 + \boldsymbol{v}_s^2 + 2\boldsymbol{u}_m \cdot \boldsymbol{u}_d - 2\boldsymbol{u}_m \cdot \boldsymbol{v}_s - 2\boldsymbol{u}_d \cdot \boldsymbol{v}_s) ds \tag{2.44}$$

which will be useful in the next section.

2.7 Force on a Moving Sphere in a Source Field

Remove the uniform flow $-\boldsymbol{v}_s$ from the previous example. The situation now represents a moving sphere with velocity $\boldsymbol{v}_s$ which can be a function of time.

As the sphere moves, the external source staying fixed, "c" will change such that

$$dc/dt = v_s \tag{2.45}$$

and the image system will adjust accordingly.

Let φ_m be the total potential due to the source and its image system while φ_d is the potential from the dipole. If d/dt represents differentiation in a coordinate system following the sphere (and the dipole),

$$\frac{d\varphi}{dt} = \frac{\partial\varphi}{\partial t} + v_s \cdot \nabla\varphi \tag{2.46}$$

rearranging and using $\varphi = \varphi_m + \varphi_d$ we have

$$\frac{\partial\varphi}{\partial t} = \frac{\partial(\varphi_m + \varphi_d)}{\partial t} = \frac{d\varphi_m}{dt} + \frac{d\varphi_d}{dt} - v_s \cdot \nabla(\varphi_m + \varphi_d) = \frac{d\varphi_m}{dt}$$

$$+ \frac{d\varphi_d}{dt} + v_s \cdot u_m + v_s \cdot u_d \tag{2.47}$$

The pressure at a point on the sphere's surface follows from Bernoulli's equation as

$$p = p_0 - \frac{1}{2}\rho u^2 + \rho\frac{\partial\varphi}{\partial t} \tag{2.48}$$

Since the flow is at rest at infinity where $p = p_0$.

Now,

$$u^2 = (u_m + u_d)^2 = u_m^2 + u_d^2 + 2u_m \cdot u_d \tag{2.49}$$

Substituting (2.47) and (2.49) into (2.48) we find the force on the sphere to be

$$F = -\int p ds = \frac{1}{2}\rho\int (u_m^2 + u_d^2 + 2u_m \cdot u_d - 2v_s \cdot u_m - 2v_s \cdot u_d) ds$$

$$- \rho\int\left(\frac{d\varphi_m}{dt} + \frac{d\varphi_d}{dt}\right) ds \tag{2.50}$$

The integral of the constant pressure, p_0, over the surface vanishes. Moreover, we also have for the constant v_s^2

$$\frac{1}{2}\rho \int v_s^2 \, ds = 0 \tag{2.51}$$

Adding (2.51) to (2.50) and comparing with (2.44) we see that the net force on the sphere is

$$F = F_1 + F_2 - \rho \int \left(\frac{d\varphi_m}{dt} + \frac{d\varphi_d}{dt}\right) ds \tag{2.52}$$

where F_1 and F_2 were found previously.

Since the sphere is a constant shape and $(d\varphi/dt)$ is evaluated following the sphere, the final two terms in (2.52) may be written as

$$F_3 = - \rho \frac{d}{dt} \int \varphi_m \, ds \tag{2.53}$$

$$F_4 = - \rho \frac{d}{dt} \int \varphi_d \, ds \tag{2.54}$$

which have an obvious connection with (A.12). In fact, (2.54) leads to a force which is exactly what would be computed for the acceleration of the sphere alone in fluid at rest

$$F_4 = - \rho C V_s \frac{dv_s}{dt} = - \rho C V_s v_s \frac{dv_s}{dx} \tag{2.55}$$

The quantity F_3 could probably be derived in some simple way. The present author used the expression given by Milne-Thomson (1949) using the nomenclature in Figure 2.1

$$\varphi_m = \frac{2m}{r} - \frac{m}{a} \ln \left(\frac{r + c + a}{r + c - a}\right) \tag{2.56}$$

After a lot of routine algebra and calculus, use of (2.56) in (2.53) gives

$$F_3 = \rho \frac{d}{dt}\left(2\,\frac{m\pi a^3}{c^2}\right) = -\,\rho\,\frac{4m\pi a^3}{c^3}\,\frac{dc}{dt} \tag{2.57}$$

$$= \rho\,\frac{4}{3}\,\pi a^3 \cdot \frac{3}{2}\left(-\,\frac{2m}{c^3}\right) v_S \tag{2.58}$$

$$= \rho V_S (1 + C)\left(\frac{dv_f}{dx}\right) v_S \tag{2.59}$$

which cancels exactly the term in (2.41). Using (2.42) and (2.55) the tot force from the fluid on the sphere - *only if the approximation implied in (2.32) is valid* - is

$$F = F_1 + F_4 = \rho V_S v_f \frac{dv_f}{dx} + C\rho V_S\left(v_f \frac{dv_f}{dx} - v_S \frac{dv_S}{dx}\right) \tag{2.60}$$

which is compatible with (2.28) since $F + F_e$ must add up to the mass of the sphere times its acceleration.

In fact, integration of (2.53) is unnecessary. Landweber and Miloh's (1980) generalization of Taylor's (1928b) and Lagally's (1922) theorem, giving the inertia terms solely in terms of the source distribution in the immersed body, shows (2.60) to be valid for any object that is constructed by means of internal and external sources and sinks located on a common axis.

Indeed, the term $2m\pi a^3/c^3$ appearing within the differential in (2.57) is exactly 4π times the negative of the dipole moment of the sources in the sphere, $(ma/c)x(a^2/2c)$ which is what Taylor's approach would give for the effective momentum of the surrounding fluid due to these sources. In fact, Taylor's theorem can be expressed in the form

$$\int \varphi ds = \sum 4\pi m_i x_i \tag{2.61}$$

where φ is the potential around an axisymmetrical body moving in the x-direction in stationary fluid and created by a set of sources m_i at locations x_i along the x-axis. The constraint $\sum m_i = 0$ applies.

The cancelling of F_2 and F_3 indicates that there should be no force exerted between a stationary sphere and a source in a uniform flow if the source is allowed to travel with the flow; this might be physically justifiable, since forces between the source and the sphere are *mutual* and the sphere will feel no force unless the source is constrained.

2.8 *General Motion*

In the general case, the motions of both the sphere and the fluid are described by vector velocities that are functions of position and time. For this situation, Drew and Lahey (1987) have deduced that the force on the sphere due to the relative motion is

$$F = \rho_f C \left[\left(\frac{\partial \mathbf{v}_f}{\partial t} + \mathbf{v}_f \cdot \nabla \mathbf{v}_f - \frac{\partial \mathbf{v}_s}{\partial t} - \mathbf{v}_s \cdot \nabla \mathbf{v}_s \right) + (\mathbf{v}_f - \mathbf{v}_s) \times (\nabla \times \mathbf{v}_f) \right] \tag{2.62}$$

The first term expresses the *relative acceleration* and is a generalization of previous results in this article, being compatible with all of them. The second terms describes a *lift force* due to relative motion in a vorticity field. The inclusion of both terms makes the total force *objective* , or invariant under a change of coordinate system, a principle emphasized in an earlier paper by Drew and Lahey (1979) that, because the lift force was omitted, gave rise to a rather peculiar form of expression that did not reduce to the *relative acceleration* and was not compatible with results describing simpler situations.

The coefficient of *virtual* or *added mass* in (2.62) is 1/2. A similar result was obtained by Auton (1983) and used by Thomas, et al. (1983) but with the coefficient for the lift force being 0.53. Experimental confirmation of the lift force is given by Beyerlein et al. (1985).

As discussed in Section 2.4, there must be some approximation implied in (2.62), which is probably valid as long as velocity changes are small on the scale of the diameter of the sphere.

2.9 **Other Properties of Single Spheres in a Fluid**

The properties developed in Sections 2.1 and 2.2 concerned the kinetic energy of the fluid associated with various motions of the sphere. It is also useful to develop further results involving momentum flux and particle stresses.

By straightforward use of (2.1) the following integrals are obtained for the fluid surrounding a sphere of volume V_2 moving with speed w in the x-direction:

$$\int u_x^2 \, dV_1 = \frac{1}{5} w^2 V_2 \tag{2.63}$$

$$\int u_y^2 \, dV_1 = \int u_z^2 dV_1 = \frac{3}{20} w^2 V_2 \tag{2.64}$$

The average x-direction stress in the sphere is obtained starting with the force balance for the spherical cap shown in Figure 2.2

$$- \sigma_x A_x = \int_{cap} p_i ds_x \tag{2.65}$$

σ_x is the average tensile stress across the flat face of area A_x and ds_x is the x-component of an element of the spherical interface. Integrating (2.65) from $x = -a$ to $x = +a$ we obtain the average x-direction stress:

$$< -\sigma_x > V_2 = - \int_{-a}^{a} \sigma_x A_x dx = \int_{-a}^{a} \int_{cap} p_i ds_x dx \tag{2.66}$$

Reversing the order of integration yields

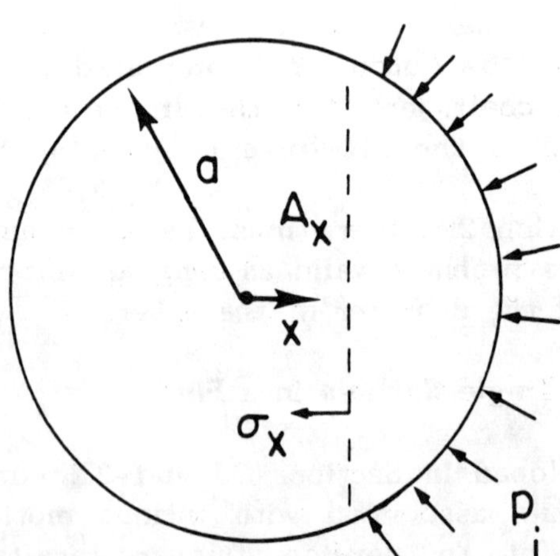

Figure 2.2 Internal x-direction stress in a particle

$$< -\sigma_x > = \frac{1}{V_2} \int_S x p_i \, ds_x \tag{2.67}$$

where the integration is over the whole surface of the sphere. Using (2.9) in (2.67) we get

$$< -\sigma_x > = p_\infty + \frac{1}{20} \rho w^2 \tag{2.68}$$

Using similar arguments, the other principal average stresses are found to be

$$< -\sigma_y > = < -\sigma_z > = p_\infty - \frac{2}{5} \rho w^2 \tag{2.69}$$

The bulk stress is the average of the three principal stresses, a result consistent with (2.11).

It might be suspected that the results in (2.63), (2.64), (2.68) and (2.69) are not independent, a consideration that will be investigated more generally in the next section.

2.10 **Properties of a General Particle in a Fluid**

For a general particle moving in the x-direction with speed w in fluid at rest at infinity, the following relationships may be derived as in Appendix A.

$$\int u_x dV_1 = C w V_2 \tag{2.70}$$

$$\int u^2 dV_1 = C w^2 V_2 \tag{2.71}$$

where C is the appropriate coefficient of added mass for rectilinear motion in the x-direction. We also have, from (2.71)

$$\int (u_x^2 + u_y^2 + u_z^2) dV_1 = C w^2 V_2 \tag{2.72}$$

The state of stress of the particles may be related to these results by considering motion at speed w relative to a stationary object oriented in the same way as above. For this situation we denote the local velocity components as w_x, w_y, w_z, which are related to the "fluctuations" in velocity above by

$$w_x = w - u_x, \qquad w_y = u_y, \qquad w_z = u_z \tag{2.73}$$

For the (large) control volume shown in Figure 2.3 we have, from continuity,

$$wA = \int w_x dA_1 + Q \tag{2.74}$$

where the integral is taken across the right-hand boundary of the control volume and Q is the volumetric flow rate out of the sides of the control volume. If the control volume is large, Q leaves with a component of velocity in the x-direction equal to w.

The momentum balance in the x-direction for the control volume is

$$A(p_\infty + \rho w^2) = \rho w Q + \int (p_1 + \rho w_x^2) dA_1 - \int \sigma_x dA_2 \tag{2.75}$$

Replacing A by $\int dA_1 + \int dA_2$ and using (2.74), (2.75) becomes

$$-\int \sigma_x dA_2 = p_\infty A_2 - \int \left[(p_1 - p_\infty) + \rho w_x^2 - \rho w_x w \right] dA_1 \tag{2.76}$$

Now, from Bernoulli's equation from infinity to a general point in the fluid,

$$p_1 - p_\infty = \frac{\rho}{2} \left[w^2 - (w_x^2 + w_y^2 + w_z^2) \right] \tag{2.77}$$

Integrating (2.76) over all x and using (2.77) we get

$$< -\sigma_x > = p_\infty - \frac{\rho}{V_2} \int \left[\frac{(w - w_x)^2}{2} - \frac{w_y^2 + w_z^2}{2} \right] dV_1 \tag{2.78}$$

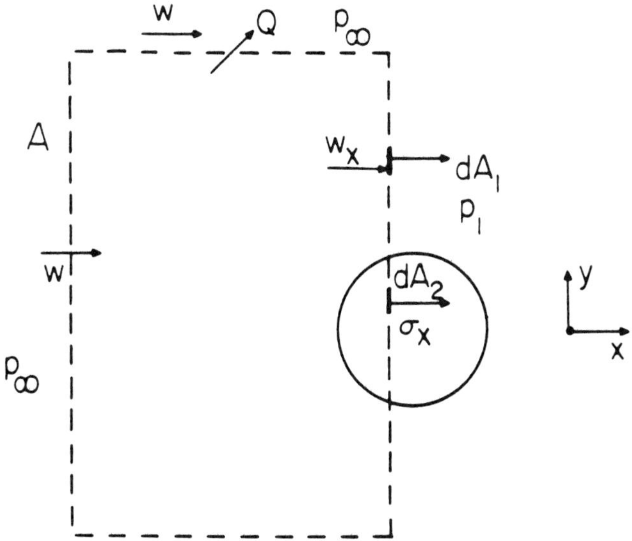

Figure 2.3 Control volume for x-direction momentum and stresses

The integrals in (2.78) involve the fluctuations in velocity that were considered in (2.70) to (2.72). Using those results we obtain

$$< -\sigma_x > = p_\infty - \rho C \frac{w^2}{2} + \frac{\rho}{V_2} \int (w_y^2 + w_z^2) dV_1 \tag{2.79}$$

or

$$< -\sigma_x > = p_\infty + \rho C \frac{w^2}{2} - \frac{\rho}{V_2} \int (w - w_x)^2 dV_1 \tag{2.80}$$

which show that the state of stress of the particles is related to the "fluctuations" in velocity in the surrounding fluid, a result that will later be derived for the more general case of a suspension.

To derive the stresses in the other directions, appeal is made to the (large) control volume in Figure 2.4. The momentum balance in the y-direction for the control volume shown is

$$p_\infty A = \rho \int (w_y^2 + p_1) dA_1 - \int \sigma_y dA_2 \tag{2.81}$$

As before, p_1 is given by (2.77) and (2.81) becomes, when integrated from $y = -\infty$ to $+\infty$,

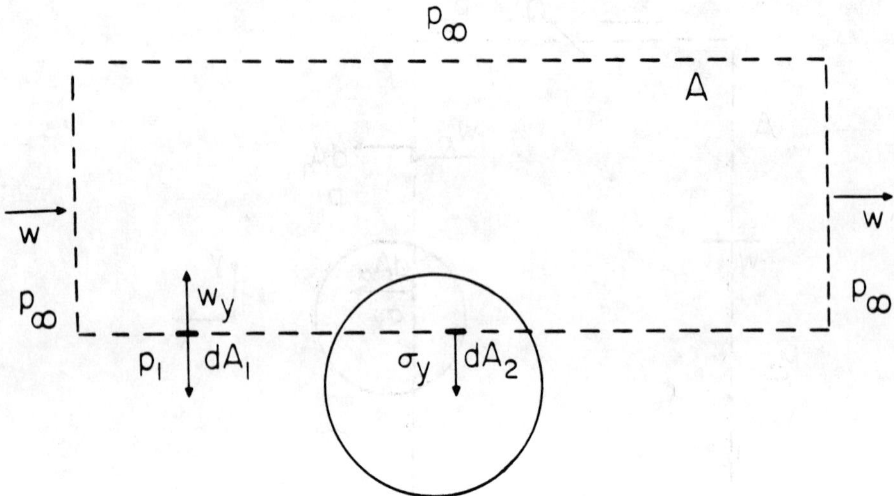

Figure 2.4 Control volume for y-direction momentum and stresses

$$\langle -\sigma_y \rangle = p_\infty - \frac{\rho}{V_2} \int \left(w_y^2 + \frac{w^2}{2} - \frac{w_x^2 + w_y^2 + w_z^2}{2} \right) dV_1 \tag{2.82}$$

now,

$$\int (w^2 - (w_x^2 + w_y^2 + w_z^2) dV_1 = \int \left[2w(w - w_x) - \left\{ (w - w_x)^2 + w_y^2 + w_z^2 \right\} \right] dV_1 = 2w^2C - Cw^2 = w^2C \tag{2.83}$$

where both (2.70) and (2.72) were used. When (2.83) is substituted into (2.82) the result is

$$\langle -\sigma_y \rangle = p_\infty - \frac{1}{2} \rho w^2 C - \frac{\rho}{V_2} \int w_y^2 dV_1 \tag{2.84}$$

Similarly, for the z-direction

$$\langle -\sigma_z \rangle = p_\infty - \frac{1}{2} \rho w^2 C - \frac{\rho}{V_2} \int w_z^2 dV_1 \tag{2.85}$$

Combining (2.79), (2.84) and (2.85) the bulk stress in the object is

$$< -\bar{\sigma} > = p_{\infty} - \frac{1}{2} \rho w^2 C \tag{2.86}$$

in agreement with (2.11). Moreover, the differences between this bulk stress, the particle "pressure," and the stresses in the three directions are given by the "Reynolds Stresses" in the surrounding fluid, a result that will be derived in a more general way in Section 3.

3. A SIMPLE MODEL OF DISPERSED FLOW

3.1 The "Base" One-Dimensional Case: Dispersed Phase at Rest

Consider an idealized potential flow past a set of dispersed objects (phase 2) held stationary in a potential fluid flow (phase 1) in a straight duct. In the simplest case the dispersed phase could consist of a row of symmetrical objects equally spaced along the axis of the pipe (Figure 3.1a). Clearly the potential will be constant across planes of symmetry that bound a "unit cell" characteristic of the flow, as shown. In a more complicated case there are many dispersed objects and the equipotentials may be interrupted by regions of phase 2 (Figure 3.1b).

In the general case the equipotentials will not be strictly planar, but deviations from a plane will be on the scale of the distances between the centers of individual particles and therefore small compared with the *averaging volume* used to obtain a quasi-continuum description of the flow. Therefore, errors due to disregard of the "wiggles" in the equipotentials can be made indefinitely small. (Those with interest in greater rigor may wish to pursue the degrees of approximation implied by this statement.) While true of the potential, this is not true of parameters such as pressure and velocity which may vary more over distances on the scale of the particles than they do, on the average, over the whole flow field; therefore, careful attention must be given to rigorous methods for averaging such variables. We denote the *macroscopic* potential appropriate for such a continuum description by Φ, while the *microscopic* potential that varies on the scale of the particles will be denoted by φ. Similarly, the microscopic pressure and velocity (i.e. the variables that would appear in the Navier-Stokes Equations) are denoted by p and $\boldsymbol{u}$, while the average values in each phase are denoted by p_1, p_2, v_1, v_2. A macroscopic pressure, P, is also used at times as a measure of the overall bulk stress in the mixture.

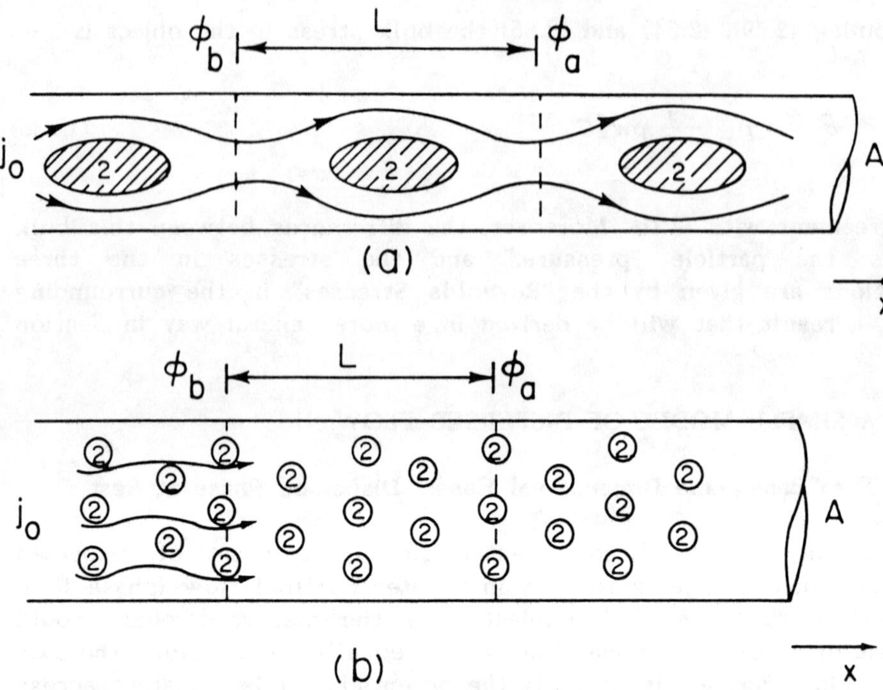

Figure 3.1 Flow past a stationary array.

The length of the unit cell is L, the duct cross-section A, the volume occupied by each phase V_1 and V_2 respectively, and the flow rate of phase 1 is

$$Q_0 = j_0 A \tag{3.1}$$

where j_0 has the usual meaning of *volumetric flux*.

The volumetric fractions of the duct occupied by the phases are

$$\alpha_1 = V_1/AL, \qquad \alpha_2 = V_2/AL \tag{3.2}$$

and, since the volumes of the phases add up to the volume of the duct,

$$\alpha_1 + \alpha_2 = 1 \tag{3.3}$$

The average velocity of phase 1 over the entire volume is

$$w = \frac{j_0}{\alpha_1} = \int_{V_1} w_X \, dV/V_1 \tag{3.4}$$

which is also the relative velocity between the phases. (The symbol w_X denotes the local or microscopic x-component of velocity in the *base case* in which the particles are at rest).

If the dispersed phase were not present, the flux of phase 1 for the same potential difference across the cell would be $(\Phi_b - \Phi_a)/L$. Let the actual flux be $(1/\beta)$ times this, then

$$j_0 = (\Phi_b - \Phi_a)/(L\beta) = -(d\Phi_0/dx)/\beta \tag{3.5}$$

where Φ_0 is the *macroscopic potential.*

The kinetic energy of the fluid in the cell follows from (A.4) as

$$K_0 = \frac{1}{2} \rho_1 j_0 A (\Phi_b - \Phi_a) \tag{3.6}$$

therefore, the kinetic energy per unit volume of phase 1 is, from (3.4), (3.5) and (3.6),

$$k_1 = \frac{1}{2} \rho_1 j_0^2 \beta/\alpha_1 = \frac{1}{2} \rho_1 w^2 (\alpha_1 \beta) \tag{3.7}$$

whereas the kinetic energy per unit of total volume is $\alpha_1 k_1$ and is $(1/\beta)$ times the kinetic energy at the same $(d\Phi/dx)$ without the phase 2 or β times the kinetic energy at the same j_0, in the absence of phase 2.

The average velocity of phase 1, or the relative velocity, is related to the potential gradient by using (3.4) and (3.5):

$$-d\Phi_0/dx = (\alpha_1 \beta) w \tag{3.8}$$

which could be expressed as

$$-d\Phi_0/dx = w + (\alpha_1\beta - 1)w \tag{3.9}$$

(3.8) and (3.9) suggest that the effect of the presence of phase 2 is to increase the potential gradient needed to produce a given average velocity, since $(\alpha_1\beta)$ is almost certainly greater than 1 in all cases (a general proof would be nice).

Now, the volumetric average pressure in phase 1 is

$$\langle p_1\rangle = \int p dV_1/V_1 = p_{01} - \int \frac{1}{2}\rho u_1^2 dV_1/V_1 \tag{3.10}$$

The final term appearing in (3.10) is the same as k_1 in (3.7), therefore

$$\langle p_1\rangle = p_{01} - \frac{1}{2}\rho_1 w^2(\alpha_1\beta) \tag{3.11}$$

which is the averaged *Bernoulli equation* for phase 1.

To use the method described in A.3.1 to get the average interfacial pressure we need to allow phase 2 to expand, keeping Φ_b and Φ_a constant. Therefore, using (3.5) and (3.7) in (A.30) written per unit total volume,

$$\langle p_i\rangle = p_{01} + \frac{d}{d\alpha_2}\left[\frac{1}{2}\rho_1 \frac{(d\Phi_0/dx)^2}{\beta}\right] \tag{3.12}$$

or

$$\langle p_i\rangle = p_{01} + \frac{1}{2}\rho_1\left(\frac{d\Phi_0}{dx}\right)^2\left(-\frac{1}{\beta^2}\right)\frac{d\beta}{d\alpha_1}\frac{d\alpha_1}{d\alpha_2} \tag{3.13}$$

Now, $d\alpha_1/d\alpha_2 = -1$ in view of (3.3) and we may also use (3.5) in (3.13) to obtain eventually

$$\langle p_i\rangle = p_{01} + \frac{1}{2}\rho_1 j_0^2 \frac{d\beta}{d\alpha_1} \tag{3.14}$$

or

$$\langle p_i\rangle = p_{01} + \frac{1}{2}\rho_1 w^2\left(\alpha_1^2 \frac{d\beta}{d\alpha_1}\right) \tag{3.15}$$

If $<p_1>$ is interpreted as the appropriate average pressure of phase 2, the difference in average pressure between the phases is

$$<p_2> - <p_1> = \frac{1}{2}\rho_1 w^2 \alpha_1^2 \left(\frac{d\beta}{d\alpha_1} + \frac{\beta}{\alpha_1}\right) = \frac{1}{2}\rho_1 w^2 \alpha_1 \left[\frac{d(\alpha_1\beta - 1)}{d\alpha_1}\right] \tag{3.16}$$

The term $(\alpha_1\beta - 1)$ will be shown to be a coefficient determining the kinetic energy due to relative motion and related to the *added mass* of the particles. Since it plays a key role in subsequent developments and represents the "external inertia" of the particles, it is proposed to name it the "exertia."

$<p_2>$ is the bulk stress within the particles but does not define their total stress field. It will be shown later (Section 3.9) that the average (principal) particle stress in the direction of relative motion differs from the stress in a perpendicular direction, as is easily demonstrated for a single sphere by using (2.9) and a force balance. In the case of bubbles, the nonuniform pressure around the interface is balanced by surface tension, implying that the surface distorts from spherical and force balances over surfaces cutting such interfaces should take this into account.

If it were possible to determine the values of the potential around the elements of dispersed phase, the average velocity of phase 1 could be determined using (A.16) (implying a sum over all the objects in the case of the pattern illustrated in Figure 3.1b) as

$$w = \frac{M_x}{\rho_1\alpha_1 AL} = \left(\frac{d\Phi_0}{dx}\right)\frac{1}{\alpha_1}\left[1 + \frac{\int\varphi ds_x}{(\Phi_b - \Phi_a)A}\right] \tag{3.17}$$

Comparing with (3.8) we see that the square bracket above is an expression for $(1/\beta)$. However, it would seem to be much simpler to determine β from (3.5). In fact, (3.5) has the same form as the equation describing conduction of electricity or heat through a matrix of insulating particles in a continuous medium and may be evaluated if such solutions are known, an approach also taken by Mercadier (1981). A classic example is the effective conductivity, c, of an array of spheres for which Maxwell (1881) obtained

$$c = c_1\left(\frac{1 - \alpha_2}{1 + \alpha_2/2}\right) \tag{3.18}$$

for the effective conductivity of a regular array when $c_2 = 0$, c_1 being the conductivity of the continuous phase. Jeffrey (1973) obtained a more accurate expansion to second order in α_2, valid for a random array. Turner (1976) showed experimentally that (3.18) is a close approximation to observed behavior of uniformly fluidized beds for the entire range of α_2 up to the packed bed value $(\alpha_2 \simeq 0.6)$. If we accept (3.18), the value of β in (3.5) is

$$\beta = (1 + \alpha_2/2)/(1 - \alpha_2) = (3 - \alpha_1)/2\alpha_1 \tag{3.19}$$

from which, using (3.3), the *exertia* is

$$(\alpha_1\beta - 1) = \alpha_2/2 \tag{3.20}$$

and

$$\frac{d}{d\alpha_1}(\alpha_1\beta - 1) = -\frac{1}{2} \tag{3.21}$$

These remarkably simple results can be used in (3.9), (3.11), (3.15) and so on. In the next section we will demonstrate a more direct relationship with the "added mass" concepts.

Since certain authors, for example Zuber (1964), have presented expressions for the added mass that are equivalent to a much larger value for β than is given in (3.19), particularly at high values of α_2, it is useful to show that the present result is reasonable. It will later be shown in Section 3.4.2 that the present results are indeed compatible with Zuber's when the various definitions of "added mass coefficient" are compared.

Figure 3.2 illustrates a very simplified view in which a typical stream tube through the particle lattice is assumed to have a length L_1 and uniform area of cross-section A_1. A straight stream tube of the same volume would have a cross section A_0 such that

$$L_1A_1 = LA_0 \tag{3.22}$$

Since the stream tube was assumed to have uniform area, the fluid speed in it is uniform and is equal to

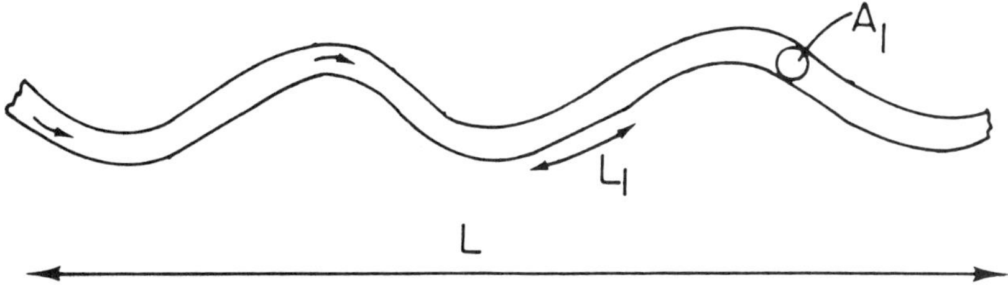

Figure 3.2 Idealized stream tube through the particle lattice.

$$u_1 = \frac{\Phi_b - \Phi_a}{L_1} \tag{3.23}$$

The flow rate through the stream tube is, using (3.22) and (3.23)

$$u_1 A_1 = \frac{\Phi_b - \Phi_a}{L_1} \frac{L}{L_1} A_0 = \frac{\Phi_b - \Phi_a}{L} \alpha_0 \left(\frac{L}{L_1}\right)^2 \tag{3.24}$$

which is $(L/L_1)^2$ times the flow that would occur in the straight tube. Therefore, the fluid flux is $\alpha_1 (L/L_1)^2$ times what it would be in the absence of particles and we have, from the definition of β,

$$\beta^{-1} = \alpha_1 \left(\frac{L}{L_1}\right)^2 \tag{3.25}$$

or

$$\alpha_1 \beta = \left(\frac{L_1}{L}\right)^2 \tag{3.26}$$

The same result can be derived from energy considerations.

Now, consider the close-packed sphere array shown in Figure 3.3. Many paths could be traced through the lattice. For the path $a - a$, (L_1/L) is $2/\sqrt{3}$ or 1.1547. For the path $b - b$, which is longer than many three-dimensional alternatives, (L_1/L) is $(1 + 2/\sqrt{3})/\sqrt{3}$ or 1.244. In the third dimension, if alternate rows are stacked as indicated, there is actually a direct path for which $L_1/L = 1$. Taking the average of these three values we have $L_1/L \approx 1.133$ giving $\alpha_1\beta = 1.28$. In an actual close-packed array, $\alpha_2 = 0.71$ and (3.20) gives $\alpha_1\beta = 1.355$, close to what would be obtained from the path $(a - a)$.

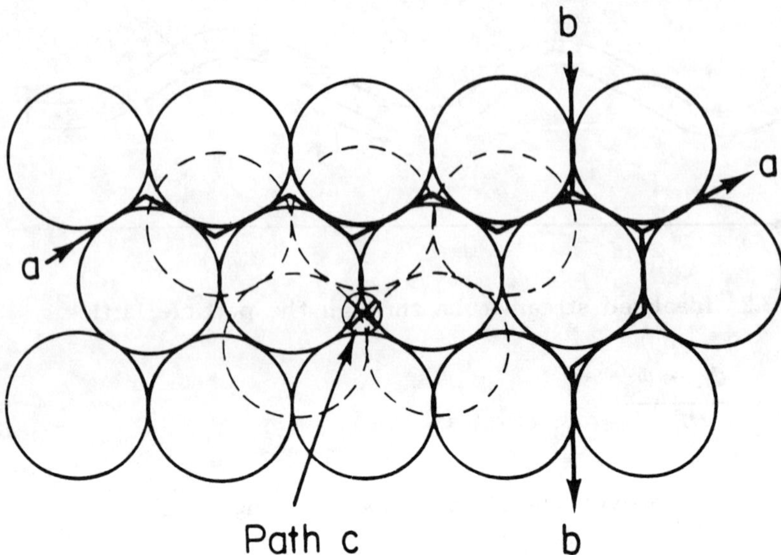

Figure 3.3 Various paths through a close-packed lattice (odd layers - solid circles; even layers - dashed circles).

3.2 One-Dimensional Motion of Both Phases

If we superimpose a potential

$$\varphi_2 = - v_2 x \tag{3.27}$$

on the motion described in the previous section, this is equivalent to adding a velocity v_2 to all the previous velocities. The average velocity of phase 2 is now v_2 and the average velocity of phase 1 is

$$v_1 = v_2 + w \tag{3.28}$$

The relative velocity remains w.

In this new motion the dispersed phase all moves *at the same velocity* v_2. Clearly additional features, not included in the present theory, will emerge if the particles have random motions about this average value and therefore have additional kinetic energy. However, if this random motion is itself scaled by the relative motion, the total "additional kinetic energy" remains scaled by w^2 and some of the present results will be unchanged, with the *exertia* merely being enhanced but having the same form.

The momentum in the unit cell, per unit volume of the entire flow, is

$$m_x = \frac{V_1}{V}\frac{\int u_{1x}\rho_1 dV_1}{V_1} + \frac{V_2}{V}\frac{\int u_{2x}\rho_x dV_2}{V_2} \tag{3.29}$$

$$= \alpha_1\rho_1 v_1 + \alpha_2\rho_2 v_2 \tag{3.30}$$

The kinetic energy of phase 1 is

$$k_1 V_1 = \frac{1}{2}\rho_1 \int_{V_1} (u_{1x}^2 + u_{1y}^2 + u_{1z}^2)\, dV_1 \tag{3.31}$$

The integral on the right-hand side can be expressed as

$$\int_{V_1} \left[(w_x + v_2)^2 + (w_y^2 + w_z^2)\right] dV_1 = V_1 v_2^2 + 2v_2 \int_{V_1} w_x dV_1$$
$$+ \int_{V_1} (w_x^2 + w_y^2 + w_z^2)\, dV_1 \tag{3.32}$$

Substituting from Section 3.1 and using (3.28), the expression on the right-hand side of (3.32) becomes, with no approximation,

$$V_1(v_2^2 + 2v_2 w + w^2\alpha_1\beta) = V_1\left[v_1^2 + w^2(\alpha_1\beta - 1)\right] \tag{3.33}$$

Using this result in (3.31) we obtain

$$k_1 = \frac{1}{2}\rho_1\left[v_1^2 + w^2(\alpha_1\beta - 1)\right] \tag{3.34}$$

The second term on the right-hand side of (3.34) represents additional kinetic energy possessed by phase 1 as a result of the relative motion.

The total kinetic energy per unit volume is

$$k = \alpha_1 k_1 + \alpha_2 k_2 = \frac{1}{2}\rho_1\alpha_1 v_1^2 + \frac{1}{2}\rho_2\alpha_2 v_2^2 + \frac{1}{2}\alpha_1(\alpha_1\beta - 1)\rho_1 w^2 \quad (3.35)$$

If (3.20) applies approximately to a suspension of spheres, the final term in (3.35) becomes remarkably simple, $\alpha_1\alpha_2\rho_1 w^2/4$, and is the same no matter which phase is dispersed, apart from the density, which corresponds to the continuous phase.

The result of imposing the potential φ_2 on the base case, i.e. adding (3.9) and (3.27), is to create an overall potential gradient across the unit cell

$$-\frac{d\Phi}{dx} = -\frac{d(\Phi_0 + \varphi_2)}{dx} = w + v_2 + (\alpha_1\beta - 1)w = v_1 + (\alpha_1\beta - 1)w$$

$$= v_2 + \alpha_1\beta w \quad (3.36)$$

A possible interpretation is that if Φ is proportional to an impulsive pressure necessary to set up the motion, as in the classical tradition (e.g. Lamb (1932), it must not only impart momentum to phase 1 but must also overcome an impulsive reaction, at the interface, from phase 2, proportional to the *exertia*. (An interesting question that arises here is the degree of approximation involved in imposing a more general non-uniform potential. Probably the results derived here, and in vector form later, are good representations as long as the average potential does not change significantly over the dimensions of one particle, a condition necessary for the continuum approximation to be valid.)

3.3 A Special Case: Added Mass

At first sight it may seem odd that there is *no correction* to the total momentum in (3.30) to account for relative motion, whereas in (A.14) and (A.15) it is found that the motion of an immersed object produces both momentum and energy in the surrounding (infinite) fluid. The explanation is that, in general, we do not have $\Phi_b = \Phi_a$ across the cell and are required to use the more complicated (A.16) instead of (A.13).

Some insight may be obtained by considering the special case in which there is no potential gradient across the unit cell. This requires, from (3.36), that

$$v_1 = -(\alpha_1\beta - 1)w \tag{3.37}$$

$$v_2 = -\alpha_1\beta w \tag{3.38}$$

Substituting these values into (3.30) and (3.35) we obtain

$$m_x = \alpha_2\rho_2 v_2 \left(1 + \frac{\rho_1}{\rho_2}\,\frac{\alpha_1\beta - 1}{\alpha_2\beta}\right) \tag{3.39}$$

$$k = \frac{1}{2}\,\alpha_2\rho_2 v_2^2 \left(1 + \frac{\rho_1}{\rho_2}\,\frac{\alpha_1\beta - 1}{\alpha_2\beta}\right) \tag{3.40}$$

The quantities in parentheses in (3.39) and (3.40) are the same. The second term in each equation could be interpreted as the effect of an *added mass*, with the same interpretation as in (A.14) and (A.15), with the coefficient of added mass given by

$$C = \frac{\alpha_1\beta - 1}{\alpha_2\beta} \tag{3.41}$$

C represents the volume of fluid effectively entrained by the moving objects, per unit volume of those objects (phase 2) - but only in the special case where there is no average potential gradient. This is easily confirmed by forming the ratio of the fluxes of the phases, $\alpha_1 v_1/\alpha_2 v_2$, and using (3.37) and (3.38).

Using (3.20) in (3.41), we find that Maxwell's theory predicts

$$C = \frac{1}{2\beta} = \frac{\alpha_1}{2 + \alpha_2} = \frac{1}{(3/\alpha_1) - 1} \tag{3.42}$$

Clearly, C decreases as α_1 decreases or the particle concentration increases.

If we rewrite (3.41) as

$$C = \frac{\alpha_1 - (1/\beta)}{\alpha_2} \tag{3.43}$$

we can identify α_2 as the volume of phase 2. $(\alpha_1 - 1/\beta)$ is the "volume of phase 1 that is entrained by phase 2," and $(1/\beta)$ the "volume of phase 1 that is not entrained," the sum being α_1, the total volume of phase 1. This gives another physical interpretation for β and forms a basis for the idealized model that will be presented in Section 4.

We may proceed further by asking how it is possible to set up this motion in the absence of a macroscopic potential gradient. In the classical view, if there is no potential difference across the unit cell, no impulsive pressure difference is applied. However, the dispersed phase has been set in motion and has entrained some fluid. One explanation could be that there was an impulsive conservative *body force* applied to the dispersed phase such that the impulsive force per unit mass of phase 2 was given by the negative of the gradient of some macroscopic potential, Ψ. Conservation of momentum then requires, by use of (3.37) and (3.38), that

$$-\rho_2 \alpha_2 \frac{d\Psi}{dx} = \alpha_2 \rho_2 (-\alpha_1 \beta w) - \alpha_1 \rho_1 (\alpha_1 \beta - 1) w \tag{3.44}$$

The motion is set up in a form such as is shown in Figure 3.4. Instantaneous equipotentials and streamlines are indicated. The net momentum of phase 1 is related to the potential differences appearing across the dispersed phase by (A.13); these may be interpreted as an impulsive reaction from phase 1 on the enclosed body (phase 2) as a result of its impulsive motion. Imposition of a *uniform* potential gradient $-d\Phi/dx$ on the situation shown in Figure 3.4 is equivalent to imposing a velocity $-d\Phi/dx$ on everything. Therefore, the mean velocities of the phases become, from (3.37) and (3.38),

$$v_1 = -\frac{d\Phi}{dx} - (\alpha_1 \beta - 1) w \tag{3.45}$$

$$v_2 = -\frac{d\Phi}{dx} - \alpha_1 \beta w \tag{3.46}$$

in agreement with (3.36). If we substitute from (3.46) in (3.44) and do some rearranging we get

$$-\frac{d(\Psi + \Phi)}{dx} = v_2 - \frac{\alpha_1 \rho_1}{\alpha_2 \rho_2} (\alpha_1 \beta - 1) w \tag{3.47}$$

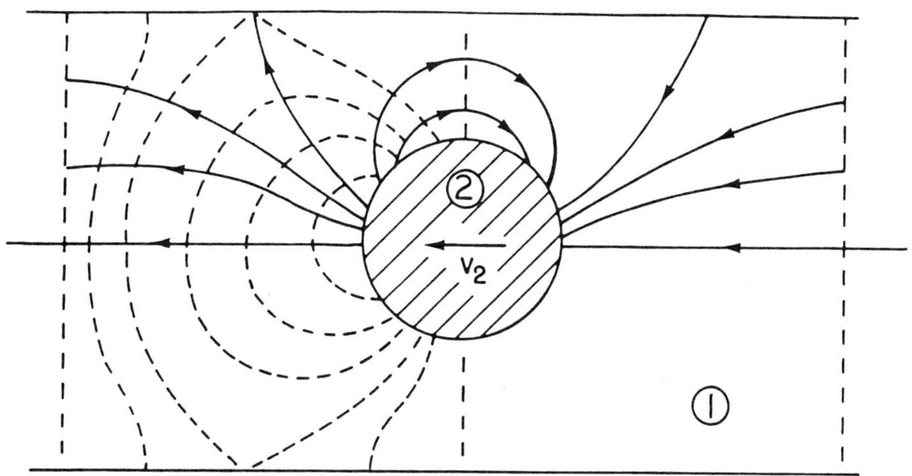

Figure 3.4 Flow around an object moving in a *unit cell* with no overall potential difference across it.

which might be interpreted as describing the result of the simultaneous imposition of gradients of Ψ and Φ on phase 2 being to produce the velocity v_2 as well as overcoming the interphase interaction resulting from the relative motion. Since Ψ and Φ are both potentials, their sum could be treated as an *effective potential* and given by the symbol "η," in which case (3.47) becomes

$$-\frac{d\eta}{dx} = v_2 - \frac{\alpha_1 \rho_1}{\alpha_2 \rho_2}(\alpha_1 \beta - 1) w \tag{3.48}$$

Since η is now the average potential imposed on phase 2 while Φ is the average potential imposed on phase 1, it is perhaps not surprising to find that the overall momentum per unit volume is given from (3.45) and (3.48) as

$$m_x = -\alpha_1 \rho_1 \frac{d\Phi}{dx} - \alpha_2 \rho_2 \frac{d\eta}{dx} = \alpha_1 \rho_1 v_1 + \alpha_2 \rho_2 v_2 \tag{3.49}$$

The relative velocity, however, is caused by the difference in these potential gradients, namely $-d\Psi/dx$, as in (3.44), and this sets up non-uniform motion at the *microscopic level*, as shown in Figure 3.4.

3.4 Derivation of "Equations of Motion"

As long as (3,.35) and (3.39) are valid, it is possible to derive separate equations of motion for each phase in the special case in which the phases are incompressible and there are no macroscopic variations in the flow pattern (i.e. α_1 and α_2 uniform along the duct).

Let a pressure difference be applied across a long duct containing a series of unit cells of the type illustrated in Figure 3.1. This is equivalent to imposing a macroscopic pressure gradient, $-\partial P/\partial x$, the fluctuations in pressure at both ends of the duct being the same and therefore cancelling each other. Let there also be body forces acting on each phase in the x-direction such that the forces on each phase per unit volume are $\rho_1 g_1$ and $\rho_2 g_2$ respectively. Considering momentum and energy conservation for the contents of each unit cell we obtain, using dots to indicate differentiation with respect to time,

$$\frac{dm_x}{dt} = \alpha_1 \rho_1 \dot{v}_1 + \alpha_2 \rho_2 \dot{v}_2 = -\frac{\partial P}{\partial x} + g_1 \rho_1 \alpha_1 + g_2 \rho_2 \alpha_2 \tag{3.50}$$

$$\frac{dk}{dt} = \alpha_1 \rho_1 v_1 \dot{v}_1 + \alpha_2 \rho_2 v_2 \dot{v}_2 + \alpha_1 (\alpha_1 \beta - 1) \rho_1 (v_1 - v_2)(\dot{v}_1 - \dot{v}_2)$$

$$= -\frac{\partial P}{\partial x}(\alpha_1 v_1 + \alpha_2 v_2) + \alpha_1 \rho_1 g_1 v_1 + \alpha_2 \rho_2 g_2 v_2 \tag{3.51}$$

The first term on the right-hand side of (3.51) arises because $(\alpha_1 v_1 + \alpha_2 v_2)$ is the total volumetric flux at both ends of the cell, no matter how distributed across the duct, and both pressure and velocity fluctuations are equal at corresponding points at opposite ends of the cell and therefore cancel in the overall balance, as does their product.

Taking v_2 times (3.50) from (3.51) and dividing the result by $\alpha_1(v_1 - v_2)$ we get

$$\rho_1 \dot{v}_1 + \rho_1 (\alpha_1 \beta - 1)(\dot{v}_1 - \dot{v}_2) = -\frac{\partial P}{\partial x} + \rho_1 g_1 \tag{3.52}$$

while taking v_1 times (3.50) from (3.51) and dividing by $\alpha_2(v_1 - v_2)$ yields

$$\rho_2 \dot{v}_2 - \rho_1 (\alpha_1/\alpha_2)(\alpha_1 \beta - 1)(\dot{v}_1 - \dot{v}_2) = -\frac{\partial P}{\partial x} + \rho_2 g_2 \tag{3.53}$$

(3.52) and (3.53) could be regarded as the equations of motion of each phase, including the effects of an *added mass force* proportional to the *relative acceleration* and the *exertia* and acting in equal and opposite directions on each phase. They reduce to a particularly simple form if (3.20) is valid.

If (3.52) and (3.53) are integrated over time for motions starting from rest, the results are, after division by the densities

$$v_1 + (\alpha_1 \beta - 1)(v_1 - v_2) = \int \left(-\frac{1}{\rho_1} \frac{\partial P}{\partial x} + g_1 \right) dt \tag{3.54}$$

$$v_2 - (\rho_1 \alpha_1 / \rho_2 \alpha_2)(\alpha_1 \beta - 1)(v_1 - v_2) = \int \left(-\frac{1}{\rho_2} \frac{\partial P}{\partial x} + g_2 \right) dt \tag{3.55}$$

Comparison with (3.45) and (3.48) gives the results (apart from constants of integration that can be set to zero):

$$\Phi = \int \left(\frac{P}{\rho_1} - g_1 x \right) dt \tag{3.56}$$

$$\eta = \int \left(\frac{P}{\rho_2} - g_2 x \right) dt \tag{3.57}$$

Thus, the *macroscopic potentials,* Φ and η, could be regarded as the result of impulses due to pressure and body forces which could be controlled independently.

3.4.1 Equations of Motion for the "Added Mass Case"

Suppose that we contrive an experiment in which there is to be no macroscopic potential gradient at any time and the motion starts from rest. Then (3.37) and (3.38) apply and may be substituted into (3.52) to obtain

$$0 = -\frac{\partial P}{\partial x} + \rho_1 g_1 \tag{3.58}$$

which indicates that the pressure variation is *hydrostatic* in phase 1. If there is no body force, there is no pressure gradient.

We may now use (3.58) in (3.53), and add an *external* force per unit volume of phase 2 for completeness, to get

$$(-\dot{w})\left[\rho_2\alpha_1\beta + \rho_1\frac{\alpha_1}{\alpha_2}(\alpha_1\beta - 1)\right] = \rho_2 g_2 - \rho_1 g_1 + f_2 \tag{3.59}$$

which shows how the relative motion could be created by the action of body forces or an external force. This result is also compatible with (3.38) and (3.39).

In the special case where there is no body force in the direction of motion (e.g. horizontal flow) an external force applied only to phase 2 will create the situation described in Section 3.3, in the absence of any externally applied pressure gradient. The rate at which the relative motion builds up will be given by (3.59)

3.4.2 *Zuber's Coefficient of Added Mass*

Zuber (1964) derived an expression for the coefficient of added mass by considering the motion of a sphere inside a spherical enclosure. His result was

$$C_Z = \left(\frac{1}{2} + \alpha_2\right)/\alpha_1 \tag{3.60}$$

This value appears to be very different from that derived in (3.42). However, we shall now show that the expressions are consistent.

While (3.42) was derived for the case in which there is no overall potential gradient, Zuber's analysis is based on a constraint of zero net flux, or in mathematical terms,

$$\alpha_1 v_1 + \alpha_2 v_2 = 0 \tag{3.61}$$

If we now consider an experiment in which an array of particles acted upon by a force per unit volume of the particles f_2 is accelerated with the above constraint, we may use (3.52) and (3.53) - including f_2 - to derive f_2. Neglecting gravity and eliminating $\partial P/\partial x$ we obtain

$$f_2 = \rho_2 \dot{v}_2 - \rho_1 \dot{v}_1 - \rho_1 (\dot{v}_1 - \dot{v}_2) \left(\frac{\alpha_1 \beta - 1}{\alpha_2} \right) \tag{3.62}$$

The first term represents the particle inertia. The other terms involving ρ_1 are components of the net additional force required as a result of inertial coupling. Using (3.61) and (3.62) we get

$$f_2 = \rho_2 \dot{v}_2 + \rho_1 \dot{v}_2 \left(\frac{\alpha_2}{\alpha_1} + \frac{\alpha_1 \beta - 1}{\alpha_2 \alpha_1} \right) \tag{3.63}$$

The factor in parentheses is the *added mass coefficient* from Zuber's point of view. For a Maxwellian suspension obeying (3.20) the value is $(\alpha_2/\alpha_1) + (1/2\alpha_1)$, which is exactly the same as (3.60). This agreement between the two approaches was pointed out by Mercadier (1981).

This example illustrates the need for clear definition of the various *coefficients* that represent inertial coupling for specific purposes.

3.5 Momentum Flux

Let the entire unit cell illustrated in Figure 3.1, but with a velocity v_2 superimposed upon it, move steadily past an imaginary plane normal to the x-direction. The average momentum flux of phase 2 in the x-direction will be $\alpha_2 \rho_2 v_2^2$; however, the average momentum flux of phase 1 is $\rho_1 \int (u_{x1})^2 dV_1/V$ which must always be less than the average kinetic energy of phase 1 in the volume, because $u^2 \geq u_{x1}^2$.

If a complete solution to the *base case* can be developed, we can evaluate the average x-direction flux of momentum and define a parameter "γ" such that

$$\int w_x^2 \, dV_1 = \gamma w^2 V_1 \tag{3.64}$$

Since (3.7) can alternatively be expressed as

$$\int (w_x^2 + w_y^2 + w_z^2) \, dV_1 = \alpha_1 \beta w^2 V_1 \tag{3.65}$$

it is clear that

$$\gamma < \alpha_1 \beta \tag{3.66}$$

When a velocity, v_2, is imposed on the base case, the momentum flux of phase 1 becomes

$$\rho_1 \int (v_2 + w_x)^2 dV_1/V = \rho_1 \int (v_2^2 + 2w_x v_2 + w_x^2) dV_1/V$$

$$= \rho_1 (v_2^2 + 2v_2 w + \gamma w^2) V_1/V = \rho_1 V_1 (v_1^2 + (\gamma - 1)w^2)/V \tag{3.67}$$

Therefore, the average momentum flux per unit area of the entire flow is

$$\Pi_x = \alpha_1 \rho_1 v_1^2 + \alpha_2 \rho_2 v_2^2 + \alpha_1 (\gamma - 1) \rho_1 w^2 \tag{3.68}$$

which is of similar form to (3.35) but lower in magnitude. The relationship between γ and $\alpha_1 \beta$ depends on the details of the flow pattern (interface geometry).

For a single sphere moving through an infinite fluid at rest we may use the potential in (2.1) to show that, for the entire fluid surrounding the sphere,

$$\int u_x^2 dV = \frac{4}{15} \pi a^3 w^2 \tag{3.69}$$

The additional fluid momentum flux for a dilute suspension of moving spheres is therefore the value given in (3.69) for each spherical volume $(4/3)\pi a^3$. The average fluid momentum flux per unit volume of the entire flow is then $\alpha_2 \rho_1 w^2/5$ and comparison with (3.68) shows that

$$\alpha_1 (\gamma - 1) = \frac{\alpha_2}{5} \tag{3.70}$$

3.6 Particle Stress

Consider an individual particle in the array sketched in Figure 3.1b. The interfacial pressure varies around the surface of the particle, as a result of the relative motion. The equation of motion for the right-hand portion of the particle shown in Figure 3.5 will be

$$- < \sigma_x > A_x - \int_x^{x_2} p_i ds_x = \rho_2 a_x \int_x^{x_2} A_x dx \tag{3.71}$$

it is being assumed that all parts of the particle have the same acceleration, a_x in the x-direction.

The average (tensile) stress, $< \sigma_x >$, throughout the particle may be computed either by averaging over the volume of the particle or by considering that a plane perpendicular to the x-direction cuts a large number of particles with every location in the particle being equally likely. The result is

$$- < \sigma_x > \int_{x_1}^{x_2} A_x dx = \int_{x_1}^{x_2} \left(\int_x^{x_2} p_i ds_x \right) dx$$

$$+ \rho_2 a_x \int_{x_1}^{x_2} \left(\int_x^{x_2} A_x dx \right) dx \tag{3.72}$$

Whence, by conventional mathematics or geometrical arguments

$$- < \sigma_x > V = \int_S x p_i ds_x + \rho_2 a_x \int_V x dV \tag{3.73}$$

If we choose the origin of coordinates at the center of gravity of the particle, $\int x dV = 0$ and we get

$$- < \sigma_x > = \frac{1}{V} \int_S x p_i ds_x \tag{3.74}$$

a result similar to Batchelor's (1970). For a single sphere in an infinite fluid we may use (2.9) in (3.74) to obtain, as in (2.68)

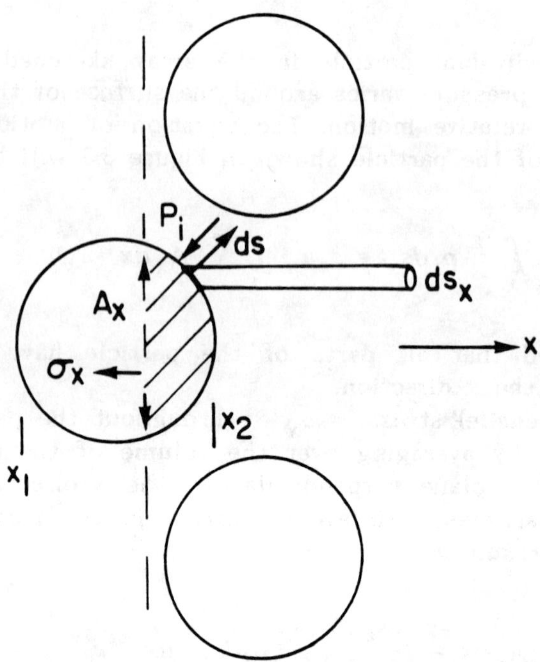

Figure 3.5 Stresses on a particle in the x-direction

$$- < \sigma_X > = p_1 + \frac{1}{20} \rho_1 w^2 \tag{3.75}$$

The combined momentum flux and pressure (stress) force over unit total area normal to the x-direction is now, using (3.68) and (3.70) for a dilute suspension,

$$P_X = \Pi_X + \alpha_1 p_1 + \alpha_2 (- < \sigma_X >)$$

$$= \alpha_1 \rho_1 v_1^2 + \alpha_2 \rho_2 v_2^2 + \frac{\alpha_2}{5} \rho_1 w^2 + \alpha_1 p_1 + \alpha_2 \left(p_1 + \frac{1}{20} \rho_1 w^2 \right)$$

$$= \alpha_1 \rho_1 v_1^2 + \alpha_2 \rho_2 v_2^2 + p_1 + \frac{1}{4} \alpha_2 \rho_1 w^2 \tag{3.76}$$

If we use (3.21) in (3.16), assuming that the average interfacial pressure defines the bulk stress in phase 2,

$$p_2 = p_1 - \frac{1}{4} \rho_1 w^2 \tag{3.77}$$

Therefore, (3.76) is equivalent to

$$P_x = \alpha_1 \rho_1 v_1^2 + \alpha_2 \rho_2 v_2^2 + \frac{1}{2} \alpha_2 \rho_1 w^2 + \alpha_1 p_1 + \alpha_2 p_2 \tag{3.78}$$

a result which could have been achieved, without rigor, by assuming (3.68) to be identical to (3.35) and adding up pressure forces in a purely one-dimensional way. Though such a derivation is erroneous, there is, as explained in Section 3.8, an overall reason why fluctuations in particle stress and fluid momentum flux should balance and lead to a compact expression such as (3.78). That such may be so is suggested by variational methods that derive momentum fluxes from energy principles without considering particle stresses at all (e.g. Geurst (1985a)).

3.7 Effects of Non-uniformity

So far we have only considered uniform suspensions of particles. Methods already developed will now be applied to a few simple examples of non-uniform suspensions. The results will later prove useful in developing the complete averaged stress and momentum flux tensors for the suspension.

3.7.1 *Lateral Variations in Concentration*

Figure 3.6 illustrates a situation in which the particles in Figure 3.1b, with initial volumetric concentration α_{20}, have been moved perpendicular to the flow to occupy a fraction f of the duct, with a concentration α_2. The fluid velocity in the region that is free of particles is v_a and the average velocity through stationary particles in the other region is v_b. The potentials at the ends of the duct remain unchanged.

It is assumed that the extent of the system is large enough compared with the scale of the particles for properties to be determined by conditions in the uniform regions, neglecting details of behavior near the interface (an assumption that bears further examination).

By symmetry, the equipotential surfaces are still normal to the duct axis. Therefore, from (3.8),

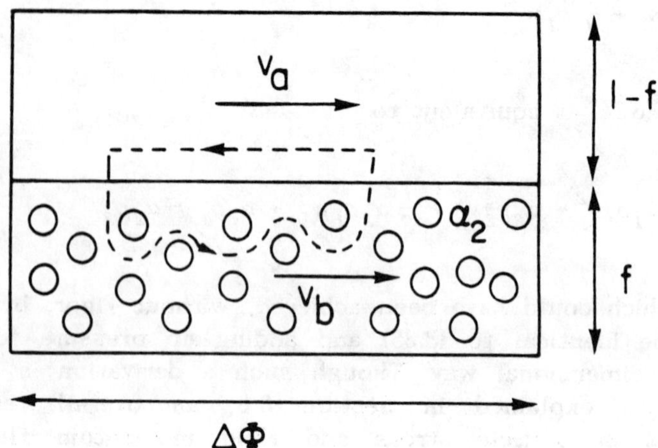

Figure 3.6 Two regions oriented parallel to the flow. There is no circulation around the dashed path.

$$\alpha_1 \beta v_b = v_a \tag{3.79}$$

where β applies to the region in which there are particles. Since $\alpha_1 \beta > 1$, $v_a > v_b$.

The flow is potential and irrotational at the microscopic level and no vorticity is contained in a circuit sketched as a dashed line in Figure 3.6. This is related to the tortuous path of a streamline, as in Figure 3.3. However, on the basis of the *average* velocities there is an *apparent vortex sheet* at the interface, since $v_a \neq v_b$. Had the concentration varied continuously, there would have appeared to be a *macroscopic vorticity,* since $\nabla \times \mathbf{v}_1$ would not have been zero, and terms containing this factor might be expected to appear in momentum and force balances.

Since the flow is reversible and comes from a common source, Bernoulli's Equation applies throughout the flow in the form

$$p_{01} = p + \frac{1}{2} \rho_1 u_1^2 \tag{3.80}$$

which may be averaged, as in (3.11), to obtain

$$p_{01} = p_1 + \frac{1}{2}\rho_1 v_1^2(\alpha_1\beta) \tag{3.81}$$

Using (3.81) in the two regions, and substituting from (3.79) we get

$$p_a = p_{01} - \frac{1}{2}\rho_1 v_a^2 \tag{3.82}$$

$$p_b = p_{01} - \frac{1}{2}\rho_1\alpha_1\beta v_b^2 = p_{01} - \frac{1}{2}\rho_1 v_a^2 \frac{1}{\alpha_1\beta} \tag{3.83}$$

Since $\alpha_1\beta > 1$, it follows that $p_b > p_a$. This might be expected to imply that there are forces on the particles at or near the interface, a hypothesis that can be investigated using the methods of Appendix A.

The kinetic energy per unit volume of the entire pipe is, from (3.7)

$$k = \frac{1}{2}\rho_1 (1 - f)v_a^2 + \frac{1}{2}\rho_1 f v_b^2 (\alpha_1\beta)\alpha_1 \tag{3.84}$$

Using (3.79) we may express this in the form

$$k = \frac{1}{2}\rho_1 v_a^2\left[1 + f\frac{1 - \beta}{\beta}\right] \tag{3.85}$$

Since v_a is given by the overall (fixed) potential gradient, it is a constant. The final term reflects the influence of f. From conservation of particles,

$$f\alpha_2 = \alpha_{20} \tag{3.86}$$

Since β depends on α_2, (3.81) and (3.82) allow the computation of k as a function of α_2. If (3.19) is used, we have

$$\frac{1 - \beta}{\beta} = -\frac{3\alpha_2}{2 + \alpha_2} \tag{3.87}$$

and (3.85) may be expressed as

$$k = \frac{1}{2} \rho_1 v_a^2 \left[1 - \frac{3\alpha_{20}}{2 + \alpha_2} \right] \tag{3.88}$$

As the particles are compressed by decreasing f, α_2 increases and so does k. Therefore, an external compressive stress must act on the surface layer. The value of this *particle pressure*, expressed as compressive force per unit total area of the interface, is

$$p_p = -\frac{dk}{df} = -\frac{dk}{d\alpha_2}\frac{d\alpha_2}{df} = -\frac{3}{2}\rho_1 v_a^2 \alpha_{20} \frac{1}{(2 + \alpha_2)^2}\left(-\frac{\alpha_{20}}{f^2}\right) \tag{3.89}$$

Using (3.86) this becomes

$$p_p = \frac{3}{2} \rho_1 v_a^2 \left(\frac{\alpha_2}{2 + \alpha_2} \right)^2 \tag{3.90}$$

This result is independent of α_{20}, as it clearly must be. Use of (3.20) and (3.79) enables the result to be expressed entirely in terms of the properties of the region containing particles,

$$p_p = \frac{3}{8} \rho_1 v_b^2 \alpha_2^2 \tag{3.91}$$

Physically, this *pressure* is a resultant of forces on particles at or near the interface and is not directly interpretable as the force on any particular one. The presence of concentration to the second power in (3.91) indicates that this is a simple interaction effect, resembling gravitational attraction or electrical space charge. This is perhaps not surprising, since objects in a perfect fluid can be represented by combinations of internal sources which effectively attract and repel each other.

Returning to (3.82) and (3.83) we may form the difference in fluid pressure between the regions as

$$p_b - p_a = \frac{1}{2} \rho_1 v_a^2 \left(\frac{\alpha_1 \beta - 1}{\alpha_1 \beta} \right) \tag{3.92}$$

Using (3.20) this may be written as

$$p_b - p_a = \frac{1}{2}\rho_1 v_a^2 \left(\frac{\alpha_2}{2 + \alpha_2}\right) \tag{3.93}$$

Now, suppose that α_2 varies continuously across the channel with v_a (or the potential gradient) remaining fixed. The velocity from (3.20) and (3.79) is, replacing v_b by v_1 which depends on y,

$$v_1 = v_a \frac{2}{(2 + \alpha_2)} \tag{3.94}$$

Using (3.94) in (3.91) we may evaluate the force in the y-direction per unit volume on the particles from the fluid as

$$-\frac{1}{\alpha_2}\frac{dp_p}{dy} = -\frac{6\rho_1 v_a^2}{(2 + \alpha_2)^3}\frac{d\alpha_2}{dy} = -\frac{3}{2(2 + \alpha_2)}\rho_1 v_1^2 \frac{d\alpha_2}{dy} \tag{3.95}$$

By further use of (3.93), (3.16) and (3.21) we obtain

$$-\frac{dp_2}{dy} = -\frac{6\rho_1 v_a^2}{(2 + \alpha_2)^3}\frac{d\alpha_2}{dy} \tag{3.96}$$

Since (3.95) is the same as (3.96), either expression is equal to the force per unit (particle) volume. In particular, (3.96) indicates that this force is *only* dependent on the gradient of the transverse average pressure in phase 2 and no further force component, due for instance to apparent macroscopic vorticity, needs to be included.

At the microscopic level, these forces are related both to the pressure distribution around the particles and their internal stresses. Indeed, a more complete theory should be able to display consistent descriptions of these phenomena from all interrelated points of view (as in the seldom realized, ideal, academic assembly or committee).

For a non-Maxwellian suspension, (3.85) is used without substitution of (3.87) to give

$$p_p = -\frac{dk}{df} = -\frac{1}{2}\rho_1 v_a^2 \left[\frac{1}{\beta} - 1 - \frac{f}{\beta^2}\frac{d\beta}{df}\right] \tag{3.97}$$

Using (3.86) we obtain

$$\frac{d\beta}{df} = \frac{d\beta}{d\alpha_2}\frac{d\alpha_2}{df} = -\frac{\alpha_{20}}{f^2}\frac{d\beta}{d\alpha_2} \tag{3.98}$$

Combining (3.98) with (3.97) and using (3.86) there results, finally,

$$p_p = -\frac{1}{2}\rho_1 v_a^2\left(\frac{1}{\beta} - 1 + \frac{\alpha_2}{\beta^2}\frac{d\beta}{d\alpha_2}\right) \tag{3.99}$$

which is generally valid for a suspension of spheres where β is some function of the void fraction.

Using (3.99), the force per unit volume of the particles from the fluid is, in place of (3.95),

$$-\frac{1}{\alpha_2}\frac{dp_p}{dy} = \frac{1}{2}\rho_1 v_a^2\frac{d}{dy}\left(\frac{1}{\beta^2}\frac{d\beta}{d\alpha_2}\right) \tag{3.100}$$

or, when (3.79) is used to express v_a in terms of the local variables,

$$-\frac{1}{\alpha_2}\frac{dp_p}{dy} = \frac{\alpha_1^2}{2}\rho_1 v_1^2\left(\frac{d^2\beta}{d\alpha_2^2} - \frac{2}{\beta}\left(\frac{d\beta}{d\alpha_2}\right)^2\right)\frac{d\alpha_2}{dy} \tag{3.101}$$

Equation (3.101) reduces to (3.95) for a Maxwellian suspension.

3.7.2 *Longitudinal Variations in Concentration*

A similar analysis may be performed for the situation illustrated in Figure 3.7, in which the particles concentrate in the direction of flow. As before, from conservation of particles

$$f\alpha_2 = \alpha_{20} \tag{3.102}$$

To use the methods in the Appendix, we keep the potential difference over the entire system fixed. The velocity in the clear region being v_a, it follows, using (3.6), that

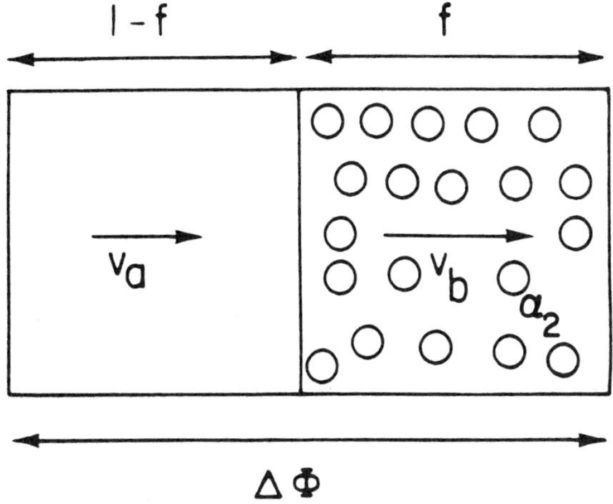

Figure 3.7 Two regions oriented normal to the flow.

$$\Delta\Phi = (1 - f)v_a + f\beta v_a \tag{3.103}$$

the kinetic energy per unit volume follows from (A.4) as

$$k = \frac{1}{2}\rho_1 v_a \Delta\Phi = \frac{1}{2}\rho_1 \Delta\Phi^2\left[1 + f(\beta - 1)\right]^{-1} \tag{3.104}$$

Using (3.102) and (3.19)

$$k = \frac{1}{2}\rho_1 \Delta\Phi^2\left(1 + \frac{3}{2}\,\frac{\alpha_{20}}{1 - \alpha_2}\right)^{-1} \tag{3.105}$$

As α_2 increases, k decreases, indicating that the particles tend to agglomerate. The particle pressure is

$$p_p = -\frac{dk}{df} = -\frac{dk}{d\alpha_2}\,\frac{d\alpha_2}{df}$$

$$= -\frac{1}{2}\rho_1 \Delta\Phi^2\,\frac{3}{2}\,\frac{\alpha_{20}}{(1 - \alpha_2)^2\left(1 + \frac{3}{2}\,\frac{\alpha_{20}}{1 - \alpha_2}\right)^2}\,\frac{\alpha_{20}}{f^2} \tag{3.106}$$

which can be expressed in the simple form, dependent only on the properties in the region containing particles,

$$p_p = -\frac{3}{4}\rho_1 \alpha_2^2 v_b^2 \tag{3.107}$$

which resembles (3.91), but with a different coefficient. This *particle pressure* acts on the surface layer(s) of particles and is in the direction that would be anticipated, tending to compress the assembly.

Of course, all of this analysis was based on potential inviscid flow and reality is usually quite different. There is some evidence (Rowe and Henwood (1961), Wallis (1962)) that drag forces on the surface layers of a uniform assembly are greater on the downstream side and smaller on the upstream side, tending to "expand" a free collection of particles, or fluidized bed, in which drag forces are balanced by gravity. However, the "forces" predicted here may be significant contributors to the momentum balances in some circumstances, such as transients, when inertia effects are more important.

Since (3.107) should be equally valid when α_2 varies continuously in the flow direction in Figure 3.7, we may write it generally, replacing v_b by v_1, as

$$p_p = -\frac{3}{4}\rho_1 \alpha_2^2 v_1^2 = -\frac{3}{4}\rho_1 \left(\frac{\alpha_2}{\alpha_1}\right)^2 j_1^2 \tag{3.108}$$

Since p_p is the force on the particles per unit area of the entire cross-section, the net force on a particle when p_p is a function of x, the coordinate in the direction of flow, is

$$f_p = -\frac{1}{\alpha_2}\frac{dp_p}{dx} \tag{3.109}$$

Substituting from (3.108) in (3.109), realizing that j_1 is constant for stationary particles and that

$$\frac{d\alpha_2}{dx} = -\frac{d\alpha_1}{dx} \tag{3.110}$$

we get

$$f_p = \frac{3}{2}\rho_1 \frac{j_1^2}{\alpha_1^3}\frac{d\alpha_2}{dx} = \frac{3}{2}\rho_1 \frac{v_1^2}{\alpha_1}\frac{d\alpha_2}{dx} \tag{3.111}$$

3.7.3 *Results for a Non-Maxwellian Suspension*

The results derived in the previous section can be generalized by keeping β as a general function of α_2 and not invoking (3.19). Starting with (3.104), the equivalent of (3.106) is

$$p_p = -\frac{dk}{df} = \frac{1}{2}\rho_1 \Delta\Phi^2 \frac{(\beta - 1) + f(d\beta/df)}{[1 + f(\beta - 1)]^{-2}} \tag{3.112}$$

which becomes, using (3.104) again,

$$p_p = \frac{1}{2}\rho_1 v_a^2 \left[(\beta - 1) + f\frac{d\beta}{df}\right] \tag{3.113}$$

Now, from (3.102), since α_{20} is a constant,

$$f\frac{d\beta}{df} = -\frac{d\beta}{d\alpha_2}\alpha_2 \tag{3.114}$$

Using (3.114) in (3.113) and replacing v_a with j_1 for more generality we get

$$p_p = \frac{1}{2}\rho_1 j_1^2 \left[(\beta - 1) - \frac{d\beta}{d\alpha_2}\alpha_2\right] \tag{3.115}$$

When (3.115) is used in (3.109), the result is

$$f_p = -\frac{\rho_1 j_1^2}{2\alpha_2}\left[\frac{d\beta}{dx} - \frac{d^2\beta}{d\alpha_2^2}\frac{d\alpha_2}{dx}\alpha_2 - \frac{d\beta}{d\alpha_2}\frac{d\alpha_2}{dx}\right] \tag{3.116}$$

The first and last terms in parentheses cancel and we are left with the simple, elegant and important result,

$$f_p = \frac{\rho_1 j_1^2}{2} \frac{d^2\beta}{d\alpha_2^2} \frac{d\alpha_2}{dx} \tag{3.117}$$

(3.117) expresses the force per unit volume on a particle in a stationary suspension in a uniform duct due to a gradient in particle concentration in the direction of fluid motion.

Equation (3.117) is sufficiently important for it to be worthwhile to deduce it by a different argument. Figure 3.8 indicates a perturbation of a uniform flow through a stationary suspension, in which a uniform concentration α_2 is slightly changed by moving each particle slightly to make the concentration vary linearly from $(\alpha_2 - \varepsilon)$ to $(\alpha_2 + \varepsilon)$ over unit length in the x-direction. We consider events in a cross section of uniform area.

The linear displacement of particles in the x-direction at a location x is, by conservation of particle volume,

$$\zeta = \frac{\varepsilon(1/4 - x^2)}{\alpha_2} \tag{3.118}$$

Therefore, the net volume displacement by all the particles is

$$V = \alpha_2 \int_{-1/2}^{1/2} \zeta dx = \frac{\varepsilon}{6} \tag{3.119}$$

If the force on the particles during this displacement is linearly dependent on ε, the average force is one half of the force at the end of the displacement, $f(\varepsilon)$. Therefore, the net work done by forces to cause the rearrangement of the particles is

$$dw = \frac{Vf(e)}{2} = \frac{\varepsilon f(\varepsilon)}{12} \tag{3.120}$$

To apply the methods in the Appendix we need to know the net change in kinetic energy in the fluid resulting from the displacement at constant $\Delta\Phi$, the macroscopic potential drop across the assembly. As in (3.6), the fluid kinetic energy is

$$k_1 = \frac{\rho_1}{2} j_1 \Delta\Phi \tag{3.121}$$

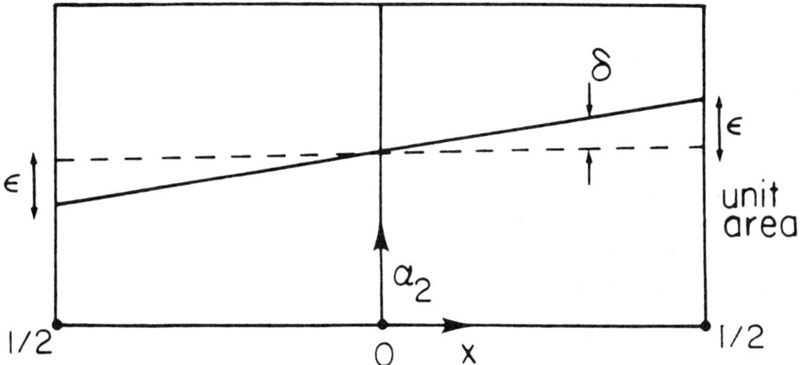

Figure 3.8 A linear perturbation in the concentration, α_2, of particles over unit length.

Therefore, the change in kinetic energy at constant $\Delta\Phi$ is

$$dk_1 = \frac{\rho_1}{2}\, dj_1 \Delta\Phi \tag{3.122}$$

dj_1 is the change in fluid flux necessary to keep the potential drop constant.

An alternative expression for the net fluid kinetic energy is, from (3.7),

$$k_1 = \int_{-1/2}^{1/2} \beta\, \frac{1}{2}\rho_1 j_1^2\, dx \tag{3.123}$$

and the change in k_1, due to changes in β and j_1, is

$$dk_1 = \int_{-1/2}^{1/2} d\beta\, \frac{1}{2}\rho_1 j_1^2\, dx + \int_{-1/2}^{1/2} \beta\rho_1 j_1\, dj_1\, dx \tag{3.124}$$

Now, the last term in (3.124) is simply $\beta\rho_1 j_1\, dj_1$ and βj_1 is the same as $\Delta\Phi$, in view of (3.5). Therefore, (3.124) is equivalent to

$$dk_1 = \int_{-1/2}^{1/2} d\beta\, \frac{1}{2}\rho_1 j_1^2\, dx + \rho_1 \Delta\Phi\, dj_1 \tag{3.125}$$

From (3.122) and (3.125),

$$dk_1 = -\int_{-1/2}^{1/2} d\beta \, \frac{1}{2} \rho_1 j_1^2 \, dx \tag{3.126}$$

We note that (3.126) is just the negative of what would have been obtained from (3.124) with the incorrect assumption that j_1 remained unchanged during the perturbation.

We now expand β in a Taylor Series in terms of the local perturbation "δ" in a_2:

$$d\beta = \frac{d\beta}{d\alpha_2}\,\delta + \frac{1}{2}\,\frac{d^2\beta}{d\alpha_2^2}\,\delta^2 + \ldots \tag{3.127}$$

Substituting (3.127) in (3.126) and evaluating the integral, with $\delta = 2x\varepsilon$, we find that the first order term vanishes and we get

$$dk_1 = -\frac{1}{12}\,\rho_1 j_1^2\,\frac{d^2\beta}{d\alpha_2^2}\,\varepsilon^2 \tag{3.128}$$

Equating (3.128) with (3.120) we find that

$$f(\varepsilon) = -\rho_1 j_1^2\,\frac{d^2\beta}{d\alpha_2^2}\,\varepsilon \tag{3.129}$$

Since

$$\varepsilon = \frac{1}{2}\,\frac{d\alpha_2}{dx} \tag{3.130}$$

(3.129) can be written as

$$f_p = -f(\varepsilon) = +\frac{1}{2}\,\rho_1 j_1^2\,\frac{d^2\beta}{d\alpha_2^2}\,\frac{d\alpha_2}{dx} \tag{3.131}$$

which agrees with (3.117). In this development $f(\varepsilon)$ was the imposed external force that is equal and opposite to the force from the fluid on the particles, f_p.

3.8 Relationship between Particle Stresses and Fluid Reynolds Stresses

In Section 2.10 it was shown that the state of stress of a single particle is related to the Reynolds stresses in the surrounding fluid. This result will now be extended to the general case of a suspension.

For the x-direction computation we use the control volume shown in Figure 3.9. The x-momentum conservation equation is

$$p_a + \rho_1 v_a^2 = -p_p + p_b \alpha_1 - \sigma_x \alpha_2 + \rho_1 \langle w_x^2 \rangle \alpha_1 \tag{3.132}$$

where w_x has the same significance as in Section 2.10.

From continuity

$$v_a = v_b \alpha_1 \tag{3.133}$$

Bernoulli's equation gives, using (3.81) and (3.133)

$$p_{01} = p_a + \frac{1}{2} \rho_1 v_a^2 = p_b + \frac{1}{2} \rho_1 \alpha_1 \beta v_b^2 = p_b + \frac{1}{2} \rho_1 \frac{\beta}{\alpha_1} v_a^2 \tag{3.134}$$

Now, p_p is given by (3.115) and we may use these together with (3.134) in (3.132) to get

$$\begin{aligned} & p_a + \rho_1 v_a^2 + \frac{1}{2} \rho_1 v_a^2 \left(\beta - 1 - \frac{d\beta}{d\alpha_2} \alpha_2 \right) \\ & = \alpha_1 \left(p_a + \frac{1}{2} \rho_1 v_a^2 - \frac{1}{2} \rho_1 v_a^2 \frac{\beta}{\alpha_1} \right) - \sigma_x \alpha_2 + \rho_1 \langle w_x^2 \rangle \alpha_1 \end{aligned} \tag{3.135}$$

We also have, from (3.14)

$$p_i = p_{01} + \frac{1}{2} \rho_1 v_a^2 \frac{d\beta}{d\alpha_1} = p_a + \frac{1}{2} \rho_1 v_a^2 \left(1 - \frac{d\beta}{d\alpha_2} \right) \tag{3.136}$$

Rearranging (3.135) and using (3.136) gives

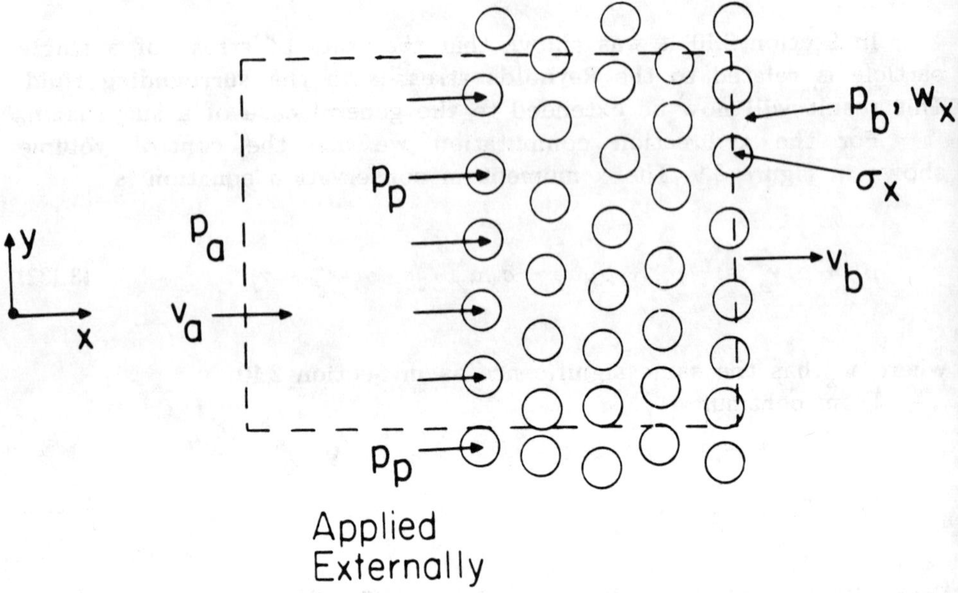

Figure 3.9 Momentum balance across a longitudinal discontinuity

$$\alpha_2 p_i + \rho_1 v_a^2 \beta = - \sigma_x \alpha_2 + \rho_1 < w_x^2 > \alpha_1 \tag{3.137}$$

where the averaging signs on σ_x have been dropped. The left-hand side of (3.137) contains quantities that can be calculated using the methods developed here. The right-hand side contains the particle stress and Reynolds stress terms that naturally appear in an overall momentum balance in the x-direction. In terms of the properties at *"b"* (3.137) may be written, using (3.133), as

$$\alpha_2 p_i + \alpha_1 \rho_1 v_b^2 + \alpha_1 \rho_1 v_b^2 (\alpha_1 \beta - 1) = - \sigma_x \alpha_2 + \rho_1 < w_x^2 > \alpha_1 \tag{3.138}$$

Now, w_x may be written as

$$w_x = v_b + \tilde{w}_x \tag{3.139}$$

where $\tilde{w}_x$ is a perturbation or fluctuating velocity with the property

$$\int \tilde{w}_x dV_1 = 0 \tag{3.140}$$

Using (3.139) and (3.140) in (3.138) we get

$$\alpha_2 p_i + \alpha_1 \rho_1 v_b^2 (\alpha_1 \beta - 1) = -\sigma_x \alpha_2 + \rho_1 \alpha_1 < \tilde{w}_x^2 > \tag{3.141}$$

When a velocity v_2 is imposed on this situation, in the x-direction, the result is, using the method used to deduce (3.34), since $v_b = w$,

$$\alpha_2 p_i + \alpha_1 \rho_1 v_1^2 + \alpha_1 \rho_1 w^2 (\alpha_1 \beta - 1) = -\sigma_x \alpha_2 + \rho_1 \alpha_1 < \tilde{w}_x^2 > + \rho_1 \alpha_1 v_1^2 \tag{3.142}$$

(3.142) describes the net momentum flux and particle pressure contributions to the overall net momentum balance in the x-direction.

For the y-direction, we use the control volume illustrated in Figure 3.10 to develop the momentum balance:

$$p_a = - p_p + p_b \alpha_1 - \sigma_y \alpha_2 + \rho_1 < w_y^2 > \alpha_1 \tag{3.143}$$

Using (3.99), (3.92), (3.79) and (3.15), (3.143) may be rearranged to give

$$\alpha_2 p_i = - \sigma_y \alpha_2 + \rho_1 < w_y^2 > \alpha_1 \tag{3.144}$$

which is related to (2.84). Similarly, for the z-direction

$$\alpha_2 p_i = - \sigma_z \alpha_2 + \rho_1 < w_z^2 > \alpha_1 \tag{3.145}$$

The bulk particle stress is obtained by averaging (3.141), (3.144) and (3.145) and it is found that

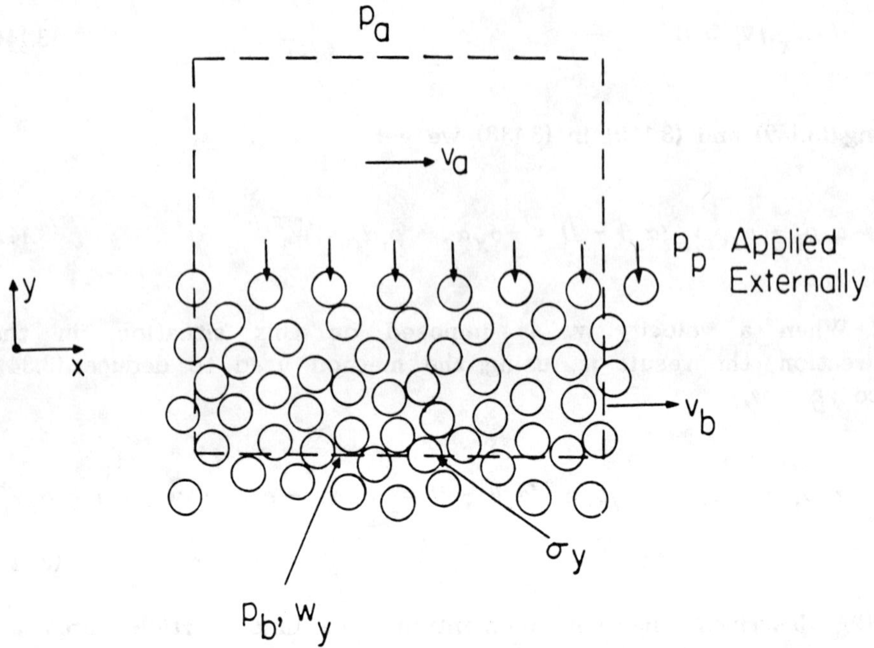

Figure 3.10 Momentum balance across a lateral discontinuity

$$p_i = -\frac{\sigma_x + \sigma_y + \sigma_z}{3} = p_2 \tag{3.146}$$

since

$$v_b^2 (\alpha_1 \beta - 1) = \langle \tilde{w}_x^2 + w_y^2 + w_z^2 \rangle \tag{3.147}$$

Therefore, the mean interface stress is indeed the same as the bulk stress in the particles, p_2.

A very important consequence of (3.142), (3.144) and (3.145) is that the components of the *overall stress tensor* for the dispersion, derived by evaluating forces and momentum fluxes across the boundaries of a rectangular control volume that cuts through the particles, are, for average motion in the x-direction

$$-\sigma_x\alpha_2 + \rho_1\alpha_1 < \tilde{w}_x^2 > + \rho_1\alpha_1 v_1^2 + p_1\alpha_1 + \rho_2\alpha_2 v_2^2$$

$$= \alpha_2 p_2 + \alpha_1 p_1 + \rho_1\alpha_1 v_1^2 + \rho_2\alpha_2 v_2^2 + \alpha_1\rho_1 w^2(\alpha_1\beta - 1) \tag{3.148}$$

$$-\sigma_y\alpha_2 + \rho_1\alpha_1 < w_y^2 > + p_1\alpha_1 = p_2\alpha_2 + p_1\alpha_1 \tag{3.149}$$

$$-\sigma_z\alpha_2 + \rho_1\alpha_1 < w_z^2 > + p_1\alpha_1 = p_2\alpha_2 + p_1\alpha_1 \tag{3.150}$$

These results suggest that the average pressure of the combined phases should be taken as

$$P = \alpha_1 p_1 + \alpha_2 p_2 \tag{3.151}$$

and that the overall momentum flux is increased above what would be accounted for by a solely one-dimensional assumption by the term $\alpha_1\rho_1 w^2(\alpha_1\beta - 1)$, a result useful for checking hypothesized general equations of two-phase flow.

These results are generalizable to an arbitrary three-dimensional motion of an isotropic suspension by imposing an overall velocity in any direction of the above pattern, yielding a combined pressure and momentum flux tensor of

$$\mathbf{P} = \alpha_1(\rho_1\mathbf{v}_1\mathbf{v}_1 + p_1\mathbf{I}) + \alpha_2(\rho_2\mathbf{v}_2\mathbf{v}_2 + p_2\mathbf{I}) + \alpha_1\rho_1\mathbf{ww}(\alpha_1\beta - 1) \tag{3.152}$$

which anticipates (3.202) for a Maxwellian suspension and the overall equation of motion of the Cook-Harlow model in Chapter 4.

Besides being important conclusions in their own right, these results reinforce the derivation of particle pressure in 3.7.1 and 3.7.2, providing evidence of an overall pattern of consistency.

3.9 Confirmatory Results

The results obtained in the previous parts of Sections 3.7 and 3.8 may be confirmed, and a few new ones derived, by using differential control volumes for situations resembling Figures 3.9 and 3.10 in which variations in particle concentration are continuous. For variations entirely in the direction of relative motion we use (3.148) to derive the momentum balance, with particles stationary.

$$\frac{d}{dx}\left[\alpha_2 p_2 + \alpha_1 p_1 + \rho_1 \alpha_1 v_1^2 (\alpha_1 \beta)\right] + f_{2x}\alpha_2 = 0 \tag{3.153}$$

where f_2 is an external force, per unit volume of the particles, that is needed to hold them in place.

Similarly, for variations entirely in the direction perpendicular to the direction of flow, (3.149) gives

$$\frac{d}{dy}(\alpha_2 \, p_2 + \alpha_1 p_1) + f_{2y}\alpha_2 = 0 \tag{3.154}$$

Now, the key step in going from (3.132) to (3.137) was to show that

$$p_a + \rho_1 v_a^2 + p_p - p_b \alpha_1 = \alpha_2 p_2 + \rho_1 v_a^2 \beta \tag{3.155}$$

Since p_a and $\rho_1 v_a^2$ are constants and $\alpha_1 v_1 = v_a$, $p_b = p_1$, use of (3.155) in (3.153) shows that

$$f_{2x} = -\frac{1}{\alpha_2}\frac{dp_p}{dx} \tag{3.156}$$

a result that was previously obtained by *assuming* that p_p was a continuous variable that described a "particle pressure" throughout the suspension.

Similarly, in the y-direction, perpendicular to the relative motion, the key step in going from (3.143) to (3.144) was to show that

$$p_a + p_p - p_b \alpha_1 = \alpha_2 p_2 \tag{3.157}$$

Using (3.157) in (3.154), putting $p_b = p_1$, we obtain

$$f_{2y} = -\frac{1}{\alpha_2}\frac{dp_p}{dy} \tag{3.158}$$

which resembles (3.156).

Now, for variations in the x-direction, the continuity equation is

$$\alpha_1 v_1 = j_1 = constant \tag{3.159}$$

The average phase pressures are related in the usual way by using (3.16) and (3.21):

$$p_2 = p_1 + \frac{1}{2}\rho_1 v_1^2 \alpha_1 \frac{d(\alpha_1\beta - 1)}{d\alpha_1} \tag{3.160}$$

Moreover, from Bernoulli's equation averaged throughout the continuous phase, as in (3.81),

$$p_1 + \frac{1}{2}\rho_1\alpha_1\beta v_1^2 = constant \tag{3.161}$$

Now, (3.153) can be arranged to the form

$$\alpha_2\frac{dp_2}{dx} + p_2\frac{d\alpha_2}{dx} - \frac{d\alpha_2}{dx}(p_1 + \rho_1 v_1^2\alpha_1\beta) + \alpha_1\frac{d}{dx}(p_1 + \rho_1 v_1^2\alpha_1\beta)$$

$$+ f_{2x}\alpha_2 = 0 \tag{3.162}$$

Using (3.160) and (3.161) in (3.162) we get

$$\alpha_2\frac{dp_2}{dx} + \frac{d\alpha_2}{dx}\left[\frac{1}{2}\rho_1 v_1^2\alpha_1\frac{d}{dx}(\alpha_1\beta - 1) - \rho_1 v_1^2\alpha_1\beta\right]$$

$$+ \alpha_1\frac{d}{dx}\left(\frac{1}{2}\rho_1\alpha_1\beta v_1^2\right) + f_{2x}\alpha_2 = 0 \tag{3.163}$$

Evaluating the terms in the square bracket and using (3.159) in the third term above, it is found that the two resulting expressions cancel and (3.163) reduces to

$$-\frac{dp_2}{dx} = f_{2x} \tag{3.164}$$

Equation (3.164) and (3.156) together confirm the conclusion reached by comparing (3.109) and (7.29) and derived for a special case in (2.24).

Similar developments for changes in α_2 entirely in the y-direction, with the flow coming from a source with a uniform stagnation pressure, using (3.160), (3.161) and the general form of (3.79), yield

$$-\frac{dp_2}{dy} = f_{2y} \tag{3.165}$$

which, together with (3.158) confirms the equivalence of (3.95) and (3.96).

3.10 Extension to Three Dimensions

The potential imposed on the base case in Section 3.2 could have had a gradient in any direction, resulting in a vector velocity for phase 2, $\mathbf{v}_2$, with components, v_{2x}, v_{2y}, v_{2z}. Equations (3.28) and (3.30) are then valid in vector form as

$$\mathbf{v}_1 = \mathbf{v}_2 + \mathbf{w} \tag{3.166}$$

$$\mathbf{m} = \alpha_1\rho_1\mathbf{v}_1 + \alpha_2\rho_2\mathbf{v}_2 \tag{3.167}$$

Since in the base case,

$$\int w_y dV_1 = \int w_z \, dV_1 = 0 \tag{3.168}$$

the equivalent of (3.34) is easily shown to be

$$k_1 = \frac{1}{2}\rho_1\left[v_1^2 + w^2(\alpha_1\beta - 1)\right] \tag{3.169}$$

Therefore, (3.35) is valid in vector form. Similarly, (3.36) becomes

$$-\nabla\Phi = \mathbf{v}_1 + (\alpha_1\beta - 1)\mathbf{w} = \mathbf{v}_2 + \alpha_1\beta\mathbf{w} \tag{3.170}$$

(3.50) and (3.51) can now be generalized to

$$\frac{dm}{dt} = \alpha_1\rho_1\dot{\mathbf{v}}_1 = + \alpha_2\rho_2\dot{\mathbf{v}}_2 - \nabla P + \mathbf{g}_1\rho_1\alpha_1 + \mathbf{g}_2\rho_2\alpha_2 \tag{3.171}$$

$$\frac{dK}{dt} = \alpha_1\rho_1\dot{\mathbf{v}}_1\cdot\dot{\mathbf{v}}_1 + \alpha_2\rho_2\mathbf{v}_2\cdot\dot{\mathbf{v}}_2 + \alpha_1(\alpha_1\beta - 1)\rho_1\mathbf{w}\cdot\dot{\mathbf{w}}$$

$$= -\nabla P\cdot(\alpha_1\mathbf{v}_1 + \alpha_2\mathbf{v}_2) + \alpha_1\rho_1\mathbf{g}_1\cdot\mathbf{v}_1 + \alpha_2\rho_2\mathbf{g}_2\cdot\mathbf{v}_2 \tag{3.172}$$

Subtracting the dot product of $\mathbf{v}_1$ or $\mathbf{v}_2$ with (3.171) from (3.172), we find that (3.52) and (3.53) are valid in vector form apart from an arbitrary vector perpendicular to $\mathbf{w}$, therefore

$$\rho_1\dot{\mathbf{v}}_1 + \rho_1(\alpha_1\beta - 1)\dot{\mathbf{w}} = -\nabla P + \rho_1\mathbf{g}_1 + \mathbf{a}_1 \times \mathbf{w} \tag{3.173}$$

$$\rho_2\dot{\mathbf{v}}_2 - \rho_1(\alpha_1/\alpha_2)(\alpha_1\beta - 1)\dot{\mathbf{w}} = -\nabla P + \rho_2\mathbf{g}_2 + \mathbf{a}_2 \times \mathbf{w} \tag{3.174}$$

where $\mathbf{a}_1$ and $\mathbf{a}_2$ are unknown at this stage. If we take α_1 times (3.173) and add it to α_2 times (3.174), we recreate (3.171) together with the condition

$$(\alpha_1\mathbf{a}_1 + \alpha_2\mathbf{a}_2) \times \mathbf{w} = 0 \tag{3.175}$$

(3.172) may also be recreated by taking the dot product of $\alpha_1\mathbf{v}_1$ with (3.173) and adding the dot product of $\alpha_2\mathbf{v}_2$ with (3.174). This gives the condition

$$\alpha_1 \mathbf{a}_1 \cdot \mathbf{v}_1 \times \mathbf{w} + \alpha_2 \mathbf{a}_2 \cdot \mathbf{v}_2 \times \mathbf{w} = 0 \tag{3.176}$$

However, (3.175) and (3.176) are not independent since (3.176) can be derived from (3.175) by using the identity

$$\alpha_2 \mathbf{a}_2 \cdot \mathbf{w} \times \mathbf{w} = 0 \tag{3.177}$$

It is unclear what physical significance $\mathbf{a}_1$ and $\mathbf{a}_2$ might have. A possible form is $\mathbf{a}_1 = \alpha_1 \nabla \times \mathbf{v}_1$, $\mathbf{a}_2 = (\alpha_1/\alpha_2)\alpha_1 \nabla \times \mathbf{v}_1$ which would have the character of a *lift* force occasioned by relative motion in an *apparent vorticity* field.

3.11 The Momentum Flux Tensor and Particle Stress in Three-Dimensions

Equation (3.68) was derived for one-dimensional flow. If we now let v_2 be a vector, we have to superpose its three components on the base case.

The velocity field in the base case could be thought of as consisting of the average velocity w in the x-direction together with the fluctuations $\tilde{w}_x = w_x - w$, w_y, and w_z which have an average value of zero. It seems reasonable for potential flow around a symmetrical arrangement (on the average) of spheres, or generally for an isotropic arrangement of any particles, that we should have*

$$< w_y^2 > = < w_z^2 > \tag{3.178}$$

$$< w_y w_z > = < w_y \tilde{w}_x > = < \tilde{w}_x w_z > = 0 \tag{3.179}$$

In view of (3.64) and (3.65),

$$< \tilde{w}_x^2 > = (\gamma - 1) w^2 \tag{3.180}$$

---

*These assumptions are justified by the form of the result already obtained in (3.152).

$$< w_y^2 > = < w_z^2 > = (\alpha_1\beta - \gamma)w^2/2 \tag{3.181}$$

The momentum flux tensor for the resulting flow has the components

$$< u_x u_x > = <(v_{2x} + w - \tilde{w}_x)^2 > = v_{1x}^2 + < \tilde{w}_x^2 > \tag{3.182}$$

$$< u_y u_y > = <(v_{2y} + w_y)^2 > = v_{1y}^2 + < w_y^2 > \tag{3.183}$$

$$< u_z u_z > = <(v_{2z} + w_z)^2 > = v_{1z}^2 + < w_z^2 > \tag{3.184}$$

$$< u_x u_y > = <(v_{2x} + w + \tilde{w}_x)(v_{2y} + w_y) > = v_{1x} v_{1y} \tag{3.185}$$

$$< u_x u_z > = <(v_{2x} + w + \tilde{w}_x)(v_{2z} + w_z) > = v_{1x} v_{1z} \tag{3.186}$$

$$< u_y u_z > = <(v_{2y} + w_y)(v_{2z} + w_z) > = v_{1y} v_{1z} \tag{3.187}$$

So, the additional momentum flux due to the relative motion only has the three components, resembling *Reynolds Stresses*, given in (3.182), (3.183) and (3.184), which can be evaluated from (3.180) and (3.181). The *principal directions* of this tensor are in the direction of the relative motion and perpendicular to it. The resulting momentum flux tensor due to the relative motion (or fluctuations) can then be expressed by using (3.180) and (3.181), as

$$\Pi_W = \rho_1 \left(\frac{\alpha_1\beta - \gamma}{2} w^2 \boldsymbol{I} + \frac{3\gamma - \alpha_1\beta - 2}{2} \boldsymbol{ww} \right) \tag{3.188}$$

where $\boldsymbol{I}$ is the unit dyadic and the second term above has the character of a dyadic product. This tensor will be completely isotropic if $3\gamma - \alpha_1\beta - 2 = 0$ which is perhaps emphasized if (3.188) is rewritten as

$$\Pi_w = \rho_1 \left(\frac{\alpha_1 \beta - 1}{3} w^2 \boldsymbol{I} + \frac{3\gamma - \alpha_1 \beta - 2}{2} \left(\boldsymbol{ww} - \frac{w^2}{3} \boldsymbol{I} \right) \right) \tag{3.189}$$

(see Lhuillier (1987), Eq. [46] for an equivalent expression in terms of two unknown functions).

In the case of a single sphere moving through an infinite fluid we may use the potential in (2.1) to show that, for the entire fluid surrounding the sphere

$$\int w_y^2 \, dV = \int w_z^2 dV = \frac{1}{5} \pi a^3 w^2 \tag{3.190}$$

The momentum flux of the fluid in a dilute suspension of spheres is then

$$\Pi_1 = \alpha_1 \rho_1 \boldsymbol{v}_1 \boldsymbol{v}_1 + \rho_1 \alpha_2 \left(\frac{3}{20} w^2 \boldsymbol{I} + \frac{1}{20} \boldsymbol{ww} \right) \tag{3.191}$$

in agreement with (3.68) and (3.70), since $\alpha_1 \approx 1$.

It would seem that, for a random dispersion of spheres, it should be possible to derive the value of γ, or any equivalent parameter determining the partition of the kinetic energy of relative motion in different directions, by invoking arguments of symmetry and invariance: a goal that has so far eluded this author. However, it is possible to make an estimate by using the approach suggested by Zuber (1964) and already used in Section 3.4.2. The idea is to use the solution developed by Lamb (1932d) for the motion of a sphere of radius "a" inside a sphere of radius "b", the space between them being filled with fluid. In this situation we have

$$\alpha_2 = \frac{a^3}{b^3}, \qquad \alpha_1 = \frac{b^3 - a^3}{b^3} \tag{3.192}$$

The total kinetic energy of the fluid per unit total volume is

$$k = \frac{1}{2} \rho_1 v_2^2 \, \frac{b^3/2 + a^3}{b^3 - a^3} \, \frac{a^3}{b^3} = \frac{1}{2} \rho_1 v_2^2 \left(\frac{1}{2} + \alpha_2 \right) \frac{\alpha_2}{\alpha_1} \tag{3.193}$$

Since we have in addition the constraint of no net volumetric flux,

$$\alpha_1 v_1 + \alpha_2 v_2 = 0 \tag{3.194}$$

(3.193) can be put into the form

$$k = \frac{1}{2}\rho_1\alpha_1 v_1^2 + \frac{1}{2}\rho_1 (v_1 - v_2)^2 \left(\frac{1}{2}\alpha_1\alpha_2\right) \tag{3.195}$$

The final term in (3.195) is just what would be predicted by (3.35) for a Maxwellian suspension.

Now, using Lamb's solution we may also compute $\int u_x^2 dV$, $\int u_y^2 dV$, and $\int u_z^2 dV$ for all of the fluid between the spheres. It is found that

$$\int u_x^2 dV = \frac{2}{3}\pi \frac{a^3}{(b^3 - a^3)}\left(2a^3 + \frac{2}{5}b^3\right) v_2^2 \tag{3.196}$$

$$\int u_y^2 dV = \int u_z^2 dV = \frac{2}{3}\pi \frac{a^3}{b^3 - a^3}\left(\frac{3}{10}b^3\right) v_2^2 \tag{3.197}$$

Therefore, the average momentum flux for the fluid is

$$\Pi_1 = \alpha_1\rho_1 \mathbf{v}_1\mathbf{v}_1 + \rho_1\alpha_1\alpha_2\left(\frac{3}{20}w^2\mathbf{I} + \frac{1}{20}\mathbf{ww}\right) \tag{3.198}$$

which differs from (3.191) only in the additional factor α_1 in the second term and is a possible representation of the fluid momentum flux for *all* void fractions.

The particle stress may also be developed as a vector quantity. By a straightforward extension of (3.76) we obtain

$$- \langle \sigma_y \rangle = \frac{1}{V}\int y p_i ds_y \tag{3.199}$$

For an isolated sphere there follows, for the average stress tensor

$$-\sigma_2 = \left(p_1 - \frac{1}{4}\rho_1 w^2\right)\mathbf{I} + \frac{9}{20}\rho_1\left(\mathbf{ww} - \frac{w^2}{3}\mathbf{I}\right) \tag{3.200}$$

The first term in (3.200) is the bulk stress which is the same as the average interfacial pressure in (3.77).

The combined momentum flux and stress tensor for a dilute suspension is then

$$\mathbf{P} = \Pi_1 + \Pi_2 + \alpha_1 p_1 \mathbf{I} - \alpha_2\sigma_2 = \alpha_1\rho_1\mathbf{v}_1\mathbf{v}_1 + \alpha_2\rho_2\mathbf{v}_2\mathbf{v}_2$$

$$+ \rho_1\alpha_2\left(\frac{\mathbf{ww}}{2} - \frac{1}{4}w^2\mathbf{I}\right) + p_1\mathbf{I} \tag{3.201}$$

Remarkably, if (3.77) is invoked, this reduces to

$$\mathbf{P} = \alpha_1(\rho_1\mathbf{v}_1\mathbf{v}_1 + p_1\mathbf{I}) + \alpha_2(\rho_2\mathbf{v}_2\mathbf{v}_2 + p_2\mathbf{I}) + \rho_1\alpha_2\frac{\mathbf{ww}}{2} \tag{3.202}$$

suggesting that the resultant is made up of the usual form of contributions from each phase together with an additional term due to the relative motion, as in (3.78). Equation (3.202) is the limit at low α_2 of (3.152).

Equations (3.191), (3.200), and (3.202) were derived by Biesheuvel and van Wijngaarden (1984).

In the quasi-continuum flow of a suspension, conservation of mass is expressed as

$$\nabla \cdot (\alpha_1\rho_1\mathbf{v}_1) = S_{12} \tag{3.203}$$

$$\nabla \cdot (\alpha_2\rho_2\mathbf{v}_2) = -S_{12} \tag{3.204}$$

where S_{12} is a rate of phase change per unit volume. If (3.202), in its more general form (3.152), is assumed to describe the momentum and stress transport (which may not be true in a non-uniform dispersion?), the overall momentum balance will involve its divergence. Performing the mathematics and using (3.203) and (3.204) we get

$$\nabla \cdot \boldsymbol{P} = \alpha_1(\rho_1 \boldsymbol{v}_1 \cdot \nabla \boldsymbol{v}_1 + \nabla p_1) + \alpha_2(\rho_2 \boldsymbol{v}_2 \cdot \nabla \boldsymbol{v}_2 + \nabla p_2)$$

$$- (p_1 - p_2) \nabla \alpha_2 + \boldsymbol{w} S_{12} + \rho_1 \alpha_1 (\alpha_1 \beta - 1)(\nabla \cdot \boldsymbol{ww})$$

$$+ \boldsymbol{ww} \cdot \left\{ \rho_1 \alpha_1 \nabla (\alpha_1 \beta - 1) + \rho_1 (\alpha_1 \beta - 1) \nabla \alpha_1 + \alpha_1 (\alpha_1 \beta - 1) \nabla \rho_1 \right\}$$

(3.205)

which might be useful for comparison with equations derived from other sources, such as postulated general three-dimensional equations of motion.

3.12 Three-Dimensional Steady Flow in a Stationary Particle Lattice

The results derived in Section 3.7 and 3.9 are special cases of a more general situation in which the phase concentrations and fluid velocity vary three-dimensionally. Since we have now established what appear to be general equations, they will be used in vector form. The continuity equation with no phase change is

$$\nabla \cdot (\rho_1 \alpha_1 \boldsymbol{v}_1) = 0 \tag{3.206}$$

With stationary particles, (3.170) is

$$- \nabla \Phi = \alpha_1 \beta \boldsymbol{v}_1 \tag{3.207}$$

The macroscopic momentum balance for a region of the flow, using (3.152) with $\boldsymbol{v}_2 = 0$ and $\boldsymbol{w} = \boldsymbol{v}_1$, is

$$\alpha_2 \boldsymbol{f}_2 = \nabla \cdot \boldsymbol{P} = \nabla p_2 + \nabla [\alpha_1 (p_1 - p_2)] + \alpha_1 \beta \boldsymbol{v}_1 \nabla \cdot (\rho_1 \alpha_1 \boldsymbol{v}_1)$$

$$+ \rho_1 \alpha_1 \boldsymbol{v}_1 \cdot \nabla (\alpha_1 \beta \boldsymbol{v}_1) \tag{3.208}$$

The third term on the right-hand side above may be discarded in view of (3.206). Bernoulli's equation for the fluid is given by (3.161) and we assume that (3.160) is also valid. We use (3.160) in the second term on the right-hand side of (3.208) and expand the final term as

$$\rho_1 \alpha_1 \mathbf{v}_1 \cdot \nabla(\alpha_1 \beta \mathbf{v}_1) = \frac{\rho_1}{\beta}\left[(\alpha_1 \beta \mathbf{v}_1)\cdot \nabla(\alpha_1 \beta \mathbf{v}_1)\right]$$

$$= \frac{\rho_1}{\beta}\left[\nabla \frac{(\alpha_1 \beta \mathbf{v}_1)^2}{2} - \alpha_1 \beta \mathbf{v}_1 \times \nabla \times (\alpha_1 \beta \mathbf{v}_1)\right] \tag{3.209}$$

In view of (3.207), the last term in (3.209) vanishes. Therefore, (3.208) becomes

$$\alpha_2 \mathbf{f}_2 = \nabla p_2 - \nabla\left[\frac{1}{2}\rho_1 v_1^2 \alpha_1^2 \frac{d}{d\alpha_1}(\alpha_1 \beta)\right] + \frac{\rho_1}{\beta}\nabla \frac{(\alpha_1 \beta \mathbf{v}_1)^2}{2} \tag{3.210}$$

We note that the last term in (3.210) is related to (3.207). Expanding the gradient terms in (3.210) and simplifying gives, eventually,

$$\alpha_2 \mathbf{f}_2 = \nabla p_2 - \frac{1}{2}\rho_1 v_1^2 \alpha_1^3 \frac{d^2\beta}{d\alpha_1^2}\nabla \alpha_1 - \frac{1}{2}\rho_1 \alpha_1 \frac{d\beta}{d\alpha_1}\nabla(v_1^2 \alpha_1^2) \tag{3.211}$$

So far, (3.161) has not been used. Combining it with (3.160) we obtain

$$p_2 - \frac{1}{2}\rho_1 v_1^2 \alpha_1^2 \frac{d\beta}{d\alpha_1} = constant \tag{3.212}$$

which may be differentiated to give

$$\nabla p_2 - \frac{1}{2}\rho_1 v_1^2 a_1^2 \frac{d^2\beta}{d\alpha_1^2}\nabla\alpha_1 - \frac{1}{2}\rho_1 \frac{d\beta}{d\alpha_1}\nabla(v_1^2 \alpha_1^2) = 0 \tag{3.213}$$

Adding α_1 times (3.213) to (3.211) we get the simple general result:

$$\mathbf{f}_1 = \nabla p_2 \tag{3.214}$$

which may be expressed in terms of gradients of α_1 and v_1 by using (3.213). Results obtained previously in Sections 3.7 to 3.9 are special cases of these expressions.

3.13 Summary

By working through a set of well-defined simple examples, a consistent overall description of the properties of an ideal suspension has been derived. The overall stress and momentum flux tensor, the difference between phase pressures, forces on particles, momentum and kinetic energy densities, are all derivable from the average velocities and void fraction in an isotropic suspension as long as the *exertia* is known as a function of particle concentration. Moreover, effective potentials for each phase can be described and equations of motion derived for some simple situations. The theory is ready to be tested in broader applications, challenged for conditions under which higher order approximations may be needed, and developed further to encompass more generally useful results, such as the complete equations of motion of each phase.

4. THE COOK-HARLOW MODEL

4.1 One-Dimensional Parallel Flow

In Section 3.3 it was found that the correct results for the overall momentum and kinetic energy could be obtained by regarding $(\alpha_1 - 1/\beta)$ as the fraction of the total volume comprised of phase 1 *entrained* in phase 2, while a fraction $1/\beta$ of the total volume consists of *unentrained* phase 1. This forms the basis of an idealized model for the flow, illustrated in Figure 4.1a, that was developed in one-dimensional form by Wallis (1978) and independently in three-dimensional form by Cook and Harlow (1984) who used the symbol "f" in place of "C" in (3.41); their nomenclature is indicated in Figure 4.1b.

In each case the flow consists of an unentrained fraction of phase 1, moving at a speed v_1' while phase 2 and the entrained part of phase 1 move as a homogeneous mixture at the same speed, v_2.

The relationship between the alternative parameters f, or C in (3.41), and β is

$$\frac{1}{\beta} = \alpha_1 - f\alpha_2 \tag{4.1}$$

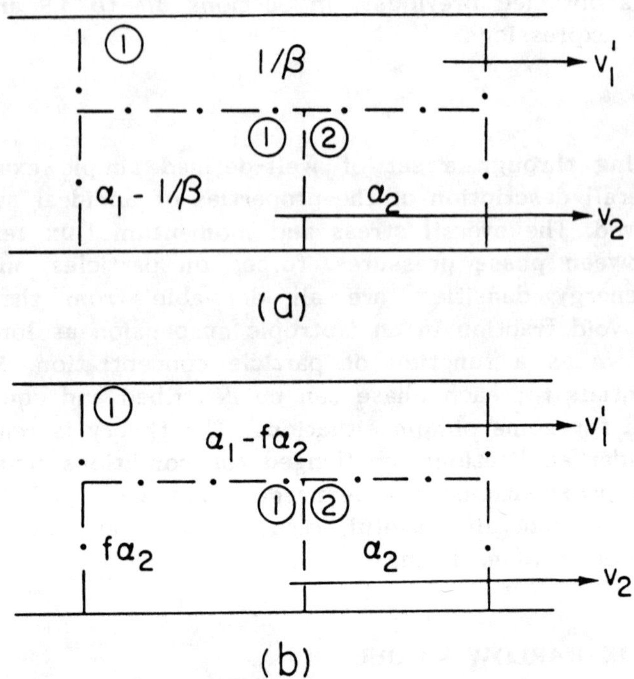

Figure 4.1 Assumed phase geometry for the Cook-Harlow model.

and formulations in terms of either are equivalent. For a Maxwellian suspension of spheres we have, from (3.20), $f = \alpha_1/(\alpha_2 + 2)$.

The volumetric average velocity of phase 1 is

$$v_1 = \left[v_1'/\beta + (\alpha_1 - 1/\beta) v_2 \right] / \alpha_1 \tag{4.2}$$

This may be rearranged to the forms

$$v_1' = v_1 + (\alpha_1 \beta - 1) w \tag{4.3}$$

$$v_1' = v_2 + \alpha_1 \beta w \tag{4.4}$$

which are compatible with (3.36) if

$$v_1' = -\, d\Phi/dx \tag{4.5}$$

which is the relationship one might expect for the unentrained phase in Figure 4.1.

The total momentum per unit volume is

$$m_x = \rho_1 v_1' (1/\beta) + \rho_1 v_2 (\alpha_1 - 1/\beta) + \rho_2 \alpha_2 v_2 = \rho_1 \alpha_1 v_1 + \rho_2 \alpha_2 v_2 \tag{4.6}$$

while the total kinetic energy per unit volume is

$$k = \frac{1}{2} \rho_1 v_1'^2 (1/\beta) + \frac{1}{2} \rho_1 v_2^2 (\alpha_1 - 1/\beta) + \frac{1}{2} \rho_2 \alpha_2 v_2^2 \tag{4.7}$$

which may be shown, by using (4.3) or (4.4) to be the same as

$$k = \frac{1}{2} \rho_1 \alpha_1 v_1^2 + \frac{1}{2} \rho_2 \alpha_2 v_2^2 + \frac{1}{2} \rho_1 \alpha_1 (\alpha_1 \beta - 1) w^2 \tag{4.8}$$

Since (4.6) and (4.8) are compatible with (3.30) and (3.35), this flow has the same momentum and kinetic energy per unit volume as the more elaborate potential flow described in Section 3.2.

To evaluate the pressure difference between phases we impose a velocity v_2 on the situation illustrated in Figure 4.1, obtaining a flow resembling the *base case* of Section 3.1. From (4.4) we find the velocity of the unentrained part of phase 1 past the stationary entrained part is

$$v_{10}' = \alpha_1 \beta w \tag{4.9}$$

Therefore, from Bernoulli's equation

$$p' = p_{01} - \frac{1}{2} \rho_1 (\alpha_1 \beta w)^2 \tag{4.10}$$

The volumetric average pressure over phase 1 is

$$p_1 = \left[\frac{1}{\beta}p' + \left(\alpha_1 - \frac{1}{\beta}\right)p_2\right]/\alpha_1 \tag{4.11}$$

where p_2 is the pressure in the entrained phase. If we now argue that p_2 should equal p_{01} and use (4.10) in (4.11) we find that

$$p_1 = p_{01} - \frac{1}{2}\rho_1\alpha_1\beta w^2 \tag{4.12}$$

which is the same result as obtained in (3.11). However, setting p_2 equal to p_{01} does not capture the influence of $(d\beta/d\alpha_1)$ in (3.15). Therefore, the models are not completely compatible in their descriptions of pressure.

Cook and Harlow (1984) assumed the pressure to be uniform throughout the volume in Figure 4.1. This is incompatible with both (3.11) and (3.14).

For certain purposes (e.g. Section 3.4) the details of pressure differences between the phases (and the interface between them) may not be important and the above inconsistencies do not matter. For example, if we impose a macroscopic pressure gradient *dP/dx*, this will be the effective pressure gradient acting on each of the *phases* in Figure 4.1, as long as the pressure difference between the phases is uniform. The equations of motion are, in gravitational fields g_1 and g_2

$$\rho_1\dot{v}_1 = -\frac{\partial P}{\partial x} + \rho_1 g_1 \tag{4.13}$$

$$\left[\rho_2\alpha_2 + \rho_1\left(\alpha_1 - \frac{1}{\beta}\right)\right]\dot{v}_2 = -\frac{\partial P}{\partial x}\left(1 - \frac{1}{\beta}\right) + \rho_1\left(\alpha - \frac{1}{\beta}\right)g_1 + \rho_2\alpha_2 g_2 \tag{4.14}$$

Using (4.3), (4.13) becomes

$$\rho_1\dot{v}_1 + \rho_1(\alpha_1\beta - 1)\dot{w} = -\frac{\partial P}{\partial x} + \rho_1 g_1 \tag{4.15}$$

in agreement with (3.46)

Adding $1/\beta$ times (4.15) to (4.14) we get

$$\rho_1 \alpha_1 \dot{v}_1 + \rho_2 \alpha_2 \dot{v}_2 = - \frac{\partial P}{\partial x} + \rho_1 \alpha_1 g_1 + \rho_2 \alpha_2 g_2 \tag{4.16}$$

which is the equation of motion for the whole mixture. Evaluating (4.16) $-\alpha_1$ x (4.15) we get, after division by $\alpha_2 = 1 - \alpha_1$,

$$\rho_2 \dot{v}_2 - (\alpha_1/\alpha_2)\rho_1 (\alpha_1 \beta - 1)\, \dot{w} = - \frac{\partial P}{\partial x} + \rho_2 g_2 \tag{4.17}$$

which is compatible with (3.53).

The momentum flux derived from this model is the same as $2k$, where k is given in (4.8), leading to an overestimate in view of the arguments leading to (3.68), but properly capturing the total stress and momentum flux in (3.152).

4.2 Three-Dimensional Flow

Cook and Harlow (1984) extended the model illustrated in Figure 4.1 to the three-dimensional case by replacing v_1' and v_2 by vector velocity components. They assumed equal pressures in the (incompressible) phases and took "f" to be a constant.

In this section a similar formulation will be derived and extended. The phases will not be considered to be incompressible but it will be assumed that there is no mass transfer between the entrained and unentrained "phases", this being equivalent to holding $(f\rho_1/\rho_2)$ constant.

Cook and Harlow introduced coefficients to account for friction and momentum transfer between the phases. These will be omitted in the present derivation in order to focus on the terms in the momentum equations that derive directly from the structure of the flow.

Omitting speculation about how this is actually possible, Cook and Harlow assumed that none of the phases illustrated in Figure 4.1 interferes with the motion of another when $\mathbf{v}_1'$ and $\mathbf{v}_2$ are not parallel. For ease in algebraic manipulation we write the *exertia* as

$$D = \alpha_1 \beta - 1 \tag{4.18}$$

and express (4.3) in vector form as

$$\mathbf{v}'_1 = \mathbf{v}_1 + D\mathbf{w} \tag{4.19}$$

If there is no mass exchange either between the phases or between the entrained and unentrained portions of phase 1, there are three independent continuity equations

$$\frac{\partial}{\partial t}\left(\frac{\rho_1}{\beta}\right) + \nabla \cdot \left(\frac{1}{\beta}\rho_1 \mathbf{v}'_1\right) = 0 \tag{4.20}$$

$$\frac{\partial}{\partial t}\left[\left(\alpha_1 - \frac{1}{\beta}\right)\rho_1\right] + \nabla \cdot \left[\left(\alpha_1 - \frac{1}{\beta}\right)\rho_1 \mathbf{v}_2\right] = 0 \tag{4.21}$$

$$\frac{\partial}{\partial t}(\alpha_2\rho_2) + \nabla \cdot (\alpha_2\rho_2\mathbf{v}_2) = 0 \tag{4.22}$$

(4.20) and (4.21) are easily combined with the vector version of (4.2) to get the overall continuity equation for phase 1:

$$\frac{\partial}{\partial t}(\alpha_1\rho_1) + \nabla \cdot (\alpha_1\rho_1\mathbf{v}_1) = 0 \tag{4.23}$$

The momentum equation for the unentrained fluid is

$$\frac{\partial}{\partial t}\left(\frac{\rho_1 \mathbf{v}'_1}{\beta}\right) + \nabla \cdot \left(\frac{1}{\beta}\rho_1 \mathbf{v}'_1 \mathbf{v}'_1\right) = -\frac{\nabla P}{\beta} \tag{4.24}$$

Expanding and combining (4.20) and (4.24) gives the equation of motion of the unentrained fluid in the usual form for single phase flow:

$$\rho_1\left(\frac{\partial \mathbf{v}'_1}{\partial t} + \mathbf{v}'_1 \cdot \nabla \mathbf{v}'_1\right) = -\nabla P \tag{4.25}$$

Substituting from (4.19)

$$\frac{\partial}{\partial t}(\mathbf{v}_1 + D\mathbf{w}) + (\mathbf{v}_1 + D\mathbf{w})\cdot\nabla(\mathbf{v}_1 + D\mathbf{w}) = -\frac{\nabla P}{\rho_1} \tag{4.26}$$

Expanding the terms in (4.26) we have

$$\frac{\partial \mathbf{v}_1}{\partial t} + \mathbf{v}_1\cdot\nabla\mathbf{v}_1 + D\left[\frac{\partial \mathbf{w}}{\partial t} + \mathbf{v}_1\cdot\nabla\mathbf{w} + \mathbf{w}\cdot\nabla\mathbf{v}_1 + \mathbf{w}\cdot\nabla(D\mathbf{w})\right]$$

$$+ \mathbf{w}\left(\frac{\partial D}{\partial t} + \mathbf{v}_1\cdot\nabla D\right) = -\frac{\nabla P}{\rho_1} \tag{4.27}$$

Now, the various terms in the square bracket in (4.27) can be written as follows:

$$\frac{\partial \mathbf{w}}{\partial t} = \frac{\partial \mathbf{v}_1}{\partial t} - \frac{\partial \mathbf{v}_2}{\partial t} \tag{4.28}$$

$$\mathbf{v}_1\cdot\nabla\mathbf{w} + \mathbf{w}\cdot\nabla\mathbf{v}_1 + \mathbf{w}\cdot\nabla(D\mathbf{w}) = \mathbf{v}_1\cdot\nabla\mathbf{v}_1 - \mathbf{v}_2\cdot\nabla\mathbf{v}_2$$

$$+ \mathbf{w}\cdot\nabla\left[(D+1)\mathbf{w}\right] \tag{4.29}$$

$\partial D/\partial t$ is evaluated by rewriting (4.21) as

$$\frac{\partial}{\partial t}\left(\frac{D\rho_1}{\beta}\right) + \nabla\cdot\left(\frac{D\rho_1\mathbf{v}_2}{\beta}\right) = 0 \tag{4.30}$$

Expanding the first differential and using (4.20) to eliminate $\partial(\rho_1/\beta)/\partial t$ we get

$$\frac{\partial D}{\partial t} = -\mathbf{v}_2\cdot\nabla D - \frac{D\beta}{\rho_1}\nabla\cdot\left(\frac{\rho_1\mathbf{v}_2 - \rho_1\mathbf{v}_1'}{\beta}\right) \tag{4.31}$$

We now use the vector version of (4.4) to simplify the final term and replace β by using (4.18). The result is

$$\frac{\partial D}{\partial t} = -\mathbf{v}_2\cdot\nabla D + \frac{D(D+1)}{\alpha_1\rho_1}\nabla\cdot(\alpha_1\rho_1\mathbf{w}) \tag{4.32}$$

When (4.28), (4.29) and (4.32) are used in (4.27) there results

$$\rho_1 \left(\frac{\partial \mathbf{v}_1}{\partial t} + \mathbf{v}_1 \cdot \nabla \mathbf{v}_1 + Z \right) + D\rho_1 \left(\frac{\partial \mathbf{v}_1}{\partial t} + \mathbf{v}_1 \cdot \nabla \mathbf{v}_1 + Z - \frac{\partial \mathbf{v}_2}{\partial t} - \mathbf{v}_2 \cdot \nabla \mathbf{v}_2 \right) = - \nabla P \tag{4.33}$$

with

$$Z = \frac{1}{\alpha_1 \rho_1} \nabla \cdot (\alpha_1 \rho_1 D \mathbf{w}\mathbf{w}) \tag{4.34}$$

(4.33) is the *equation of motion of phase 1* in which the effective acceleration is modified by introducing an additional term Z, due to the relative motion, in *both* the direct inertia term *and* the *added mass* term. One might anticipate such a modification since, in a rigorous averaging procedure,

$$\langle \mathbf{u}_1 \cdot \nabla \mathbf{u}_1 \rangle \neq \langle \mathbf{u}_1 \rangle \cdot \nabla \langle \mathbf{u}_1 \rangle \tag{4.35}$$

The interested reader may wish to show that these results are compatible with Cook and Harlow's (1984) equation (21), which can be derived directly from their (17) without using their (16) at all, thus emphasizing that phase 2 is not involved.

To obtain the equation of motion for the combined phases we make a momentum balance on the entire contents of a unit cell. The momentum flux tensor is

$$\Pi = \alpha_2 \rho_2 \mathbf{v}_2 \mathbf{v}_2 + \rho_1 \left(\frac{1}{\beta} \mathbf{v}_1' \mathbf{v}_1' + \frac{D}{\beta} \mathbf{v}_2 \mathbf{v}_2 \right) \tag{4.36}$$

It is straightforward to show that this is the same as

$$\Pi = \alpha_1 \rho_1 \mathbf{v}_1 \mathbf{v}_1 + \alpha_2 \rho_2 \mathbf{v}_2 \mathbf{v}_2 + \alpha_1 \rho_1 D \mathbf{w}\mathbf{w} \tag{4.37}$$

The overall momentum balance equation is

$$\frac{\partial \boldsymbol{m}}{\partial t} + \nabla \cdot \Pi = - \nabla P \tag{4.38}$$

where $\boldsymbol{m}$ is the vector equivalent of (4.6).

When (4.37) is used in (4.38) and (4.22) and (4.23) are invoked, the *overall equation of motion* is found to be

$$\alpha_1 \rho_1 \left(\frac{\partial \mathbf{v}_1}{\partial t} + \mathbf{v}_1 \cdot \nabla \mathbf{v}_1 + \mathbf{Z} \right) + \alpha_2 \rho_2 \left(\frac{\partial \mathbf{v}_2}{\partial t} + \mathbf{v}_2 \cdot \nabla \mathbf{v}_2 \right) = - \nabla P \tag{4.39}$$

If we now subtract α_1 times (4.33) from this, the result is the *equation of motion of phase 2:*

$$\rho_2 \left(\frac{\partial \mathbf{v}_2}{\partial t} + \mathbf{v}_2 \cdot \nabla \mathbf{v}_2 \right) - \bar{D} \frac{\alpha_1 \rho_1}{\alpha_2} \left(\frac{\partial \mathbf{v}_1}{\partial t} + \mathbf{v}_1 \cdot \nabla \mathbf{v}_1 + \mathbf{Z} - \frac{\partial \mathbf{v}_2}{\partial t} - \mathbf{v}_2 \cdot \nabla \mathbf{v}_2 \right) = - \nabla P \tag{4.40}$$

The form of (4.23) and (4.40) emphasizes the *mutual nature* of the *added mass* force, that acts in equal and opposite directions on the phases.

Of course, all of the above is merely a self-consistent derivation of the consequences of a certain simplifying assumption. While the form of the results is enlightening, they do not necessarily clarify the nature of the rigorously-derived momentum flux and the appropriate phase pressures for a real suspension.

Clearly, equations (4.33) and (4.39) could be combined to give explicit expressions for the quantities

$$\left(\frac{\partial \mathbf{v}_2}{\partial t} + \mathbf{v}_2 \cdot \nabla \mathbf{v}_2 \right) \quad \textit{and} \quad \left(\frac{\partial \mathbf{v}_1}{\partial t} + \mathbf{v}_1 \cdot \nabla \mathbf{v}_1 + \mathbf{Z} \right)$$

should be this advantageous.

If it is desired to reduce these equations to one-dimensional form, care must be taken over the correct reduction of (4.34), using continuity equations to evaluate divergences.

5. A SIMPLE EXAMPLE: SLUG FLOW

Many of the results obtained in Section 3, and several new ones, may be illustrated by analyzing the simple idealized flow pattern shown in Figure 5.1. By considering a specific geometry of the phases it is possible to determine the actual coefficients in the general theory and to check the compatibility of its parts. The overall equations can be related to *averages* of more fundamental formulations. Moreover, difficulties in evaluating all of the terms under some circumstances identify aspects of the theory that need better understanding.

It is assumed that phase 2 moves in the form of long cylindrical slugs in a pipe. The fraction of the length of each unit cell that contains these slugs is "b" and they occupy a fraction "α_s" of the cross-section. The (uniform) velocities of phase 1 in the regions between and around the slugs are respectively v_{1f} and v_{1s}. The slugs and the gaps between them are so long that *end effects* can be ignored at this stage; they will be considered in more detail later. It is assumed that the transitions between v_{1f} and v_{1s} at both ends of each slug are reversible.

From volume averaging, the mean velocity of phase 1 is

$$v_1 \left[(1 - \alpha_s)bv_{1s} + (1 - b)v_{1f}\right]/(1 - \alpha_s b) \tag{5.1}$$

The term in the denominator of (5.1) is the average volumetric concentration of phase 1:

$$\alpha_1 = 1 - \alpha_s b \tag{5.2}$$

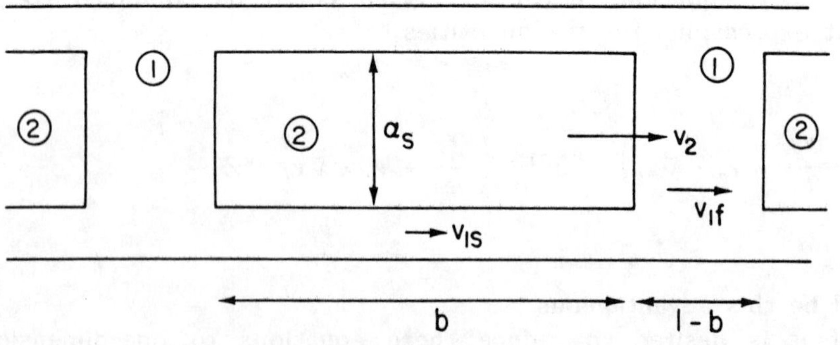

Figure 5.1 Idealized slug flow.

Similarly,

$$\alpha_2 = \alpha_s b \tag{5.3}$$

If the phases are incompressible, the overall volumetric flux is constant, therefore:

$$\alpha_s v_2 + (1 - \alpha_s) v_{1s} = v_{1f} \tag{5.4}$$

From (5.1) and (5.4), v_{1f} and v_{1s} may be expressed in terms of the phase average velocities:

$$v_{1f} = v_1 - b\alpha_s (v_1 - v_2) \tag{5.5}$$

$$v_{1s} = v_1 + \frac{\alpha_s}{1 - \alpha_s} (1 - b)(v_1 - v_2) \tag{5.6}$$

If the flow is *potential*, the macroscopic gradient across an entire unit cell is

$$-\frac{d\Phi}{dx} = b v_{1s} + (1 - b) v_{1f} \tag{5.7}$$

Using (5.5) and (5.6) in (5.7) we get

$$-\frac{d\Phi}{dx} = v_1 + (v_1 - v_2) \frac{b(1 - b)\alpha_s^2}{1 - \alpha_s} \tag{5.8}$$

Therefore, by comparison with (3.36) we have, for the *exertia*,

$$(\alpha_1 \beta - 1) = \frac{b(1 - b)\alpha_s^2}{1 - \alpha_s} \tag{5.9}$$

If the flow is accelerated in the x-direction, the pressure gradients across the two regions are

$$\left(-\frac{dp}{dx}\right)_f = \rho_1 \dot{v}_{1f} \tag{5.10}$$

$$\left(-\frac{dp}{dx}\right)_s = \rho_1 \dot{v}_{1s} = \rho_2 \dot{v}_2 \tag{5.11}$$

The overall pressure gradient is

$$-\frac{dP}{dx} = (1-b)\left(-\frac{dp}{dx}\right)_f + b\left(-\frac{dp}{dx}\right)_s = \rho_1\left[(1-b)\dot{v}_{1f} + b\dot{v}_{1s}\right] \tag{5.12}$$

The term in square brackets is just the time derivative of (5.7), therefore we may use (5.8) to obtain

$$-\frac{dP}{dx} = \rho_1 \dot{v}_1 + \rho_1(\dot{v}_1 - \dot{v}_2)\frac{b(1-b)\alpha_s^2}{1-\alpha_s} \tag{5.13}$$

which is equivalent to (3.52) and consistent with (5.9).

The equation of motion of phase 2 may be derived by using the second part of (5.11) in (5.12) to get

$$-\frac{dP}{dx} = \rho_1(1-b)\dot{v}_{1f} + \rho_2 b\dot{v}_2 \tag{5.14}$$

we now use (5.5) in (5.14), divide the result by b and add $(1 - 1/b)$ times (5.13) to get

$$-\frac{dP}{dx} = \rho_2 \dot{v}_2 - \rho_1(\dot{v}_1 - \dot{v}_2)\frac{(1-b)\alpha_s}{1-\alpha_s}(1 - b\alpha_s) \tag{5.15}$$

which is compatible with (3.53) when (5.2), (5.3) and (5.9) are used.

The pressure difference between the regions "*f*" and "*s*" may be determined by imposing a velocity $(-v_2)$ on Figure 5.1 and using Bernoulli's equation:

$$p_f - p_s = \frac{1}{2}\rho_1 \left[(v_2 - v_{1s})^2 - (v_2 - v_{1f})^2\right] \tag{5.16}$$

Using (5.5) and (5.6) this becomes

$$p_f - p_s = \frac{1}{2}\rho_1 (v_1 - v_2)^2 \frac{a_s(2 - \alpha_s)}{(1 - \alpha_s)^2} (1 - \alpha_s b)^2 \tag{5.17}$$

The mean pressure in phase 1 is

$$p_1 = \frac{(1 - b)p_f + (1 - \alpha_s)bp_s}{(1 - \alpha_s b)} \tag{5.18}$$

while

$$p_2 = p_s \tag{5.19}$$

Combining (5.18) and (5.19) leads to

$$p_1 - p_2 = \frac{1 - b}{1 - \alpha_s b} (p_f - p_s) \tag{5.20}$$

Using this in (5.17) gives an expression for the difference between the mean phase pressures

$$p_1 - p_2 = \frac{1}{2}\rho_1 (v_1 - v_2)^2 \frac{\alpha_s(2 - \alpha_s)}{(1 - \alpha_s)^2} (1 - b)(1 - \alpha_s b) \tag{5.21}$$

To compare this with (3.16) we choose it in the form

$$p_1 - p_2 = -\frac{1}{2}\rho_1 (v_1 - v_2)^2 \alpha_1 \frac{\partial}{\partial \alpha_1} (\alpha_1 \beta - 1) \tag{5.22}$$

Since end effects are ignored, the only way α_1 can change in a uniform expansion of phase 2 is by α_s changing. Therefore, b should be regarded as constant and, from (5.2), we have

$$\frac{\partial \alpha_1}{\partial \alpha_s} = -\, b \tag{5.23}$$

Using (5.9) and (5.23) in (5.22),

$$(p_1 - p_2) = -\frac{1}{2}\rho_1 (v_1 - v_2)^2 \alpha_1 \frac{\partial}{\partial \alpha_s}\left[\frac{b(1-b)\alpha_s^2}{1-\alpha_s}\right] / (-b)$$

$$= \frac{1}{2}\rho_1 (v_1 - v_2)^2 \alpha_1 (1-b)\,\frac{\alpha_s(2-\alpha_s)}{(1-\alpha_s)^2} \tag{5.24}$$

which agrees with (5.21) in view of (5.2).

The kinetic energy per unit volume of the entire flow is

$$k = \frac{1}{2}\rho_2 b \alpha_s v_2^2 + \frac{1}{2}\rho_1 \left[(1-b)v_{1f}^2 + b(1-\alpha_s)v_{1s}^2\right] \tag{5.25}$$

Using (5.2), (5.3), (5.5) and (5.6) this can be transformed to

$$k = \frac{1}{2}\rho_2 b \alpha_s v_2^2 + \frac{1}{2}\rho_1 \alpha_1 \left[v_1^2 + (v_1 - v_2)^2 \frac{b(1-b)\alpha_s^2}{1-\alpha_s}\right] \tag{5.26}$$

which is compatible with (3.35) in view of (5.9).

5.1 Macroscopic and Average Pressure Gradients

The mean pressure gradient in phase 1 is not necessarily the same as the macroscopic pressure gradient. We may illustrate this idea for the present example by averaging the *local* equation of motion of phase 1:

$$\left\langle -\frac{dp}{dx}\right\rangle_1 = \rho_1 \langle \dot{v}_1 \rangle \tag{5.27}$$

The average value of $\dot{v}_1$ is merely the time derivative of the average of v_1. However

$$< -\frac{dp}{dx} >_1 = \frac{(1-b)\left(-\frac{dp}{dx}\right)_f + b(1-\alpha_s)\left(-\frac{dp}{dx}\right)_s}{1 - b\alpha_s} \tag{5.28}$$

whereas the macroscopic pressure gradient is given by

$$-\frac{dP}{dx} = (1-b)\left(-\frac{dp}{dx}\right)_f + b\left(-\frac{dp}{dx}\right)_s \tag{5.29}$$

From (5.28) and (5.29) we get

$$-\frac{dP}{dx} = < -\frac{dp}{dx} >_1 - \frac{b(1-b)\alpha_s}{1-b\alpha_s}\left[\left(-\frac{dp}{dx}\right)_f - \left(-\frac{dp}{dx}\right)_s\right] \tag{5.30}$$

Using (5.10) and (5.11), as well as (5.5) and (5.6) it is found that

$$-\left(\frac{dp}{dx}\right)_f - \left(-\frac{dp}{dx}\right)_s = \rho_1(\dot{v}_{1f} - \dot{v}_{1s}) = -\rho_1(\dot{v}_1 - \dot{v}_2)\frac{\alpha_s}{1-\alpha_s}(1 - b\alpha_s) \tag{5.31}$$

Using (5.27) and (5.31) in (5.30) recreates (5.13) and shows that it is the interphase *added mass force* that accounts for the difference between the macroscopic and average pressure gradients, as might be deduced from an appropriate *momentum balance*.

5.2 Convective Acceleration

Suppose that the pattern shown in Figure 5.1 is introduced into a *nozzle* with converging walls. By suitable design it is possible to keep α_s constant and arrange for constant acceleration in each region. An observer at a given location will see region "*f*" (pure phase 1) for a fraction *(1 - b)* of the time and the composite stratified, region "*s*", for a fraction *(b)* of the time. The pressure gradients in each of these regions, which are essentially in steady flow, are $-\rho_1 v_{1f}(dv_{1f}/dx)$ and $-\rho_1 v_{1s}(dv_{1s}/dx)$. The pressure jumps between the regions, given by (5.17) do not play a role in determining the macroscopic pressure gradient which is

$$- \frac{dP}{dx} = \rho_1 \left((1 - b)v_{1f} \frac{dv_{1f}}{dx} + bv_{1s} \frac{dv_{1s}}{dx} \right) = \frac{\rho_1}{2} \frac{d}{dx} \left[(1 - b)v_{1f}^2 + bv_{1s}^2 \right] \tag{5.32}$$

using (5.5) and (5.6), the term in square brackets above can be arranged to give

$$- \frac{dP}{dx} = \frac{\rho_1}{2} \frac{d}{dx} \left[v_1^2 + \frac{\alpha_s^2 b (1 - b)}{1 - \alpha_s} (v_1^2 - v_2^2) + \right.$$

$$\left. (v_1 - v_2)^2 \frac{\alpha_s^2 b}{(1 - \alpha_s)} (1 - b)(1 - b\alpha_s)(2 - \alpha_s) \right] \tag{5.33}$$

when the differentiation is performed, the first term gives the *single-phase* contribution to the pressure gradient, the second term gives an *added mass* term proportional to $v_1(dv_1/dz) - v_2(dv_2/dz)$, and the third *Bernoulli* term is α_2 times the factor multiplying $\rho_1/2$ in (5.24); a term of this form but with different coefficient was introduced by Drew and Wood (1985). Equation (5.24), however, cannot be used to evaluate $(p_1 - p_2)$ because pressures are not uniform in the various regions.

If we persist in *neglecting "end effects"* and assume that the pressure gradient along phase 2 is the same as that in the fluid between it and the wall, we may write (based on time-averaging),

$$- \left\langle \frac{dp_1}{dx} \right\rangle = \frac{\rho_1}{2(1 - \alpha_s b)} \frac{d}{dx} \left[b(1 - \alpha_s)v_{1s}^2 + (1 - b)v_{1f}^2 \right] \tag{5.34}$$

$$- \left\langle \frac{dp_2}{dx} \right\rangle = \frac{\rho_1}{2} \frac{d}{dx} v_{1s}^2 \tag{5.35}$$

(5.33) through (5.35) are consistent with the statement

$$\frac{dP}{dx} = \alpha_1 \left\langle \frac{dp_1}{dx} \right\rangle + \alpha_2 \left\langle \frac{dp_2}{dx} \right\rangle \tag{5.36}$$

To emphasize the three basic forms of terms contributing to effective accelerations we can express v_{1f}^2 and v_{1s}^2, using (5.5) and (5.6), as

$$v_{1s}^2 = v_1^2 + (v_1^2 - v_2^2)\frac{\alpha_s(1 - b)}{1 - \alpha_s} + (v_1 - v_2)^2 \frac{\alpha_s(1 - b)(1 - \alpha_s b)}{(1 - \alpha_s)^2} \tag{5.37}$$

$$v_{1f}^2 = v_1^2 - (v_1^2 - v_2^2)\alpha_s b - (v_1 - v_2)^2\alpha_s b(1 - \alpha_s b) \tag{5.38}$$

Using (5.37) and (5.38) in (5.34) we obtain

$$-\left\langle \frac{dp_1}{dx} \right\rangle = \frac{\rho_1}{2}\frac{d}{dx}\left[v_1^2 + (v_1 - v_2)^2 \frac{\alpha_s^2 b(1 - b)}{1 - \alpha_s}\right] \tag{5.39}$$

which looks like an effective Bernoulli equation for phase 1 with the kinetic energy due to relative motion included.

Again *neglecting end effects* on the slug of phase 2, we may write

$$-\left\langle \frac{dp_2}{dx} \right\rangle = \frac{\rho_2}{2}\frac{d}{dx}(v_2^2) \tag{5.40}$$

When this is combined with (5.35), (5.32) can be put in the form

$$-\frac{dP}{dx} = \frac{\rho_2}{2}\frac{dv_2^2}{dx} + \frac{\rho_1}{2}\frac{d}{dx}(1 - b)(v_{1f}^2 - v_{1s}^2) = \frac{\rho_2}{2}\frac{dv_2^2}{dx}$$

$$-\frac{\rho_1}{2}\frac{d}{dx}\left[(v_1^2 - v_2^2)\frac{\alpha_s(1 - b)}{1 - \alpha_s}(1 - \alpha_s b)\right.$$

$$\left. + (v_1 - v_2)^2 \frac{\alpha_s(1 - b)(1 - \alpha_s b)(1 - \alpha_s b(2 - \alpha_s))}{(1 - \alpha_s)^2}\right] \tag{5.41}$$

(5.33) and (5.41) have the pleasing characteristic that the added mass term is a *mutual force* that disappears when we take α_1 times (5.33) and add it to α_2 times (5.41) to get the *equation of motion for the mixture*:

$$-\frac{dP}{dx} = \frac{1}{2}\frac{d}{dx}\left[\alpha_1\rho_1 v_1^2 + \alpha_2\rho_2 v_2^2 + \rho_1\alpha_1 \frac{\alpha_s^2 b(1-b)}{1-\alpha_s}(v_1 - v_2)^2\right] \tag{5.42}$$

which is consistent with (5.36), (5.39) and (5.40).

Of course, all of the preceding derivations were contingent upon the *disregard of end effects.* In the case of the pipe of uniform diameter this was justifiable since pressure drop and force from the fluid on the *nose* of the slug was balanced by an equal and opposite effect at the *tail.* In the converging pipe, however, it is not clear that this is so and a more thorough accounting for these effects must be made.

It is notable that the term in square brackets in (5.42) is identical with $2k$ in (5.26), suggesting that (5.42) has the character of an energy-conserving *two-phase Bernoulli equation.*

5.3 End Effects

In the previous section, changes occurring around the ends of the *slug* of phase 2 were ignored. It will now be shown that when they are considered, the coefficients of some of the $(v_1 - v_2)^2$ terms in the former analysis should be changed.

Consider first the steady flow past the nose of the slug, brought to rest by imposing a velocity $-v_2$ on the slug. The velocities in this *base* case are v_{fo} and v_{so} as shown in Figure 5.2.

From continuity

$$v_{fo} = (1 - \alpha_s)v_{so} \tag{5.43}$$

From Bernoulli's equation

$$p_f - p_s = \frac{1}{2}\rho_1(v_{so}^2 - v_{fo}^2) \tag{5.44}$$

From a momentum balance for the control volume in Figure 5.2, if p_2 is the average pressure in phase 2,

$$p_f + \rho v_{fo}^2 = p_s(1 - \alpha_s) + p_2\alpha_s + \rho v_{fo} v_{so} \tag{5.45}$$

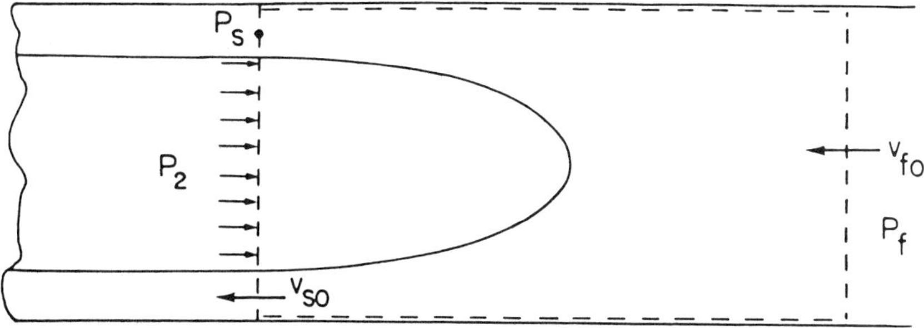

Figure 5.2 Flow around the nose of the slug.

Combining (5.43) to (5.45) we get

$$p_2 = p_s + \frac{1}{2} \rho_1 \alpha_s v_{so}^2 \tag{5.46}$$

Therefore, p_2 and p_s are not equal, as assumed previously. Since

$$v_{so} = v_2 - v_{1s} \tag{5.47}$$

(5.46) may be transformed with the help of (5.6) to

$$p_2 = p_s + \frac{1}{2} \rho_1 \frac{\alpha_s}{(1 - \alpha_s)^2} (1 - \alpha_s b)^2 (v_1 - v_2)^2 \tag{5.48}$$

These results are still valid to first order if the pipe is slightly tapered.

When the flow is accelerating, these pressure changes are superimposed on the ends of the slug (Figure 5.3).

Since the slug is long, its acceleration can be computed by neglecting the inertia of the small masses contained in the nose and tail. The p_s contribution around the whole slug is the same as it would be in a uniform accelerated field $v_{1s}(dv_{1s}/dz)$ in phase 1. The *additional* contribution in (5.48) will be compatible with a continuous additional pressure gradient

$$\rho_1 \frac{\alpha_s}{(1 - \alpha_s)^2} (1 - \alpha_s b)^2 (v_1 - v_2) \frac{d(v_1 - v_2)}{dx}$$

along the entire length of the interior of the slug as long as there is an interfacial hoop stress of this amount everywhere; it is needed to maintain the geometry of the slug. Therefore, the equation of motion of every particle of phase 2 in the slug is

$$-\frac{dp_2}{dx} = \rho_2 v_2 \frac{dv_2}{dx} = \frac{\rho_1}{2}\frac{d}{dx}\left[v_{1s}^2 - \frac{\alpha_s}{(1-\alpha_s)^2}(1-\alpha_s b)(v_1 - v_2)^2\right] \tag{5.49}$$

which differs from (5.35) because of the effect of end-conditions. If we substitute for v_{1s}^2 in (5.49) from (5.37), the result is

$$\rho_2 v_2 \frac{dv_2}{dx} = \frac{\rho_1}{2}\frac{d}{dx}\left[v_1^2 + (v_1^2 - v_2^2)\frac{\alpha_s(1-b)}{1-\alpha_s} - (v_1 - v_2^2)\frac{\alpha_s b(1-\alpha_s b)}{1-\alpha_s}\right] \tag{5.50}$$

The added mass term in this equation is the same as would occur if (5.37) were combined with (5.35). However, the *Bernoulli* term is different and has even changed sign.

The need to invoke a hoop stress in order to maintain the interfacial geometry is awkward since it introduces a new force,

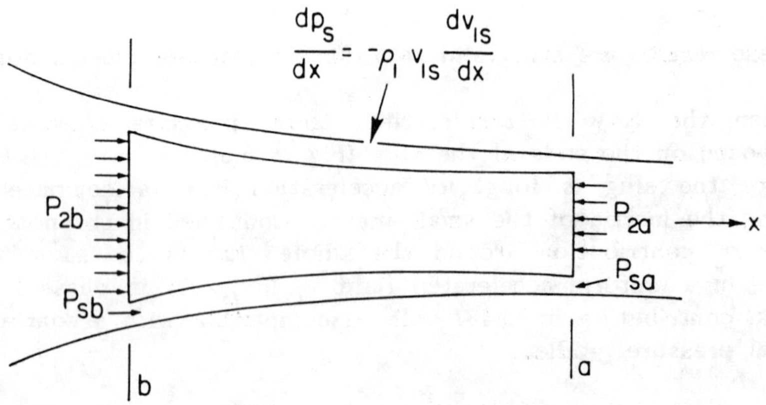

Figure 5.3 Slug flow through a nozzle

capable of doing work, that will play a role in the conservation laws. Moreover, the attendant jump condition at the interface complicates the evaluation of certain average properties (see Drew and Wood (1985)).

The conclusion from this more thorough, but still incomplete, investigation appears to be to uphold the added mass formulation in its usual form while pointing to the need for a better understanding of the *Bernoulli* term that produces apparent forces proportional to $(v_1 - v_2)(d/dx)(v_1 - v_2)$.

6. A SIMPLE EXAMPLE: ONE-DIMENSIONAL INCOMPRESSIBLE REVERSIBLE STEADY NOZZLE FLOW

Figure 6.1 illustrates one of the simplest examples in which convective accelerations play a role. An incompressible steady two-phase flow with known flow rates is accelerated reversibly in a nozzle from stagnation conditions. This case might serve as a *standard problem* for testing any hypothesized *two-fluid model* of two phase flow.

At any location, x, the volumetric fluxes of the phases, j_1 and j_2, are determined by dividing the individual volumetric flow rates by the total cross-section area. The average velocities of the phases are then

$$v_1 = \frac{j_1}{\alpha_1} \tag{6.1}$$

$$v_2 = \frac{j_2}{\alpha_2} \tag{6.2}$$

with

$$\alpha_1 + \alpha_2 = 1 \tag{6.3}$$

If we assume that (3.35) defines the kinetic energy density everywhere, then

$$k = \frac{1}{2}\rho_1 \frac{j_1^2}{\alpha_1} + \frac{1}{2}\rho_2 \frac{j_2^2}{\alpha_2} + \frac{1}{2}\rho_1\alpha_1(\alpha_1\beta - 1)\left(\frac{j_1}{\alpha_1} - \frac{j_2}{\alpha_2}\right)^2 \tag{6.4}$$

Since β is a function of α_1 (or α_2), the only free parameter in (6.4) is α_1 (or α_2). Moreover, j_1 and j_2 are always known at any section, for given flow rates. Therefore, if we minimize (6.4) with respect to α_1 at any section, it will be minimized at all sections and *the total kinetic energy will be a minimum.* It is uncertain whether the flow must satisfy this condition but it is interesting to pursue the consequences of making such an assumption. To do this we first define B by the equation

$$(\alpha_1\beta - 1) = \alpha_2 B \tag{6.5}$$

and then differentiate (6.4) and set the result to zero to obtain, after reuse of (6.1) and (6.2),

$$\frac{dk}{d\alpha_1} = -\rho_1 v_1^2 (1 + B) + \rho_2 v_2^2 \left(1 + \frac{\rho_1 B}{\rho_2}\right) + \rho_1 \alpha_1 \alpha_2 (v_1 - v_2)^2 \frac{dB}{d\alpha_1} = 0 \tag{6.6}$$

Since (j_1/j_2) is known and B is a function of α_1, use of (6.1) and (6.2) shows that (6.6) is an equation for α_1 which is constant throughout the nozzle.

While one might speculate about the conditions for real roots to (6.6), the simplest case is obtained when B is independent of α_2. We then obtain

$$n = \left(\frac{1 + B}{\frac{\rho_2}{\rho_1} + B}\right)^{1/2} \tag{6.7}$$

where "n" is the *slip ratio*, defined as

$$n = v_2/v_1 \tag{6.8}$$

For example, if Maxwell's prediction for spheres, equation (3.20), is used, we have $B = 1/2$. This would imply that in the reversible acceleration of a suspension of light $(\rho_2/\rho_1 \ll 1)$ spheres, or bubbles, from rest the slip ratio should be $\sqrt{3}$. Of course, this prediction rests on the hypothesis that kinetic energy should be minimized.

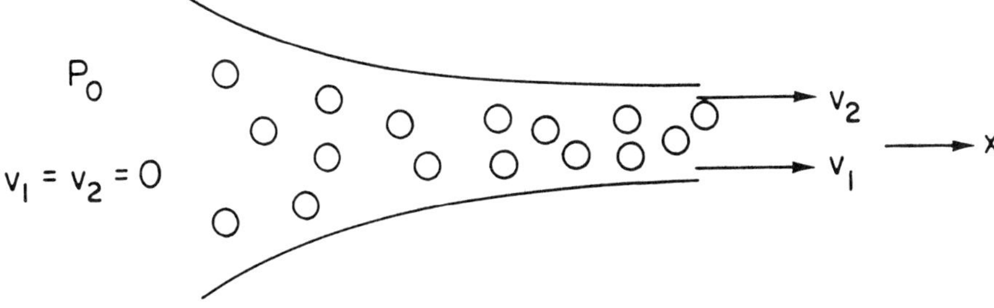

Figure 6.1 Dispersed flow through a nozzle

Now, if (6.6) is valid throughout the nozzle, we may differentiate it to obtain

$$-\rho_1 v_1 \frac{dv_1}{dx} + \rho_2 v_2 \frac{dv_2}{dx} + B\rho_1 \left(v_2 \frac{dv_2}{dx} - v_1 \frac{dv_1}{dx} \right)$$

$$+ \rho_1 \alpha_1 \alpha_2 \frac{dB}{d\alpha_1} (v_1 - v_2) \left(\frac{dv_1}{dx} - \frac{dv_2}{dx} \right) = 0 \tag{6.9}$$

B, α_1 α_2 and $dB/d\alpha_1$ have been treated as constants that are determined from (6.5) and (6.6).

Equation (6.9) can be rearranged in several ways. For instance, we could write it as

$$\rho_1 v_1 \frac{dv_1}{dx} + \alpha_2 B\rho_1 \left(v_1 \frac{dv_1}{dx} - v_2 \frac{dv_2}{dx} \right)$$

$$+ b\alpha_1 \alpha_2 \frac{dB}{d\alpha_1} \rho_1 (v_1 - v_2) \frac{d}{dx} (v_1 - v_2) = \rho_2 v_2 \frac{dv_2}{dx}$$

$$- \alpha_1 B\rho_1 \left(v_1 \frac{dv_1}{dx} - v_2 \frac{dv_2}{dx} \right) + (1+b) \alpha_1 \alpha_2 \frac{dB}{d\alpha_1} \rho_1 (v_1 - v_2) \frac{d}{dx} (v_1 - v_2) \tag{6.10}$$

Moreover, both sides of (6.10) could be set equal to $(-dP/dx)$, without specifying exactly what "P" is, though it has the dimensions of pressure. The results would be two *equations of motion* with the added mass term resembling what was obtained in (3.52) and (3.53).

"b" is an arbitrary, or unknown, parameter at this stage. Adding α_1 times the equation for phase *1* to α_2 times the equation for phase 2 we get

$$\alpha_1\rho_1 v_1 \frac{dv_1}{dx} + \alpha_2\rho_2 v_2 \frac{dv_2}{dx} + (b + \alpha_2)\alpha_1\alpha_2 \frac{dB}{d\alpha_1} \rho_1 (v_1 - v_2) \frac{d(v_1 - v_2)}{dx}$$

$$= - \frac{dP}{dx} \tag{6.11}$$

which could be a reasonable form for the *equation of motion for the mixture.* The term involving the relative velocity will be zero if $dB/d\alpha_1 = 0$, as it is for a Maxwellian suspension of spheres, or if $b = - \alpha_2$. If we choose $b = (\alpha_1/\alpha_2) - \alpha_2$, the term resembles the *pressure correction* in (3.15). However, there is not enough information in these derivations to determine what the value of "b" should be or whether "P" is some easily definable *pressure* or a composite including terms representing momentum flux.

Perhaps the most interesting feature of this short derivation is the way in which the *added mass* and *Bernoulli* terms occur naturally in (6.9) and (6.10) as a result of differentiating the kinetic energy density, suggesting that the true equations of motion might be derivable from a suitably-formulated variational principle.

7. COMPARISON WITH HYPOTHESIZED GENERAL EQUATIONS

Up to now the discussion of *general equations of motion* has been avoided, the emphasis being on establishing a set of fundamental results that might eventually be used to test or infer a more comprehensive theory.

We will now make some comparisons with a few equations selected from those that have appeared in the literature. This will serve to illustrate how the present developments might be related to more global approaches, and also some of the difficulties that arise when this is done.

7.1 Pauchon-Banerjee Equations

The equations of motion for the two phases developed by Pauchon and Banerjee (1986) are, in one-dimensional form using the present nomenclature,

$$\rho_1 \left(\frac{\partial v_1}{\partial t} + v_1 \frac{\partial v_1}{\partial x} \right) = - \frac{\partial p_1}{\partial x} - \frac{1}{4} \rho_1 \frac{w^2}{\alpha_1} \frac{\partial \alpha_1}{\partial x}$$

$$- \frac{1}{2} \rho_1 \frac{\alpha_2}{\alpha_1} \left[\left(\frac{\partial v_1}{\partial t} + v_1 \frac{\partial v_1}{\partial x} \right) - \left(\frac{\partial v_2}{\partial t} + v_2 \frac{\partial v_2}{\partial x} \right) \right] \tag{7.1}$$

$$\rho_2 \left(\frac{\partial v_2}{\partial t} + v_2 \frac{\partial v_2}{\partial x} \right) = - \frac{\partial p_2}{\partial x} + \frac{1}{2} \rho_1 \left[\frac{\partial v_1}{\partial t} + v_1 \frac{\partial v_1}{\partial x} - \left(\frac{\partial v_2}{\partial t} + v_2 \frac{\partial v_2}{\partial x} \right) \right] \tag{7.2}$$

$$p_2 - p_1 = - \frac{1}{4} \rho_1 w^2 \tag{7.3}$$

Equation (7.3) is compatible with (3.16) and (3.21) in the limit where $\alpha_1 \approx 1$.

Adding α_1 x (7.1) to α_2 x (7.2), the *added mass* terms cancel and we get the *equation of motion of the mixture* that can be rearranged, using (7.3) into the form

$$\alpha_1 \rho_1 \left(\frac{\partial v_1}{\partial t} + v_1 \frac{\partial v_1}{\partial x} \right) + \alpha_2 \rho_2 \left(\frac{\partial v_2}{\partial t} + v_2 \frac{\partial v_2}{\partial x} \right) = - \frac{\partial}{\partial x} (\alpha_1 p_1 + \alpha_2 p_2) \tag{7.4}$$

The final *added mass term* in (7.1) and (7.2) contains the *relative acceleration*, as in (2.62), resembling a more general version of (3.52) and (3.53) but with an additional α_1 in the denominator of (3.20). The rest of the equations reduce to (3.52) and (3.53) for the example being considered there, since α_1 and w are independent of x.

Equation (7.4) resembles (3.205) but without the final square bracket that arises from the divergence of the final term in (3.202). Apparently, in the Pauchon-Banerjee formulation, there is no additional overall momentum flux due to the relative motion.

In the special case of a very dilute suspension of spheres at rest in a steady converging flow we have

$$v_2 = 0, \quad \frac{\partial}{\partial t} = 0, \quad \alpha_2 = 0 \tag{7.5}$$

(7.1) to (7.3) then reduce to

$$\rho_1 v_1 \frac{dv_1}{dx} = - \frac{dp_1}{dx} \tag{7.6}$$

$$0 = - \frac{dp_2}{dx} + \frac{1}{2} \rho_1 v_1 \frac{dv_1}{dx} + f_2 \tag{7.7}$$

$$p_2 - p_1 = - \frac{1}{4} \rho_1 v_1^2 \tag{7.8}$$

where an external force f_2 per unit volume has been added to keep the sphere in position.

Combining (7.6) through (7.8) by eliminating the pressures, we get

$$f_2 = \rho_1 v_1 \frac{dv_1}{dx} \left(1 + \frac{1}{2} + \frac{2}{4}\right) \tag{7.9}$$

which has the effect of *counting the added mass force twice* compared with (2.23). Clearly, something is not compatible between this formulation and G.I. Taylor's classical result for the force on a single sphere in a convergent flow (2.23).

7.2 Drew-Wood Equations

The one-dimensional equations of motion presented by Drew and Wood (1985) resemble (7.1) and (7.2) but with an additional term. In the present nomenclature, after correcting what appears to be an error in sign in the additional term, the equations are

$$\rho_1 \left(\frac{\partial v_1}{\partial t} + v_1 \frac{\partial v_1}{\partial x}\right) = - \frac{\partial p_1}{\partial x} - \xi \rho_1 \frac{w^2}{\alpha_1} \frac{\partial \alpha_1}{\partial x} - \frac{\alpha_2}{\alpha_1} c_{vm} \rho_1 \left[\left(\frac{\partial v_1}{\partial t}\right.\right.$$

$$\left.\left. + v_1 \frac{\partial v_1}{\partial x}\right) - \left(\frac{\partial v_2}{\partial t} + v_2 \frac{\partial v_2}{\partial x}\right)\right] + \frac{1}{\alpha_1} \frac{\partial}{\partial x} \left[\alpha_1 \alpha_2 (a_1 + b_1) \rho_1 w^2\right] \tag{7.10}$$

$$\rho_2 \left(\frac{\partial v_2}{\partial t} + v_2 \frac{\partial v_2}{\partial x} \right) = - \frac{\partial p_1}{\partial x} + 2\xi \rho_1 w \frac{\partial w}{\partial x} + c_{vm} \rho_1 \left[\left(\frac{\partial v_1}{\partial t} + v_1 \frac{\partial v_1}{\partial x} \right) \right.$$

$$\left. - \left(\frac{\partial v_2}{\partial t} + v_2 \frac{\partial v_2}{\partial x} \right) \right] \qquad (7.11)$$

ξ plays the same role as the factor (1/4) in (7.3) as long as it is constant. c_{vm} is a *coefficient of virtual mass* equivalent to the present results as long as

$$c_{vm} = (\alpha_1 \beta - 1) \frac{\alpha_1}{\alpha_2} \qquad (7.12)$$

(i.e. different from *all* of $\tilde{B}$, *C* and *D*!). a_1 and b_1 are coefficients in the (Reynolds) momentum flux tensor resulting from the relative motion. Drew and Wood mention that Garipov (1973) used the values

$$\alpha_1 = - \frac{3}{20}, \qquad b_1 = - \frac{1}{20} \qquad (7.13)$$

which are consistent with the coefficients in (3.191), their origin being the same. However, their sum is (-1/5), not the same as the coefficient in the final term of (3.202), presumably because Drew and Wood do not consider the particle stress tensor. Therefore, the additional term that arises in the equivalent of (7.4) by adding α_1 x (7.10) to α_2 x (7.11) differs (in several ways) from the final term in (3.205).

When (7.10) and (7.11) are used to solve for the force on a dilute suspension of spheres in a converging flow the result is

$$f_2 = \rho_1 v_1 \frac{dv_1}{dx} (1 + c_{vm} + 2\xi) \qquad (7.14)$$

leading to the same inconsistency with (2.23) that occurred with the Pauchon-Banerjee equations.

We may also analyze the limiting case of *steady flow through a porous medium* at rest. Equation (7.10) becomes

$$\rho_1 v_1 \frac{\partial v_1}{\partial x} = -\frac{\partial p_1}{\partial x} - \xi \rho_1 \frac{v_1^2}{\alpha_1}\frac{d\alpha_1}{dx} - \frac{\alpha_2}{\alpha_1} c_{vm}\rho_1 v_1 \frac{dv_1}{dx}$$

$$+ \frac{1}{\alpha_1}\frac{d}{dx}\left[\alpha_1 \alpha_2 (a_1 + b_1)\rho_1 v_1^2\right] \tag{7.15}$$

Now, (3.81) should be valid for this situation, being the averaged *Bernoulli Equation* for the continuous phase. We can rewrite it as

$$\frac{1}{2}\rho_1 v_1^2 = p_{01} - p_1 - \frac{1}{2}\rho_1 v_1^2 (\alpha_1\beta - 1) \tag{7.16}$$

Differentiating (7.16) we get, along the flow path with incompressible flow,

$$\rho_1 v_1 \frac{dv_1}{dx} = \frac{dp_1}{dx} - (\alpha_1\beta - 1)\rho_1 v_1 \frac{dv_1}{dx} - \frac{1}{2}\rho_1 v_1^2 \frac{d}{dx}(\alpha_1\beta - 1) \tag{7.17}$$

The term involving c_{vm} on the right-hand side of (7.15) is the same as the second term in (7.17) by virtue of (7.12). Moreover, by virtue of (3.16) and the definition of ξ

$$\frac{d(\alpha_1\beta - 1)}{dx} = \frac{d(\alpha_1\beta - 1)}{d\alpha_1}\frac{d\alpha_1}{dx} = -\frac{2\xi}{\alpha_1}\frac{d\alpha_1}{dx} \tag{7.18}$$

When (7.18) is used in (7.17) it is found that the resulting term has the opposite sign to the one in (7.15). Moreover, the final term in (7.15) is missing entirely. Therefore, (7.15) is not compatible with the simpler (7.16).

The above comparison was between two forms of the "equation of motion for the fluid." We may also use (3.117) to check the momentum equation for phase 2, (7.11).

Firstly, it is clear that the second derivative of β appearing in (3.117) must result from ξ being a function of α_2. If we replace (7.3) by (3.16) and use (7.18), the additional term that appears in (7.11) is $\rho_1 w^2 (\partial\xi/\partial x)$.

For the situation illustrated in figure (3.8), we have

$$v_1 = \frac{j_1}{\alpha_1} \tag{7.19}$$

Therefore,

$$\frac{dv_1}{dx} = -\frac{j_1}{\alpha_1^2}\frac{d\alpha_1}{dx} = \frac{j_1}{\alpha_1^2}\frac{d\alpha_2}{dx} \tag{7.20}$$

Because dv_1/dx and $d\alpha_2/dx$ are related, it may not be possible to separate the "effect of each" unequivocally. We proceed, nevertheless, to try to create (3.117) from combinations of identifiable terms. Expanding (7.18),

$$\beta + \alpha_1 \frac{d\beta}{d\alpha_1} = -\frac{2\xi}{\alpha_1} \tag{7.21}$$

Multiplying (7.21) by α_1 and differentiating with respect to α_1 gives

$$\alpha_1 \frac{d\beta}{d\alpha_1} + \beta + 2\alpha_1 \frac{d\beta}{d\alpha_1} + \alpha_1^2 \frac{d^2\beta}{d\alpha_1^2} = -2\frac{d\xi}{d\alpha_1} \tag{7.22}$$

We now multiply (7.22) by (α_1) and substitute from (7.21) to get

$$-6\xi - 2(\alpha_1\beta - 1) - 2 + \alpha_1^3 \frac{d^2\beta}{d\alpha_1^2} = -2\alpha_1 \frac{d\xi}{d\alpha_1} \tag{7.23}$$

Now, the term we want on the right-hand side of (7.11) is, from (3.117),

$$f_p = \frac{\rho_1 j_1^2}{2} \frac{d^2\beta}{d\alpha_2^2} v_1 \frac{d\alpha_2}{dx} \tag{7.24}$$

From (7.20), this is equivalent to

$$f_p = \frac{\rho_1}{2}\alpha_1^3 \frac{d^2\beta}{d\alpha_2^2} v_1 \frac{dv_1}{dx} \tag{7.25}$$

Substituting from (7.23) in (7.25) we get, since $d^2\beta/d\alpha_2^2 = d^2\beta/d\alpha_1^2$,

$$f_p = \frac{1}{2}\rho_1 v_1 \frac{dv_1}{dx}\left[\left\{2 + 2(\alpha_1\beta - 1) + 2\xi\right\} + 4\xi - 2\alpha_1 \frac{d\xi}{d\alpha_1}\right] \quad (7.26)$$

Now, by using (7.20) we can show that the term involving the curly brackets is the same as $-dp_1/dx$ in (7.17), while the final term becomes $\rho_1 v_1^2 d\xi/dx$ and (7.26) can be written as

$$f_p = -\frac{dp_1}{dx} + 2\xi\rho_1 v_1 \frac{dv_1}{dx} + \rho_1 v_1^2 \frac{d\xi}{dx} \quad (7.27)$$

Since v_1 is the same as w in this case, (7.27) can be written as

$$f_p = -\frac{d}{dx}\left[p_1 - \rho_1 w^2 \xi\right] \quad (7.28)$$

$$= -\frac{d}{dx}p_2 \quad (7.29)$$

This extremely simple result was *deduced* from (7.24) and appears to indicate either that the added mass term does not belong in (7.11) or that there is some additional term that cancels it.

It is interesting to speculate about how (7.10) and (7.11) might be modified to achieve consistency with (2.23), (3.52), (3.53), (3.81) and (3.117). The most promising candidate, in the author's experience, is the set of equations postulated by Geurst (1986). They satisfy not only the above "tests" (Wallis (1988)), but appear to be consistent with every development in the present article that permits direct comparison.

7.3 Lhuillier's Equations

Lhuillier (1985) used thermodynamic arguments and averaging to derive relations between several properties of particle suspensions. His function *"f"* is the same as the *exertia* $(\alpha_1\beta - 1)$ or *"D"* appearing in the present work and his equation (29) is equivalent to (3.16), being one of the few instances in the literature in which the added mass coefficient is directly related to the difference in pressure between the phases. He introduces functions $g(\alpha_2)$ and $h(\alpha_2)$ that represent coefficients in the momentum flux and particle stress due to relative motion, resembling some of the developments in 3.9. However, he only evaluates these functions in the dilute limit, using the same coefficients as obtained in 3.9. His expression for the *interphase momentum exchange* contains the same added mass forces proportional to relative acceleration as appear here; however, the additional terms that he derives contain unknown functions and do not appear to resolve the inconsistencies pointed out in the previous two sections in any obvious way.

7.4 Nigmatulin's Equations

Nigmatulin (1979) discusses the process of averaging the microscopic equations in each phase to obtain the macroscopic conservation laws. For an inviscid potential flow he obtains

$$- \nabla\Phi = \boldsymbol{v}_1 + \frac{\alpha_2}{2} \boldsymbol{w} \tag{7.30}$$

which is the vector form of (3.36) combined with (3.20). The form of his added mass term is the same as described previously, being proportional to the relative acceleration of the phases, and the coefficient is (1/2) for spheres. In the equations of motion of each phase there appears the divergence of a *nonhydrostatic stress tensor* proportional to $(\boldsymbol{ww} - w^2\boldsymbol{I}/3)$ which appears simpler than the developments in 3.9.

7.5 Variational Methods

Pursuing what may be a very promising approach, Geurst ((1985a) (1985b) (1986)) used variational methods to derive equations of motion of the individual phases. The equations contain terms that can be identified as *added mass effects* originating from kinetic energy resulting from the relative motion. Similar approaches had been taken earlier by Drumheller and Bedford (1980).

The basis of Geurst's approach is the assertion that an equation of the form of (3.30) is generally valid, not only in the uniform flow case, but when flows are expanding, converging or completely three-dimensional:

$$k = \frac{1}{2}\rho_1\alpha_1 v_1^2 + \frac{1}{2}\rho_2\alpha_2 v_2^2 + \frac{1}{2}\rho_1 m(\alpha) w^2 \tag{7.31}$$

$m(\alpha)$ is a dimensionless coefficient that is a function of α (which we have been calling α_2 up to now). Comparing with (3.35) we have yet another version of the *added mass coefficient.*

$$m = \alpha_1(\alpha_1\beta - 1) \qquad (7.32)$$

The justification for (7.31) is that it is a suitable quadratic form that is invariant under coordinate transformations.

Geurst's variational principle is

$$\delta\int_{t_0}^{t_1} dt \int_{x_0}^{x_1} Ldx = 0 \qquad (7.33)$$

Where L is the difference between the kinetic energy density and the free energy density

$$L = k - F \qquad (7.34)$$

F is defined as

$$F = \alpha_1\rho_1(u_1 - Ts_1) + \alpha_2\rho_2(u_2 - Ts_2) \qquad (7.35)$$

where u and s denote specific internal energy and entropy respectively. Use is also made of *thermodynamic potentials.*

$$\mu_1 = u_1 - Ts_1 + p_2/\rho_1 \qquad (7.36)$$

$$\mu_2 = u_2 - Ts_2 + p_2/\rho_2 \qquad (7.37)$$

Why p_2 should appear in (7.36), rather than p_1, is not clear to this author. A possible justification is that one should use the *interface* pressure, p_i, in both (7.36) and (7.37), since it is the pressure that is *externally imposed* on both phases. If phase 2 is the dispersed phase, one might argue that $p_i = p_2$.

Pursuing the formalism of the calculus of variations, Geurst introduces Lagrange multipliers, that play the roles of potentials, whose gradients in the one-dimensional case are related to the velocities by

$$\frac{\partial \Phi}{\partial x} = v_1 - \frac{m}{\alpha_1} w \tag{7.38}$$

$$\frac{\partial \eta}{\partial x} = v_2 + \frac{\rho_1}{\rho_2} \frac{m}{\alpha_2} w \tag{7.39}$$

(In Geurst's notation ρ_1 and ρ_2 are the mass densities per unit total volume. The densities used here are the phase densities, which Geurst calls ρ_l and ρ_g.)

Using (7.32) in (7.38) and (7.39) we obtain (3.36) and (3.48) but for a change in sign convention. Thus, the potentials introduced by Geurst are compatible with those discussed in Section 3.

Geurst also derives an expression for the *hydrodynamic pressure, P,* as

$$P = p_2 + \frac{1}{2} \rho_1 (m + \alpha_1 m')w^2 \tag{7.40}$$

where the prime denotes differentiation with respect to α_2. Substituting from (7.32) we find that (7.40) is compatible with (3.16) as long as P is taken as the volumetric averge pressure:

$$P = \alpha_1 p_1 + \alpha_2 p_2 \tag{7.41}$$

as in (3.151). In his earlier publications (Geurst (1985a), (Geurst (1985b)), Geurst goes on to deduce equations of motion of the phases which are not obviously related to the results derived in this article. He also develops a *momentum flux* (including the pressure)

$$\pi = \alpha_1 \rho_1 v_1^2 + \alpha_2 \rho_2 v_2^2 + \rho_1 m w^2 + P \tag{7.42}$$

which is the one-dimensional equivalent of (3.152).

In a more recent publication (Geurst (1986)), Geurst compares his appraoch with the traditional forms of the phase momentum equations such as those described by Ishii (1975) and Drew (1983). Some forms, e.g. his (5.7), of his *equation of motion of phase 1* contain additional terms resembling $\boldsymbol{Z}$ in (4.33). Others, e.g. his (6.1), contain *Bernoulli* terms proportional to $\nabla(w^2/2)$, and a *lift force*, proportional to $\mathbf{w} \times \nabla \times \mathbf{v}_2$.

In a direct comparison with Drew's form of equations of motion, Geurst ((1986), Section 7) introduces the divergence of a stress tensor which is zero for the dispersed phase but equal to

$$\sigma_1 = -\frac{1}{2}\rho_1(\alpha_1\beta - 1)w^2\boldsymbol{I} - \rho_1(\alpha_1\beta - 1)\mathbf{ww} \tag{7.43}$$

for the continuous phase. This is coupled with a pressure difference between the phases of

$$p_1 - p_2 = \frac{1}{2}\rho_1 m' w^2 \tag{7.44}$$

which is not the same as (3.16). However, he points out that alternative definitions giving equivalent equations of motion are, in the present notation,

$$p_1 - p_2 = \frac{1}{2}\rho_1\alpha_1 w^2 \frac{d}{d\alpha_2}(\alpha_1\beta - 1) \tag{7.45}$$

$$\sigma_1 = -\rho_1(\alpha_1\beta - 1)\mathbf{ww} \tag{7.46}$$

(7.45) is compatible with (3.16) and (7.46) is compatible with the final term in (3.152) as long as an appropriate definition of pressure is chosen, as in (7.41).

Rather surprisingly, his (5.9) that expresses the *added mass terms* is not directly related to the relative acceleration. However, it can be shown to be equivalent apart from an additional term of the Bernoulli type and one involving the convective derivative of the added mass coefficient, to the results reported here.

As a simple test of this approach we may compare Geurst's predictions with (7.16) and (7.29) for the case of one-dimensional steady flow in a porous medium at rest in which there is a void fraction gradient. With the present notation, his equation of motion of the continuous phase, (2.24) of Geurst (1986), becomes, for incompressible phases,

$$v_1 \frac{d}{dx}\left(v_1 + \frac{mv_1}{\alpha_1}\right) + \frac{mv_1}{\alpha_1}\frac{dv_1}{dx} + \frac{d}{dx}\left(\frac{p_2}{\rho_1} + \frac{1}{2}m'v_1^2\right) = 0 \qquad (7.47)$$

now, from (7.18) and (7.32)

$$(p_2 - p_1) = -\xi\rho_1 v_1^2 = +\frac{\alpha_1}{2}\frac{d(\alpha_1\beta - 1)}{d\alpha_1}\rho_1 v_1^2 = -\frac{(m + \alpha_1 m')}{2\alpha_1}\rho_1 v_1^2 \qquad (7.48)$$

Substituting (7.48) in (7.47) and rearranging we get

$$v_1 \frac{dv_1}{dx} + \frac{d}{dx}\left(\frac{mv_1^2}{\alpha_1}\right) + \frac{d}{dx}\left(\frac{p_1}{\rho_1} - \frac{mv_1^2}{2\alpha_1}\right) = 0 \qquad (7.49)$$

whence,

$$\frac{d}{dx}\left(\frac{v_1^2}{2} + \frac{mv_1^2}{2\alpha_1} + \frac{p_1}{\rho_1}\right) = 0 \qquad (7.50)$$

Using (7.32), (7.50) is seen to agree with (7.16).

For this case, Geurst's (1986) equation of motion for the discontinuous phase, (2.25), predicts that the only force per unit volume from the fluid on the suspension is $-\partial p_2/\partial x$, in agreement with (7.29). Thus, Geurst's equations pass both "tests" for this special case.

Many other comparisons can be made with Geurst's theory. Some are given by Wallis (1988). One interesting result is that Geurst's equations of motion for a steady nozzle flow lead to (6.6) and (6.7) Wallis (1989). Generally, his equations contain more terms than are included in the usual versions of the two-fluid theory, as outlined in 7.1 and 7.2.

7.6 Summary

The comparisons in this section have presented some tantalizing similarities between general equations suggested by various authors and the anlaysis presented earlier in this paper. The implication might well be that a universally valid theory, within the scope of the inviscid flow assumptions, should soon emerge. However, there remain incompatibilities and unresolved differences between all of the formulations that must be resolved by further work.

ACKNOWLEDGEMENTS

Significant parts of this work were developed in Spring, 1986 while the author was a Visiting Fellow at Wolfson College and the Department of Engineering Sciences at Oxford University as well as enjoying a sabbatical appointment in the Thermal Hydraulics Division of Harwell Laboratory, Oxfordshire, England. Further derivations were supported by the U.S. Department of Energy, Contract #DE-FG02-86-ER13528, administered by Dr. Oscar Manley.

NOMENCLATURE

a	Radius of a sphere
a_x	Acceleration in x-direction
a_1	Coefficient in (7.10)
A	Area (usually of cross-section perpendicular to flow)
b	Factor in (6.10), fraction of length occupied by slugs
b_1	Coefficient in (7.10)
B	Variant of added mass coefficient, (6.5)
c	Distance of source from centre of sphere, conductivity
c_{vm}	Variant of added mass coefficient, (7.12)
C	Added mass coefficient
C_d	Drag coefficient

d	Dipole strength
D	Variant of added mass coefficient, or *exertia*, (4.18)
f	Fraction of overall volume; same as C in Cook-Harlow model
F	Force; Free energy density
f_1, f_2	External forces per unit volume on each phase
g	Acceleration due to gravity
j	Volumetric flux
k	Kinetic energy per unit total volume
k_1, k_2	Kinetic energy of each phase per unit volume of that phase
K	Total kinetic energy
L	Length; difference between kinetic energy density and free energy density
m	Source strength; Geurst's added mass coefficient (7.32)
m_X	Momentum in x-direction per unit total volume
M_X	Total momentum in x-direction
n	Slip ratio, v_2/v_1
p	Microscopic fluid pressure (e.g., as in Navier-Stokes equations)
p_0	Stagnation pressure
P	Macroscopic pressure
p_1, p_2	Average pressure in phase 1, 2
p_i	Interfacial pressure

P_x	Combined resultant stress and momentum flux in x-direction
q	Volumetric flow rate in stream tube
Q	Volumetric flow rate in a duct
r	Spherical coordinate
s	Entropy
ds_x	Surface element normal to x-direction
t	Time
T	Temperature
u_x, u_y, u_z	Components of microscopic velocity
u	Internal energy
v_1, v_2	Average velocity of phases (one-dimensional)
V	Volume
w	Relative velocity $(v_1 - v_2)$
w_x, w_y, w_z	Components of microscopic fluid velocity with stationary particles
x	One-dimensional coordinate
α_1, α_2	Fraction of total volume occupied by each phase
β	Defined in (3.5)
γ	Defined in (3.64)
δ	Perturbation in void fraction
ε	Perturbation in void fraction
ζ	Particle displacement

η	Macroscopic potential
ϑ	Angle in spherical coordinates
μ	Thermodynamic potential
ξ	Coefficient in (7.10), (7.11)
ρ	Density
σ	Particle stress (tensile)
φ	Microscopic potential
$\mathbf{\Phi}$, $\mathbf{\Psi}$	Macroscopic (average potentials)

Subscripts

a	Due to added mass
d	Due to drag, due to dipole
e	External
f	Fluid
i	Summation variable; interface
m	Due to source
o	Object, base case, original value
s	Sphere
x, y, z	Components in Cartesian directions

Superscripts

.	Rate of change with time (i.e. d/dt)
~	Fluctuation
'	Differentiated

Vectors and Tensors

$\boldsymbol{I}$	Unit dyadic
$\boldsymbol{M}$	Momentum
$\boldsymbol{m}$	Momentum per unit volume
$\boldsymbol{P}$	Combined momentum flux and stress tensor
$d\boldsymbol{s}$	Surface element
$\boldsymbol{u}$	Microscopic velocity
$\boldsymbol{v}$	Macroscopic (average) velocity
$\boldsymbol{w}$	Relative velocity
$\hat{\boldsymbol{x}}$	Unit vector in x-direction
$\boldsymbol{Z}$	Defined in (4.34)
$\boldsymbol{\Pi}$	Momentum flux
$\boldsymbol{\sigma}$	Particle stress (tensile)

APPENDIX A. THEOREMS FOR POTENTIAL INCOMPRESSIBLE FLOWS

A.1 Fluid Kinetic Energy in a Sourceless Flow

Denote the total fluid kinetic energy in a volume V by "K". Then

$$\frac{2K}{\rho} = \int u^2 dV = \int \nabla\varphi \cdot \nabla\varphi dV \tag{A.1}$$

The following vector identity is easily proved

$$\nabla \cdot (\varphi \nabla\varphi) = \nabla\varphi \cdot \nabla\varphi + \varphi \nabla^2 \varphi \tag{A.2}$$

Since

$$\nabla^2 \varphi = 0 \tag{A.3}$$

in a sourceless flow, the second term on the right-hand side of (A.2) may be removed. Using (A.2) in (A.1) and applying Gauss' theorem we obtain

$$\frac{2K}{\rho} = \int \nabla \cdot (\varphi \nabla\varphi) dV = \int \varphi \nabla\varphi \cdot \mathbf{ds} = - \int \varphi \mathbf{u} \cdot \mathbf{ds} \tag{A.4}$$

Therefore, the kinetic energy in a region may be evaluated from the surface integral of the potential times the volumetric flow rate into the region through the surface element $\mathbf{ds}$. This flow could result either from flow crossing the boundary or from the displacement of fluid by moving boundaries.

This result is a classical one (e.g. Lamb (1932), Milne-Thomson (1949)). Some examples follow.

A1.1 *Kinetic Energy in a Stream Tube*

Consider a stream tube carrying a flow rate q and having potentials φ_b and φ_a at its upstream and downstream ends (Figure A.1).

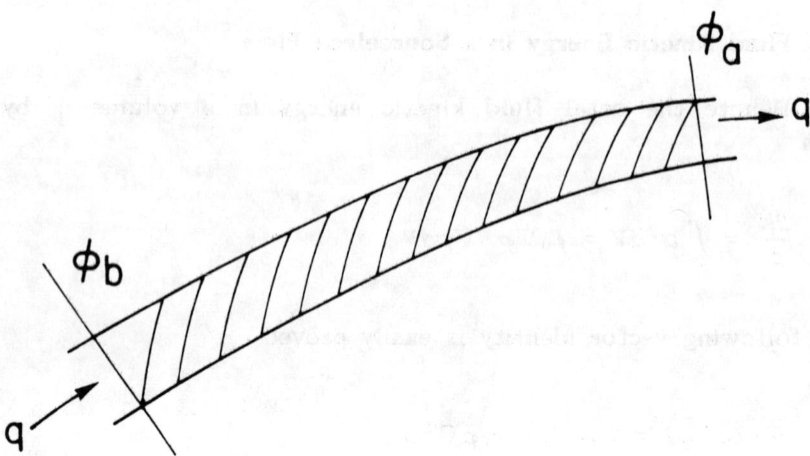

Figure A.1 A stream tube

$\int \boldsymbol{u} \cdot \boldsymbol{ds}$ is equal to q at the downstream end and $-q$ at the upstream end. Therefore, from (A.4), the kinetic energy in the stream tube is

$$K = \frac{1}{2} \rho q (\varphi_b - \varphi_a) \tag{A.5}$$

A.1.2 *Kinetic Energy of Fluid Surrounding a Translating Object, with Fluid at rest at Infinity*

Let the object have speed w in the x-direction (Figure A.2).

The fluid velocity on the surface of the object must satisfy the condition

$$(\boldsymbol{u} - w\hat{\boldsymbol{x}}) \cdot \boldsymbol{ds} = 0 \tag{A.6}$$

Therefore,

$$\boldsymbol{u} \cdot \boldsymbol{ds} = w\hat{\boldsymbol{x}} \cdot \boldsymbol{ds} = -w ds_x \tag{A.7}$$

where ds_x represents the projection of the area element ds in the x-direction

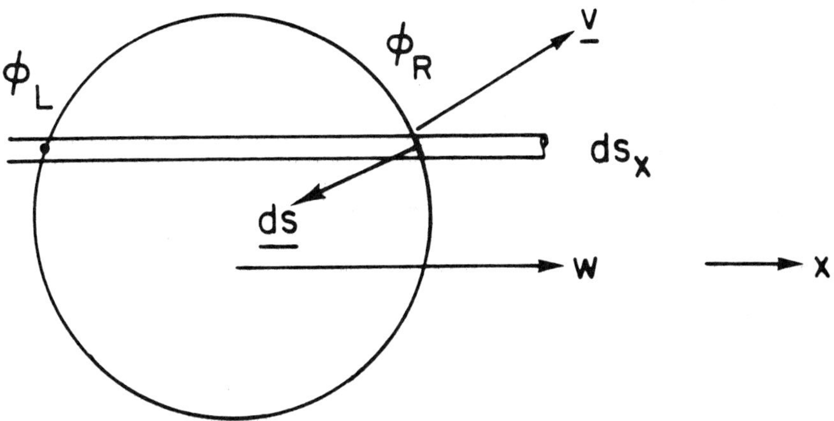

Figure A.2 A submerged object with speed w.

The kinetic energy of the fluid follows from (A.4) and (A.7) as

$$K = \frac{1}{2} \rho w \int \varphi ds_X = \frac{1}{2} \rho w \int (\varphi_R - \varphi_L) | ds_X | \tag{A.8}$$

where φ_R and φ_L are the potentials on the right and left of the body in Figure A.2 at the location corresponding to ds_X. $\mathbf{ds}$ is a vector pointing out of the fluid and into the object, accounting for the signs in (A.4) and (A.7). It is necessary here that streamlines begin and end on the body and that there are not contributions from conditions at infinity.

In the case of several objects moving in a fluid, a set of integrals such as that given in (A.8) would have to be added to obtain the kinetic energy.

The case of a more general motion involving translation in any direction together with rotation is discussed by Lamb (1932). For unsymmetrical objects in pure translation the formulation requires six coefficients and is tensorial in nature (see Tam (1977), Yih (1977)).

A.1.3 *Kinetic Energy of Fluid Surrounding an Object Moving in a Duct*

$\mathbf{u} \cdot \mathbf{ds}$ is zero over the walls of the duct. The kinetic energy is, from (A.4),

$$K = \frac{\rho}{2} \left[v_2 \int \varphi ds_X - \int_a \varphi_a \mathbf{u} \cdot \mathbf{ds} - \int_b \varphi_b \mathbf{u} \cdot \mathbf{ds} \right] \tag{A.9}$$

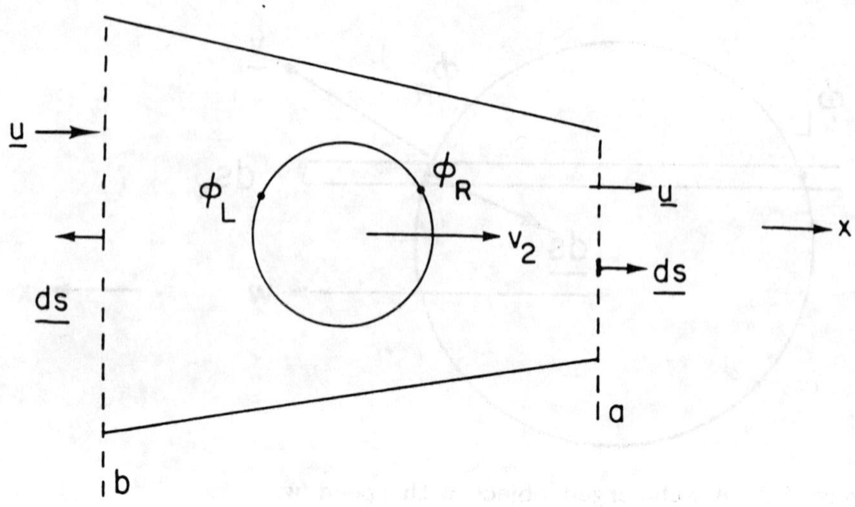

Figure A.3 An object moving in a duct.

In particular, if the ends of the flow are chosen as the equipotentials, φ_a = constant, φ_b = constant and the volume of the object is constant with time

$$K = \frac{\rho}{2}\left[v_2 \int \varphi ds_x - Q(\varphi_a - \varphi_b)\right] \tag{A.10}$$

where

$$Q = \int \mathbf{u} \cdot \mathbf{ds} \tag{A.11}$$

is the total flow rate of the fluid in the duct. This formulation is more realistic for practical purposes than (A.8) and avoids possible difficulties with boundary conditions at infinity.

A.2 Momentum in Sourceless Potential Flow

The net momentum in an incompressible fluid is

$$\mathbf{m} = \rho \int \mathbf{u} dV = -\rho \int \nabla\varphi dV = -\rho \int \varphi \mathbf{ds} \tag{A.12}$$

where, again, $\boldsymbol{ds}$ is an element of area pointing away from the fluid.

The net x-direction momentum associated with the motion in Figure A.2 is then

$$M_x = \rho \int \varphi ds_x \tag{A.13}$$

Since φ is proportional to w and the scale of the object and the same integrals appear in (A.8) and (A.13), we may represent the momentum and kinetic energy of the fluid by

$$M_x = \rho C V_0 w \tag{A.14}$$

$$K = \frac{1}{2} \rho C V_0 w^2 \tag{A.15}$$

where V_0 is the volume of the object and C is a *coefficient of added mass.*

In the case of the general motion of an unsymmetrical object, C must be treated as a tensor (e.g. Tam (1977), Yih (1977)).

We note that in order to obtain (A.8) and (A.13) it was necessary for the potential and $\int \varphi \boldsymbol{u} \cdot \boldsymbol{ds}$ to vanish at infinity. (A discussion of some related problems of convergence at infinity is given by Darwin (1952)). Otherwise, $\int \varphi \boldsymbol{ds}$ in (A.12) must be taken over all boundaries. For example, if the walls of the duct in Figure A.3 are parallel to the x-direction, so that they contribute no x-component in (A.12), the x-direction momentum for the fluid is

$$M_x = \rho \left[\int \varphi \, ds_x + (\varphi_b - \varphi_a) A \right] \tag{A.16}$$

where A is the cross-section area of the duct. It is only when $\varphi_b = \varphi_a$ that results as simple as (A.14) and (A.15) emerge.

A.3 Average Pressure on a Boundary

Consider a region V_0, containing an incompressible potential flow, bounded at its end by specified potentials φ_A and φ_B (not necessarily uniform) and at its sides by streamlines (Figure A.4).

Let there be perturbations, denoted by primed quantities, about an original flow described by potential φ_0 and velocity $\boldsymbol{u}_0$. For the new flow

$$\varphi = \varphi_0 + \varphi' \tag{A.17}$$

$$\boldsymbol{u} = \boldsymbol{u}_0 + \boldsymbol{u}' \tag{A.18}$$

while we also have

$$\boldsymbol{u} = -\nabla\varphi, \qquad \boldsymbol{u}' = -\nabla\varphi' \tag{A.19}$$

$$\nabla^2\varphi = \nabla^2\varphi' = \nabla^2\varphi_0 = 0 \tag{A.20}$$

As a result of the perturbation, the streamlines will change, being locally displaced from the old streamlines at the edge by an amount that changes the enclosed volume by dV'.

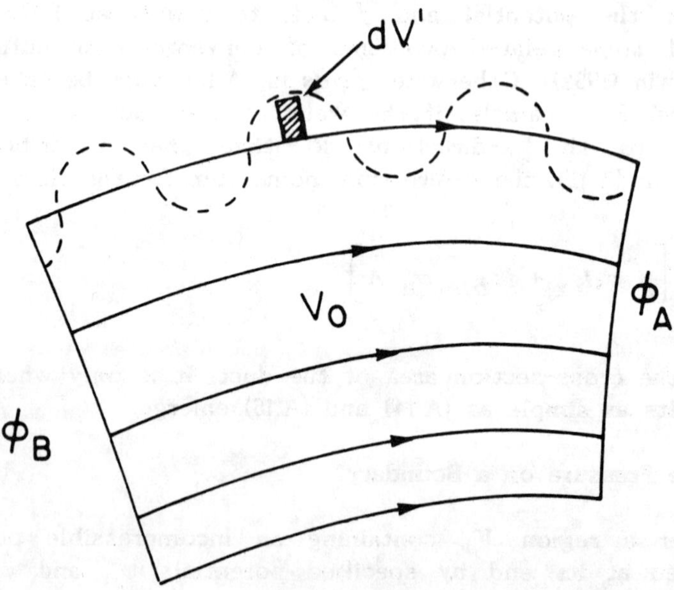

Figure A.4 A region of potential flow with perturbed boundaries.

The kinetic energy in the original volume after the perturbation is, to first order,

$$\frac{2K}{\rho} = \int_{V_0} u^2 dV = \int_{V_0} (\mathbf{u}_0 + \mathbf{v}')^2 dV = \int_{V_0} (u_0^2 + 2\nabla\varphi_0 \cdot \nabla\varphi')dV \tag{A.21}$$

Now

$$\nabla\varphi_0 \cdot \nabla\varphi' = \nabla\varphi' \cdot \nabla\varphi_0 = \nabla \cdot (\varphi'\nabla\varphi_0) - \varphi'\nabla^2\varphi_0 \tag{A.22}$$

and the last term in (A.22) vanishes because of (A.20). Using Gauss' theorem on the remaining term we get

$$\int_{V_0} u^2 dV = \int_{V_0} u_0^2 dV + 2\int_{s_0} \varphi'\nabla\varphi_0 \cdot \mathbf{ds} \tag{A.23}$$

where s_0 is the surface containing V_0.

Now φ' is zero on the ends of the volume and $\nabla\varphi_0 \cdot \mathbf{ds}$ is zero on the sides. Therefore, the final integral in (A.23) vanishes and we get

$$\int_{V_0} u^2 dV = \int_{V_0} u_0^2 dv \tag{A.24}$$

(which is related to the minimization of kinetic energy in potential flow).

Now, $\int_{V_0} u^2 dV$ may be considered to be composed of an integral over the volume, V, contained by the new bounding streamlines, less the integral over the perturbations in volume at the boundaries, i.e.

$$\int_{V_0} u^2 dV = \int_{V} u^2 dV - \int_{V'} u^2 dV' \tag{A.25}$$

To first order we may replace u^2 by u_0^2 in the final integral in (A.25), and express u_0^2 as $2(p_0 - p)/\rho$ by using Bernoulli's equation. The use of (A.24) to substitute on the left-hand side finally leads to

$$\frac{1}{2}\rho\left[\int_V u^2 dV - \int_{V_0} u_0^2 dV\right] = \int_{V'} (p_0 - p)\,dV' \qquad (A.26)$$

which may be interpreted as an expression of energy conservation, in which work done by pressures on the flexible side walls changes the total kinetic energy of the enclosed fluid. Note that this result depends on the invariance of the potential distributions on the ends of the volume.

If the integrals on the left-hand side of (A.26) can be evaluated for specific perturbations of the boundary, the *average pressure* on the boundary can be determined, with weighting in proportion to the relative size of the displacements dV'.

A.3.1 *Average Pressure over a Submerged Body*

Figure A.5 shows a submerged body (phase 2) around which phase 1 flows, confined between known potentials at the ends and fixed streamlines on the sides.

Imagine the body to change its volume, each incremental change around the surface being denoted by dV_2, which equals $-dV_1$.

From (A.26) we have

$$dK_1 = \int (p_{01} - p_i)\,dV_1 = -\int (p_{01} - p_i)\,dV_2 \qquad (A.27)$$

where

$$K_1 = \frac{1}{2}\rho_1 \int u_1^2\, dV_1 \qquad (A.28)$$

is the total kinetic energy of phase 1 contained in the volume and p_i is the pressure at the interface between the phases. Body forces have been neglected.

Define an average interfacial pressure as

$$\langle p_i \rangle = \int p_i\, dV_2 \,/ \int dV_2 \qquad (A.29)$$

Then, from (A.27) and (A.29)

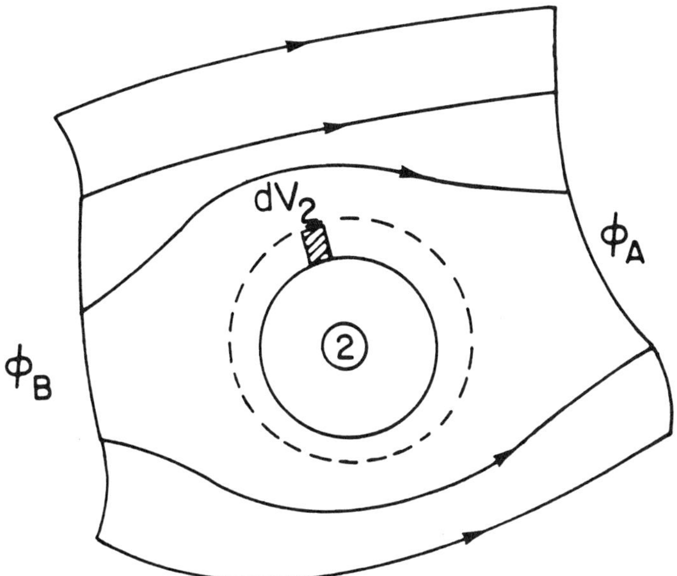

Figure A.5 Changes in volume of the dispersed phase 2.

$$< p_i > = p_{01} + \frac{dK_1}{\int dV_2} \tag{A.30}$$

Since the body can change its volume in an infinite number of ways, $< p_i >$ is a particular average interfacial pressure that is *weighted* by the volume displacements dV_2 (which could even arise from the creation of separate small disconnected regions). For the particular case of a sphere that grows uniformly radially, $< p_i >$ is equal to the area-averaged pressure $\int p_i ds / \int ds$ where ds is an element of the surface area of the sphere. Another special case is the displacement of the body by a small distance without change of shape, in which case the component of the force exerted by the fluid on the object in the direction of the displacement can be evaluated. A more complicated situation would be the rearrangement of positions of several objects in a flow, a special case being the creation of a gradient in concentration (or volume fraction) in a uniform dispersion. Analysis of the accompanying changes in kinetic energy of the fluid leads to the prediction of the *compressibility* of the dispersion, as discussed in Section 3.7.

REFERENCES

Auton, T. R. (1983) Ph.D. Dissertation, Cambridge University.

Banerjee, S. and Chan, A. M. C. (1980) Separated Flow Models - 1. Analysis of the Averaged and Local Instantaneous Formulations. *Int. J. Multiphase Flow,* **6**, 1.

Batchelor, G. K. (1970) The Stress System in a Suspension of Fource-Free Particles, *J. Fluid Mech.*, **41**, 545.

Beyerlein, S. W., Cossmann, R. K. and Richter, H. J. (1985) Prediction of Bubble Concentration Profiles in Vertical Turbulent Two-Phase Flow. *Int. J. Multiphase Flow,* **11**, 629.

Biesheuvel, A. and van Wijngaarden, L. (1984) Two-Phase Flow Equations for a Dilute Dispersion of Gas Bubbles in Liquid, *J. Fluid Mech.*, **148**, 301.

Bourè, J. A. (1987) Two-Phase Flow Models: The Closure Issue. *Multiphase Science and Technology,* **3**, Ch. 1, Hemisphere.

Bourè, J. A. (1975) Mathematical Modelling and the Two-Phase Constitutive Equations. European Two-Phase Flow Group Meeting, Haifa.

Cook, T. L. and Harlow, F. H. (1984) Virtual Mass in Multiphase Flow, *Int. J. Multiphase Flow,* **10**, 691.

Darwin, C. (1952) Note on Hydrodynamics. *Proc. Camb. Phil. Soc.,* **49**, 342.

Delhaye, J. M. (1981a) Basic Equations for Two-Phase Flow Modelling. Chapter 2 in *Two-Phase Flow and Heat Transfer in the Power and Process Industries,* Bergles, A. E., Collier, J. G., Delhaye, J. M., Hewitt, G. F. and Mayinger, F., Hemisphere & McGraw-Hill

Delhaye, J. M. (1981b) *Thermohydraulics of Two-Phase Systems for Industrial Design and Nuclear Engineering.* Chapters 7, 8 and 9, Eds. Delhaye, J. M., Giot, M. and Riethmuller, M. K., McGraw-Hill.

Delhaye, J. M. and Achard, J. L. (1978) On the Averaging Operators Introduced in Two-Phase Flow Modelling. *Transient Two-Phase Flow,* Eds. Banerjee, S. and Weaver, AECL, K. R., 1, 5.

Delhaye, J. M. (1968) Equations Fondamentales des Ecoulements Diphasiques. *Commissariat á l'Energie Atomique,* Paris (France) CEA-R-3429.

Drew, D. A. and Lahey, R. T., Jr. (1987) The Virtual Mass and Lift Force on a Sphere in Rotating and Straining Inviscid Flow. *Int. J. Multiphase Flow,* **13**, 113.

Drew, D. A. and Wood, R. T. (1985) Overview and Taxonomy of Models and Methods. *International Workshop on Two-Phase Flow Fundamentals,* National Bureau of Standards, Maryland, September.

Drew, D. A. (1983) Mathematical Modelling of Two-Phase Flow. *Ann. Rev. Fluid Mech.*, **15**, 261.

Drew, D., Cheng, L. and Lahey, R. T., Jr. (1979) The Analysis of Virtual Mass Effects in Two-Phase Flow. *Int. J. Multiphase Flow*, 5, 233.

Drew, D. A. (1971) Averaged Equations for Two-Phase Flow. *Studies in Appl. Math.*, **50**, 205.

Drumheller, D. S. and Bedford, A. (1980) A Thermomechanical Theory for Reacting Immiscible Mixtures. *Arch. Rational Mech. Anal.*, **73**, 257.

Garipov, R. M. (1973) Closed System of Equations for the Motion of a Liquid with Gas Bubbles, *Z. Prik. Mekh. i tech. Fiz.*, **6**, 3.

Geurst, J. A. (1986) Variational Principles and Two-Fluid Hydrodynamics of Bubbly Liquid/Gas Mixtures. *Physica*, **135A**, 455.

Geurst, J. A. (1985a) Virtual Mass in Two-Phase Bubbly Flow, *Physica*, **129A**, 233.

Geurst, J. A. (1985b) Two-Fluid Hydrodynamics of Bubbly Liquid/Vapour Mixture Including Phase Change. *Philips J. Res.* **40**, 352.

Harlow, F. H. and Amsden, A. A. (1975) Numerical Calculation of Multiphase Fluid Flow. *J. of Comp. Phys.*,. 17, 19.

Ishii, M. (1987) Two-Fluid Models for Two-Phase Flow. Second International Workshop on Two-Phase Flow Fundamentals, RPI, Troy, New York, March.

Ishii, M. and Mishima, K. (1984) Two-Fluid Model and Hydrodynamic Constitutive Relations. *Nuc. Engng. and Design*, **82**, 107.

Ishii, M. (1975) *Thermo-Fluid Dynamic Theory of Two-Phase Flow*, Eyrolles, Paris (France).

Jeffrey, D. J. (1973) Conduction Through a Random Suspension of Spheres. *Proc. Roy. Soc.*, **A335**, 335.

Lagally, M. (1922) Berechnung der Kräfte und Momente die Stromende Flüssigkeiten auf ihre Begrenzung ausüben. *Z. Angew. Math. Mech.*, 2, 409.

Lamb, H. (1932) *Hydrodynamics*, 6th ed., Cambridge University Press, (a) page 46; (b) Chapter 6; (c) page 166; (d) page 125.

Landweber, L. and Miloh, T. (1980) Unsteady Lagally Theorem for Multipoles and Deformable Bodies. *J. Fluid Mech.*, **96**, 33-46.

Lhuillier, D. (1985) Phenomenology of Inertia Effects in a Dispersed Solid-fluid Mixture. *Int. J. Multiphase Flow*, **11**, 427.

Maxwell, J. C. (1881) *A Treatise on Electricity and Magnetism*. 2nd ed., **1**, Clarendon Press, Oxford, 398.

Mercadier, Y. (1981) Contribution à l'étude des propagations de Perturbations de Taux de Vide dans les Ecoulements Diphasiques

Eau-Air à Bulles. D. Ing. Thesis. Univ. Sci. Medicale et Institut National Polytechnique de Grenoble.

Milne-Thomson, L. M. (1949) *Theoretical Hydrodynamics.* 2nd ed. MacMillan & Co., London, (a) page 420, paragraph 15-43 - note error in denominator in book; (b) page 417; (c) page 433, problem 39; (d) page 92.

Nigmatulin, R. I. (1979) Spatial Averaging in the Mechanics of Heterogeneous and Dispersed Systems. *Int. J. Multiphase Flow,* 5, 353.

Pauchon, C. and Banerjee, S. (1986) Interphase Momentum Interaction Effects in the Averaged Multifluid Model, Part 1, Void Propagation in Bubbly Flow. *Int. J. Multiphase Flow,* **12**, 559.

Prosperetti, A. and Jones, A. V. (1984) Pressure Forces in Disperse Two-Phase *Flow,* **10**, 425-440.

Rowe, P. N. and Henwood, G. A. (1961) Drag Forces in a Hydraulic Model of a Fluidized Bed - Part 1. *Trans. Inst. Chem. Engrs.,* **39**, 43.

Stewart, H. B. and Wendroff, B. (1984) Two-Phase Flow: Models and Methods. *J. Comp. Phys.,* **56**, 363.

Tam, P. D. (1977) Quelques Rappels sur la Notion de Masse Ajoutée en Mécanique des Fluides. CEA-R-4855.

Taylor, G. I. (1928a) The Forces on a Body Placed in a Curved or Converging Stream of Fluid. *Proc. Roy. Soc.,* **A120,** 260-283.

Taylor, G. I. (1928b) The Energy of a Body Moving in an Infinite Fluid with an Application to Airships. *Proc. Roy. Soc.,* **A120**, 13, (see also Reference 18).

Thomas, N. H., Auton, T. R., Sene, K., and Hunt, J. C. R. (1983) Entrapment and Transport of Bubbles by Transient Large Eddies in Multiphase Turbulent Shear Flows. *Physical Modelling of Multiphase Flow, BHRA,* 169.

Turner, J. C. R. (1976) Two-Phase Conductivity: The Electrical Conductance of Liquid-Fluidized Beds of Spheres. *Chem. Eng. Sci.,* **31**, *487.*

Wallis, G. B. (1989) On Geurst's Equations for Inertial Coupling in Two-Phase Flow. Inst. Math. Appl. Workshop "Two Phase Waves in Fluidized Beds, Flowing Composites and Granular Media", U. Minnesota, to be published by Springer Verlag.

Wallis, G. B. (1988) Some Tests of Two-Fluid Models for Two-Phase Flow. U.S.-Japan Seminar on Two-Phase Flow Dynamics, Ohtsu, Japan, July.

Wallis, G. B. (1978) Internal Memoranda, Thayer School of Engineering, Dartmouth College, January-February.

Wallis, G. B. (1962) One-Dimensional Waves in Two-Component Flow (With Particular Reference to the Stability of Fluidized Beds). UKAEA, AEEW-R162.

Yadigaroglu, G. and Lahey, R. T., Jr. (1976) On the Various Forms of the Conservation Equations in Two-Phase Flow. *Int. J. Multiphase Flow*, 2, 477.

Yih, C. S. (1977) *Fluid Mechanics*. West River Press, Ann Arbor, pp. 99-115.

Zuber, N. (1964) On the Dispersed Two-Phase Flow in the Laminar Flow Regime. *Chem. Eng. Sci.*, **19**, 897.

INDEX